Advances in Phase Transitions

Pergamon Titles of Related Interest

ASHBY & HIRTH
Perspectives in Hydrogen in Metals

ASHWORTH
Corrosion: Industrial Problems, Treatment and Control Techniques

BEE & GARRETT
Materials Engineering

CRAWFORD
Plastics Engineering, 2nd Edition

FARLEY & NICHOLS
Non-Destructive Testing (4-volume set)

GIBSON & ASHBY
Cellular Solids

LANSDOWN
Materials to Resist Wear

LIU & NICHOLS
Pressure Vessel Technology (2-volume set)

MICHIELS & DE HERDT
Molecular Sieve Catalysts

NOWAKI
Theory of Asymmetric Elasticity, 2nd edition
Thermoelasticity, 2nd edition

YAN et al
Mechanical Behaviour of Materials V

Pergamon Related Journals *(free sample copy gladly sent on request)*

Acta Metallurgica

Corrosion Science

European Polymer Journal

International Journal of Plasticity

International Journal of Solids and Structures

Journal of the Mechanics and Physics of Solids

Materials Research Bulletin

Scripta Metallurgica

Solid State Communications

The Journal of Physics and Chemistry of Solids

Advances in Phase Transitions

Proceedings of the International Symposium
held at McMaster University
Ontario, Canada, 22-23 October 1987

Edited by

J. D. EMBURY

and

G. R. PURDY

*Department of Materials Science and Engineering
McMaster University, Ontario, Canada*

PERGAMON PRESS
OXFORD · NEW YORK · BEIJING · FRANKFURT
SÃO PAULO · SYDNEY · TOKYO · TORONTO

U.K.	Pergamon Press plc, Headington Hill Hall, Oxford OX3 0BW, England
U.S.A.	Pergamon Press, Inc., Maxwell House, Fairview Park, Elmsford, New York 10523, U.S.A.
PEOPLE'S REPUBLIC OF CHINA	Pergamon Press, Room 4037, Qianmen Hotel, Beijing, People's Republic of China
FEDERAL REPUBLIC OF GERMANY	Pergamon Press GmbH, Hammerweg 6, D-6242 Kronberg, Federal Republic of Germany
BRAZIL	Pergamon Editora Ltda, Rua Eça de Queiros, 346, CEP 04011, Paraiso, São Paulo, Brazil
AUSTRALIA	Pergamon Press Australia Pty Ltd., P.O. Box 544, Potts Point, N.S.W. 2011, Australia
JAPAN	Pergamon Press, 5th Floor, Matsuoka Central Building, 1-7-1 Nishishinjuku, Shinjuku-ku, Tokyo 160, Japan
CANADA	Pergamon Press Canada Ltd., Suite No. 271, 253 College Street, Toronto, Ontario, Canada M5T 1R5

Copyright © 1988 Pergamon Press plc

All Rights Reserved. No part of this publication may be reproduced, stored in a retrieval system or transmitted in any form or by any means: electronic, electrostatic, magnetic tape, mechanical, photocopying, recording or otherwise, without permission in writing from the publishers.

First edition 1988

Library of Congress Cataloging in Publication Data
Advances in phase transitions: proceedings of the international symposium held at McMaster University, Ontario, Canada, 22-23 October 1987/edited by J. D. Embury and G. R. Purdy. – 1st ed. p. cm.
1. Phase rule and equilibrium–Congresses.
2. Materials–Congresses. I. Embury, J. D.
II. Purdy, Gary R.
QD503.A28 1988 88-19609
669'.94—dc19

British Library Cataloguing in Publication Data
Advances in phase transitions.
1. Phase transitions
I. Embury, J. D. II. Purdy, Gary R.
536'.401

ISBN 0-08-036234-6

In order to make this volume available as economically and as rapidly as possible the author's typescript has been reproduced in its original form. This method unfortunately has its typographical limitations but it is hoped that they in no way distract the reader.

Printed in Great Britain by A. Wheaton & Co. Ltd., Exeter

Contents

Preface		vii
Mats Hillert and John Ågren	Diffusional Transformations under Local Equilibrium in Fe-C-M Alloys	1
H.I. Aaronson, M. Enomoto and W.T. Reynolds	Influence of the Chemistry and Structure of Austenite: Ferrite Allotriomorphs in Fe-C-X Alloys	20
K. Balasubramanian and J.S. Kirkaldy	Thermodynamics of Fe-Ti-C and Fe-Nb-C Austenites and Non-stoichiometric Titanium and Niobium Carbides	37
A.D. Pelton, W.T. Thompson, C.W. Bale, and N.G. Eriksson	Phase Equilibrium Calculations in Multicomponent Systems	52
J.E. Morral and M.S. Thompson	Phase Transitions and the Square Root Diffusivity	68
Maruti Bhandarkar and J.C.M. Li	Molecular Dynamics Simulation of Melting	78
D. Venugopalan	Cellular Solidification	90
J.R. Sarazin and A. Hellawell	Channel Flow in Partly Solidified Alloy Systems	101
D.J. Young	Dealloying Reactions as Cellular Phase Transformations	116
W.W. Smeltzer	Modelling of Oxidation and Sulfidation Reactions	131
Ichiro Arakawa, Yoshikata Koga, and James A. Morrison	Wetting Transitions of Ethylene Absorbed on Graphite	145
Yoshikata Koga	Metastability/Unstability: Kinetics of Polymorphic Transition from Phase II to Phase III of C_2Cl_6	151
Alfred R. Cooper	Freezing in Transitions	154
Ricardo B. Schwarz	Recent Advances in the Synthesis of Amorphous Metallic Materials	166
S. Radelaar	Structural Relaxation and Phase Transformations in Silicon/Transition Metal Multilayers	178

F. Larché	The Role of Stresses on Phase Transformations	193
Gary R. Purdy and John R. Dryden	On the Role of Elastic Energy in Diffusional Phase Transformations	204
J.P. Hirth	The Role of Creep in Diffusional Transformations	221
Elias C. Aifantis	Dislocation Patterns and Deformation Bands	231
G. Burger, E. Koken, D.S. Wilkinson, J.D. Embury	The Influence of Spatial Distributions on Metallurgical Processes	247
J.S. Kirkaldy	Thermologistics; A Science of Pattern Formation	263
Keywords		297

Preface

It is fitting that as part of McMaster University's Centennial celebrations, the members of the Department of Materials Science and Engineering should honour their distinguished colleague, John Samuel Kirkaldy.

It is appropriate, in this context, to consider areas of materials science where the framework of thermodynamics and kinetics has provided a basic structure for the detailed exploration of a broad range of phase transitions, which encompasses solidification, reactions in the solid state, gas-metal reactions and amorphous-crystalline transitions.

In the past three decades, Jack Kirkaldy has contributed much to both the basic understanding of diffusive transport and phase transitions, and to the application of this understanding to the control and optimisation of phase transitions in the heat treatment of steel and other areas of technological import. Thus, in the fall of 1987, the Department of Materials Science and Engineering at McMaster organised a Symposium on Advances in Phase Transitions. The program attempted to cover a broad range of fundamental issues and attracted a number of speakers of international repute. The present volume records the presentations and indicates the range of theoretical and experimental approaches which have been developed in the area of phase transformations. It is clear that the basic understanding of phase transformations will continue to play a central role in the understanding and control of the structures of advanced materials.

This volume is dedicated by his colleagues at McMaster to John Samuel Kirkaldy in recognition of the intellectual leadership and stimulation he has brought to McMaster over the past 30 years.

David Embury

Gary Purdy

McMaster University
March 1988.

Diffusional Transformations Under Local Equilibrium in Fe-C-M Systems

MATS HILLERT AND JOHN ÅGREN

Division of Physical Metallurgy
Royal Institute of Technology
S-100 44 Stockholm, Sweden

ABSTRACT

The special features of diffusional transformations in ternary systems, where one component is much more mobile than the others, are discussed. It is shown how the understanding of these features developed independently from two quite different starting points. The features are discussed with reference to a number of calculations.

The most recent development is based upon numerical solutions of the diffusion equation written with gradients of chemical potentials as driving forces. By reference to a number of calculations it is demonstrated that many of the earlier limitations have thus been removed.

KEYWORDS

Alloy partition; alloyed steels; chemical potential gradient; diffusional transformations; impingement; no-partition; paraequilibrium; ternary diffusion; up-hill diffusion.

INTRODUCTION

When considering diffusional phase transformations in alloys it is often a very useful approximation to assume that there is local equilibrium at a phase interface even if it is moving. This approximation will be accepted in the present paper and a discussion will be given of the special features of diffusional phase transformations in ternary systems with one interstitial and one substitutional solute. The classical example of this type of system is Fe-C-M where M represents such alloying elements as Mn, Ni, Cr, Si etc. The special features are caused by the fact that interstitial diffusion is usually many orders of magnitude faster than substitutional diffusion. The study of these features was carried out independently over a number of years in Hamilton and Stockholm. In reviewing the developments in the two groups it will be shown that the starting points and the objectives were quite different but the results were nearly the same. Finally, a more recent development will be discussed.

SINGLE-PHASE DIFFUSION

Darken (1951) in his classical paper on a formal basis of diffusion theory mentioned one of the special features found in Fe-C-M systems but finally he concluded that "diffusion in a multicomponent system cannot be adequately described by a simple extension of the method currently used for binary systems." J.S. Kirkaldy took this as a challenge and in the period 1957-1969 he published a series of 11 papers with the common title "Diffusion in Multicomponent Metallic Systems"

(Kirkaldy, 1957, 1958a, 1958b, 1958c, 1959; Kirkaldy and Purdy, 1962; Kirkaldy, Weichert and Zia-Ul-Haq, 1963; Kirkaldy, Lane and Mason, 1963; Lane and Kirkaldy, 1964; Kirkaldy and Lane, 1966; Kirkaldy and Purdy, 1969) in which he gradually developed the theory. His starting point was Onsager's treatment of multicomponent diffusion (Onsager, 1945-6) which gives

$$-J_i = \Sigma D_{ij} \frac{\partial c_j}{\partial z} \qquad (1)$$

z is here the length coordinate in which diffusion occurs. Already in the first paper Kirkaldy considered the interstitial-substitutional case and identified C with component 1, a substitutional alloying element M with component 2 and Fe with component 3. Onsager thus gave 3 equations for the ternary case and 3 terms in each, making a total of 9 diffusion coefficients. Kirkaldy's first problem was to reduce that number. A review of his work will now be given in simple terms and a special concentration variable will be used defined as follows

$$u_i = x_i/(x_{Fe}+x_M) \qquad (2)$$

As a simplifying approximation it will be assumed that the lattice parameter is independent of composition, which results in

$$V_m/(x_{Fe}+x_M) = \text{constant } V_o \qquad (3)$$

One can then easily change from c to u in eq. 1 because the following transformation is linear.

$$c_i = x_i/V_m = \frac{x_i}{x_{Fe}+x_M} / \frac{V_m}{x_{Fe}+x_M} = u_i/V_o \qquad (4)$$

With a volume-fixed frame of reference the result will be

$$J_1 V_o = -D_{11}\frac{\partial u_1}{\partial z} - D_{12}\frac{\partial u_2}{\partial z} \qquad (5)$$

$$J_2 V_o = -D_{21}\frac{\partial u_1}{\partial z} - D_{22}\frac{\partial u_2}{\partial z} = -J_3 V_o \qquad (6)$$

because component 1 (C) will not affect the volume under the assumption of constant lattice parameter. Kirkaldy had thus reduced the number of diffusion coefficients from 9 to 4. In order to reduce it further he made use of Onsager's principle of microscopic reversibility (Onsager, 1931, 1932). However, in order to apply it one must use potential gradients as the driving force for diffusion. For the present case Kirkaldy thus wrote

$$J_1 = -L_{11}\frac{\partial \mu_1}{\partial z} - L_{12}\frac{\partial (\mu_2-\mu_3)}{\partial z} \qquad (7)$$

$$J_2 = -L_{21}\frac{\partial \mu_1}{\partial z} - L_{22}\frac{\partial (\mu_2-\mu_3)}{\partial z} = -J_3 \qquad (8)$$

The difference in form between the two terms is caused by the fact that component 1 diffuses by jumping from one interstitial site to the next but component 2 must change position with component 3 if viewed from a volume-fixed frame of reference.

By applying the principle of microscopic reversibility, Onsager found that the off-diagonal elements must be equal, making the L matrix symmetric,

$$L_{21} = L_{12} \qquad (9)$$

This is usually called Onsager's reciprocal relation. Kirkaldy further argued that this coefficient, which represents a cross-effect similar to the thermoelectric effect, should be negligible because the two diffusion processes occur on different sublattices and their coupling should be very weak. He thus introduced

$$L_{21} = L_{12} = 0 \qquad (10)$$

Only 2 diffusion coefficients thus remained.

When switching back to concentration gradients Kirkaldy chose first to eliminate μ_3 using the Gibbs-Duhem relation which we may here write in the following form

$$u_1 d\mu_1 + u_2 d\mu_2 + u_3 d\mu_3 = 0 \qquad (11)$$

This gives

$$J_1 = -L_{11}\frac{\partial \mu_1}{\partial z} \qquad (12)$$

$$J_2 = -\frac{L_{22}}{u_3}\left[u_1\frac{\partial \mu_1}{\partial z} + \frac{\partial \mu_2}{\partial z}\right] \qquad (13)$$

and finally

$$J_1 = -L_{11}\frac{\partial \mu_1}{\partial u_1}\cdot\frac{\partial u_1}{\partial z} - L_{11}\frac{\partial \mu_1}{\partial u_2}\cdot\frac{\partial u_2}{\partial z} \qquad (14)$$

$$J_2 = -\frac{L_{22}}{u_3}\left[u_1\frac{\partial \mu_1}{\partial u_1} + \frac{\partial \mu_2}{\partial u_1}\right]\cdot\frac{\partial u_1}{\partial z} - \frac{L_{22}}{u_3}\left[u_1\frac{\partial \mu_1}{\partial u_2} + \frac{\partial \mu_2}{\partial u_2}\right]\cdot\frac{\partial u_2}{\partial z} \qquad (15)$$

By comparing with eqs. 5 and 6 one can now express the four D coefficients in terms of L_{11} and L_{22}, but it is then necessary to know how μ_1 and μ_2 vary with the composition as expressed by u_1 and u_2. Using Wagner's (1952) dilute solution formalism Kirkaldy obtained

$$D_{12} = D_{11}\frac{\varepsilon_{12}x_1}{1+\varepsilon_{11}x_1} \qquad (16)$$

$$D_{21} = D_{22}\frac{(1+\varepsilon_{12})x_2}{1+\varepsilon_{22}x_2} \qquad (17)$$

where the ε:s are so-called interaction coefficients. Knowing them one can thus evaluate D_{11} from the Fe-C system, D_{22} from the Fe-M system and then calculate D_{12} and D_{21}. In this way Kirkaldy was able to find an analytical solution to the famous up-hill diffusion by Darken (1949) and to test the data with calculations. Good agreement was obtained already in the first paper of the series. See Fig. 1. However, it should be emphasized that the analytical solution was valid for constant coefficients only and average values for D_{12} and D_{21} had to be used.

PHASE TRANSFORMATIONS

When considering the rate of diffusion controlled phase transformations Kirkaldy (1958b) applied the same type of analytical solution and obtained two equations, one for carbon and one for the alloying element. From practical point of view one may ask how they can give the same growth rate in spite of the fact that carbon is much more mobile than the alloying element. In order to explain the solution to this problem and its consequences we shall here give Kirkaldy's results in a greatly simplified form

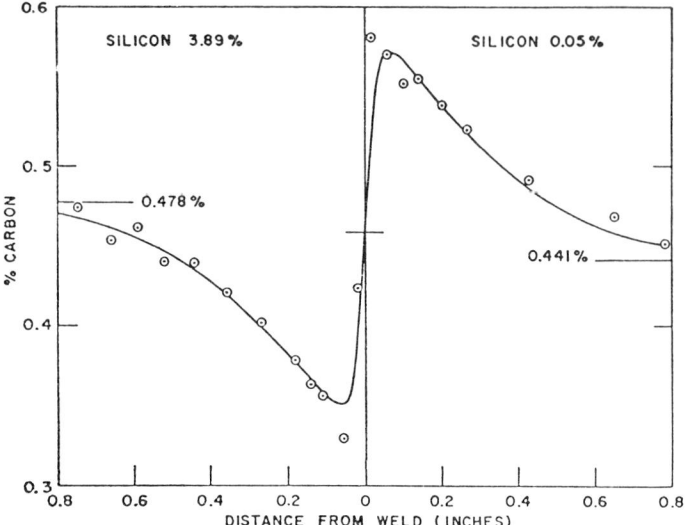

Fig. 1   Darken's up-hill diffusion experiment explained by Kirkaldy's ternary diffusion calculations. From Kirkaldy (1957).

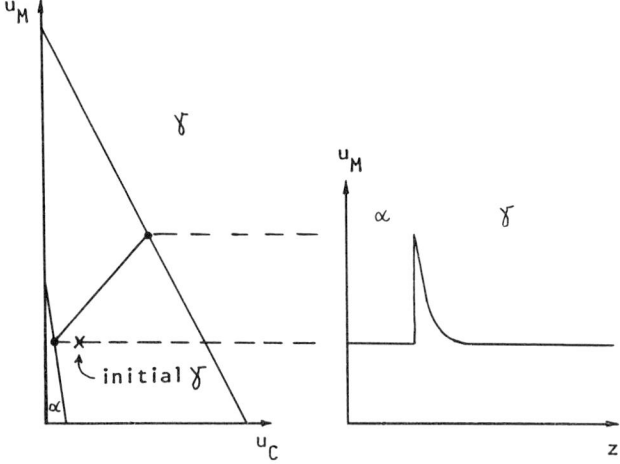

Fig. 2   Relation between the position of the alloy composition in the Fe-C-M phase diagram and the alloy concentration profile. The profile shows a very thin spike if α forms under rapid, no-partition growth.

$$v_M = \frac{D_M^\gamma}{\ell^\alpha}\left(\frac{\Delta u_M^\gamma}{u_M^{\gamma o}-u_M^\alpha}\right)^2 \tag{18}$$

$$v_C = \frac{D_C^\gamma}{\ell^\alpha}\left(\frac{\Delta u_C^\gamma}{u_C^{\gamma o}-u_C^\alpha}\right)^2 \tag{19}$$

These equations apply to the planar precipitation of α from γ. The solution is obtained by finding the appropriate tie-line in the isothermal phase diagram. The tie-line may be regarded as the second unknown, in addition to the growth rate. Kirkaldy and his collaborators found two limiting cases (Purdy, Weichert and Kirkaldy, 1964). The solution for a rapid reaction, controlled by diffusion of carbon is illustrated in Fig. 2. In that solution v_M is made to give a high growth rate by choosing a tie-line for which u_M^α is very nearly equal to the initial alloy content, $u_M^{\gamma 0}$. The denominator in eq. 18 can thus be made as small as necessary. The tie-line can be used to construct the concentration profile for the alloying element, as shown to the right in Fig. 2. A thin spike (positive or negative) of the alloying element is thus obtained in front of the growing α phase.

The other limiting case describes a slow reaction controlled by diffusion of the alloying element. See Fig. 3. In that solution v_C is made to give a low growth rate by choosing a tie-line for which u_C^γ is very nearly equal to the initial carbon content, $u_C^{\gamma 0}$. Their difference, which is denoted by Δu_C^γ in eq. 19, can thus be made as small as necessary. The construction based upon this tie-line is shown to the right in Fig. 3. Here a thick pile-up of the alloying element is found.

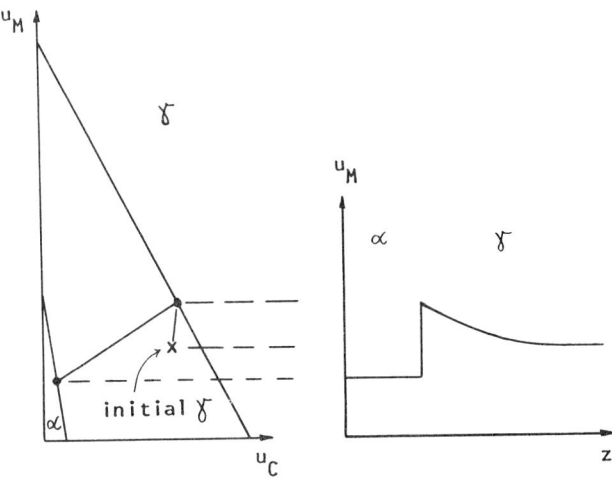

Fig. 3 Relation between the position of the alloy composition in the Fe-C-M phase diagram and the alloy concentration profile. The profile shows a thick pile-up if α forms under slow, partition growth.

Thanks to the fact that D_C is many orders of magnitude larger than D_M in austenite, most of the cases of precipitation from austenite fall very close to one or the other of the two limiting solutions and it was pointed out by Purdy (Purdy, Weichert and Kirkaldy, 1964) that one can draw a critical line in the phase diagrams, dividing the two-phase region in two portions. See Fig. 4. If the initial austenite falls below the line then a rapid reaction would occur and ferrite would form without partitioning of the alloying element between ferrite and austenite. If it falls above the line the slow reaction would occur and the alloying element would partition between ferrite and austenite. Intermediate cases would only occur very close to the critical line.

By taking into account two spikes, one in front of ferrite and one in front of cementite, Kirkaldy (1958b) was able to explain the strong effect some alloying elements have on the growth of pearlite. Later on, that explanation was greatly simplified by the application of two critical lines, one in the $\gamma+\alpha$ two-phase

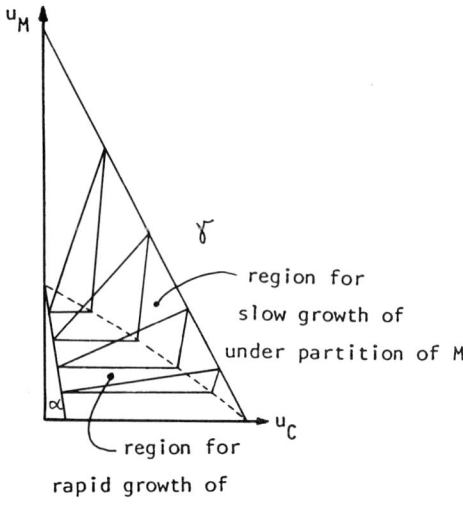

Fig. 4 Construction to find the critical line separating regions for slow growth of α under partition of the alloy element and rapid growth of α under no partition of the alloying element. The steep lines are isoactivity lines for carbon in γ.

field, the other in the γ+cementite two-phase field (Puls and Kirkaldy, 1972).

In order to simplify the discussion, approximations were here introduced in deriving eq. 18. In Kirkaldy's expression for the growth rate the slow growth rate solution is not obtained when u_C^γ has the same value at the interface, i.e. according to the tie-line, and in the initial alloy but when the carbon activity has the same value. As a consequence, isoactivity lines for carbon rather than vertical lines should be used in the constructions. This was actually done in Fig. 14.

It is interesting to note that Kirkaldy was primarily interested in the mathematics of diffusion and his main goal was reached when he had found the analytical solution and shown that it applies not only to diffusion in a single phase but also to a phase transformation. He found the spike when he studied the limiting cases and without having any previous knowledge about partition and no-partition. He seems to have felt that he had fulfilled his obligations when he in 1964 (Purdy, Weichert and Kirkaldy, 1964) was able to show that experimental data on the rapid growth of ferrite from austenite agreed very well with his predictions. He did not present the same kind of test for the slow reaction until 1972 (Gilmour, Purdy and Kirkaldy) and for the pearlite transformation until 1979 (Sharma, Purdy and Kirkaldy). In Fig. 5 we reproduce a diagram from the slow growth of ferrite. The diagram is reported to show a 3 μm thick ferrite layer surrounded by austenite on both sides. However, from the concentration profile it seems that the right-hand part of the diagram is actually ferrite formed under no-partition conditions, i.e. by a rapid reaction. This diagram may be a beautiful example of both reactions. Other examples will be discussed in the next section.

In this connection it should be mentioned that Darken also arrived at the prediction of a spike and attempted to apply it to the case of pearlite (Darken and Fisher, 1962). However, his discussion was not very quantitative and he did not pursue this field of research further.

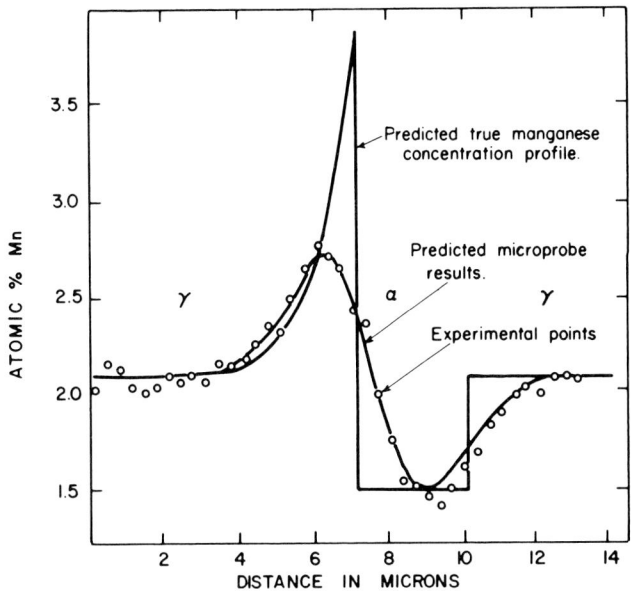

Fig. 5 Observed and predicted alloy concentration profile in
α formed from γ by rapid growth (down to 10 microns)
and slow growth (from 10 to 7 microns). α in the right
part mistakenly identified as γ. From Gilmour, Purdy
and Kirkaldy (1972).

THE THERMODYNAMIC APPROACH

In Stockholm Hultgren studied various reactions in alloyed steels with careful metallography. For example, in specimens carburized in the austenitic condition he found a surface layer containing cementite if the steel was alloyed with chromium but not if it was alloyed with nickel (Hultgren and Hägglund, 1948). He explained the absence of cementite in the latter case by referring to the isothermal Fe-C-Ni phase diagram, Fig. 6. On carburization one would expect that the surface of a steel, which is originally in B_o, will move along the arrow towards the C corner. At B_1 the two-phase filed would be entered and the remaining austenite in the surface layer would thus move to point B_2 or B_3. However, B_3 has a lower carbon content than B_2 which in turn has a lower carbon content than B_1. Hultgren suggested that it should not be possible for carbon to diffuse from a surface layer with B_3 and to an inner layer with B_2 and further into a layer with B_1. He concluded that further carburization would thus stop and the cementite would dissolve again if it were ever formed.

As a chemical engineer one of the present authors took Hultgren's explanation as a challenge of the common understanding of how thermodynamics works. By considering the slopes of the tie-lines in the two-phase field γ+cementite (given as $(Fe,Ni)_3C$ in Fig. 6) he was able to show that the carbon activity along the γ phase boundary increases with the nickel content in spite of the fact that the carbon content decreases (Hillert, 1952). Provided that the carbon diffusion is driven by the gradient of carbon activity rather than carbon content, it should thus be possible to obtain the same reaction in the Fe-C-Ni system as in the Fe-C-Cr system provided that the carbon activity is sufficiently high. Together with Hultgren he was able to prove this experimentally (Hillert and Hultgren, 1953).

The idea that carbon diffusion is governed by activity differences rather than con-

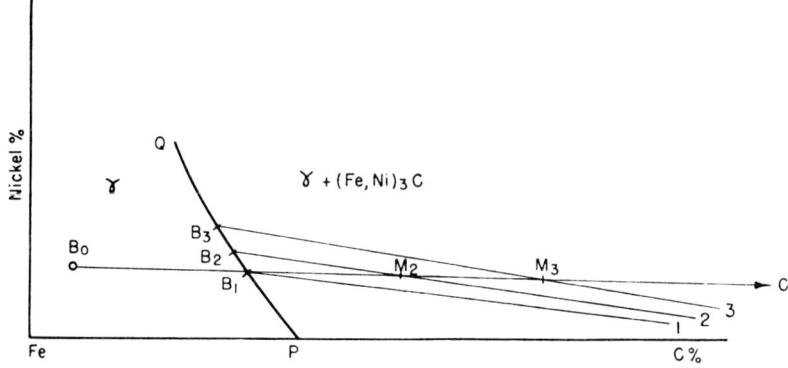

Fig. 6 Fe-C-Ni phase diagram used to illustrate that the austenite in a surface layer containing cementite has less carbon than the next layer. From Hultgren and Hägglund (1948).

centration differences was advanced by Darken in 1942. Evidently, this was a new idea to established physical metallurgists and Darken's thesis was opposed by Mehl and Seltz (1942). Hultgren's explanation was thus in line with general thinking in his field. Finally, Darken decided to prove his thesis experimentally by his famous up-hill diffusion experiment in 1949.

By inspiration from Darken's up-hill diffusion experiment the work in Stockholm continued by replacing $D_C dc_C$ in Fick's law for diffusion by $D_C^a da_C$ (Hillert and Sharp, 1953) and treating the "activity diffusion constant", D_C^a, as independent of the alloy content. The application to a phase transformation controlled by the rate of carbon diffusion in austenite between two phase boundaries, γ/α and γ/β, is demonstrated in Fig. 7. In the binary case the rate would be proportional to

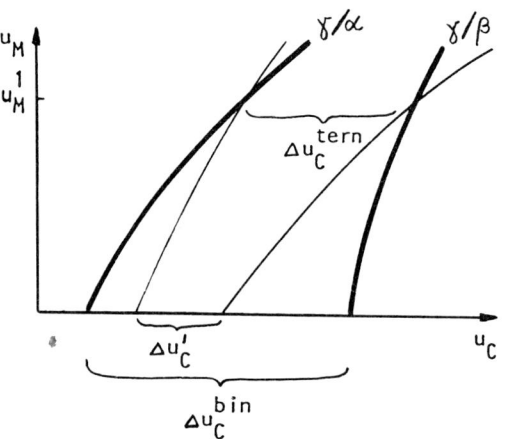

Fig. 7 Fe-C-M phase diagram with phase boundaries of γ in equilibrium with α and β, respectively. Δu_C^{bin} and Δu_C^{tern} represent the driving force for a three-phase reaction in a binary and in a ternary case. The thin lines are isoactivity lines of carbon in austenite.

$D_C^{bin} \cdot \Delta u_C^{bin}$. In the ternary case at the alloy content of u_M^1 the rate would be proportional to $D_C^{tern} \cdot \Delta u_C^{tern}$. However, in order to evaluate D_C^{tern} and Δu_C^{tern} it would be necessary to have detailed information on the slopes of the phase boundaries and also of how the carbon activity changes with composition in the austenite phase field. A better method was found by using the isoactivity lines in austenite going through the two points on the phase boundaries. If D_C^a is really independent of the alloy content then one should get the same rate of diffusion between the two isoactivity lines at any level of u_M and even down at the binary Fe-C side. $D_C^{tern} \cdot \Delta u_C^{tern}$ would then be qual to $D_C^{binM} \cdot \Delta u_C^1$ and that could even be evaluated from the value for the binary case by simply multiplying with the ratio of the driving activity differences in the two cases because

$$u_C^{bin} \cdot \frac{a_C^{tern}}{a_C^{bin}} = u_C^1 \tag{20}$$

according to Henry's law. It would thus be sufficient to start from the difference in carbon activity in the binary system and evaluate the change in carbon activity along each one of the phase boundaries. It should not even be necessary that the two points in the alloyed case fall on the same level of u_M. As a first step in the further development, the carbon activity of two-phase equilibria in the Fe-C system were studied (Hillert, 1954).

When looking for a method of evaluating the change of carbon activity in a two-phase field it was finally found that an answer would be given by applying the Gibbs-Duhem equation to each one of the two coexisting phases (Hillert, 1955). After some approximations an equation was found which can be written in the following convenient form

$$\ln(a_C^{tern}/a_C^{bin}) = \frac{1-K_M^{\alpha/\beta}}{u_C^\alpha - u_C^\beta} \cdot u_M^\beta \tag{21}$$

α and β can be identified with any one of the two phases. It may be mentioned that Kirkaldy and his collaborators first evaluated the change in carbon activity by comparing the slopes of isoactivity lines and phase boundaries, a method which requires more information and may be subject to a larger uncertainty. Eq. 21 only requires information on the distribution constant of the alloying element between the two phases $K_M^{\alpha/\beta}$.

The combination of eqs. 19 and 20 proved to be a very simple and powerful tool for explaining various phenomena known from alloyed steels and cast irons (Hillert, 1971). An example is shown in Fig. 8 which demonstrates that 0.5% Mo has a very slight influence on the rate of growth of graphite in a steel in ferritic condition but 0.5% Cr has a very strong effect from the A1 temperature and down to about 600°C (Hillert, 1957). The left-hand curves were calculated for ordinary carbon diffusion and the right-hand curves were calculated under the assumption that space must be provided for the growing graphite by diffusion of iron, a mechanism making use of Nabarro-Herring creep. The results presented in Fig. 8 give an explanation of the practical experience that molybdenum alloyed steels have a strong tendency to graphitize after long-time service in power plants, a tendency which can be reduced considerably by the addition of chromium.

PARA- AND ORTHOEQUILIBRIUM

Hultgren also studied the decomposition of austenite on cooling, finding that alloying elements sometimes partition between two growing phases or a growing phase and the parent phase, sometimes they do not partition (Hultgren, 1947, 1951, 1953). He was a strong believer in the assumption of local equilibrium at the

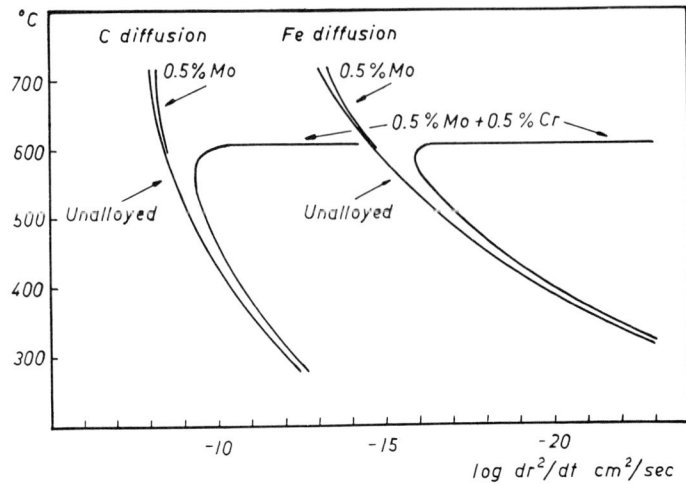

Fig. 8   Theoretical calculation of the effect of 0.5% Mo and 0.5% Cr on the rate of growth of graphite in steel. The left-hand set of curves is for rate control by carbon diffusion and the right-hand set of curves is for rate control by stress induced diffusion of iron. From Hillert (1957).

phase interfaces during transformations and already in 1920 he had proposed that the growth conditions in Fe-C alloys could be obtained by extrapolation of the phase boundaries below the eutectoid point (the so-called Hultgren extrapolation) (Hultgren, 1920). He thus suggested that partition would occur as a result of full equilibrium at the phase interfaces but for no-partition he was led to suggest that there would be equilibrium for carbon only. He suggested that, due to the much lower diffusivity of the alloying elements they would not be able to move at all. The growing phase would thus inherit the alloy content of the parent phase. That kind of local equilibrium he called para-equilibrium and he suggested that full equilibrium in this context should be called orthoequilibrium.

Hultgren even penetrated the case of paraequilibrium in enough detail to suggest that the phase boundaries for paraequilibrium should fall inside the ordinary two-phase fields in the diagram. One of the present authors (M.H.) also took this suggestion as a challenge to his understanding of thermodynamics and by the application of isoactivity lines for carbon he was able to prove that Hultgren was right (Hillert 1952, 1955). In working with the proof he realized that, due to the build-up of a very local change in alloy content in front of the growing phase, it may grow with the same alloy content as the initial alloy content even if there is full local equilibrium (orthequilibrium) at the interface (Hillert, 1953). He called this false paraequilibrium. It is interesting to note that he thus arrived at the prediction of the thin spike without being confronted with the question how two growth equations can give the same result. He did not consider any problem of ternary diffusion mathematics but he applied binary diffusion to the spike and found that its form can be described by a function $\exp-(z-vt)v/D_M$. The width of the spike may thus be estimated as D_M/v.

Using the concept of false paraequilibrium the author was able to predict the rate of various rapid reactions, for instance the rate of growth of a ferritic surface layer during decarburization of a steel in austenitic condition below 912°C (Hillert, 1971). He also applied the method to predict the effect of alloying elements on the growth of pearlite (Hillert, 1953). The construction is reproduced in Fig. 9 and it shows that the rate should be decreased by a factor obtained as the ratio of the difference in carbon activity between γ_α^e and γ_c^e and the difference between γ_α^o and γ_c^o. The index c here stands for cementite and the superscript e represents the level of the alloy content in the initial γ. It is evident that the driving force for carbon diffusion would vanish at a slightly higher alloy content and then the growth can only proceed with a very much slower rate, controlled by alloy diffusion.

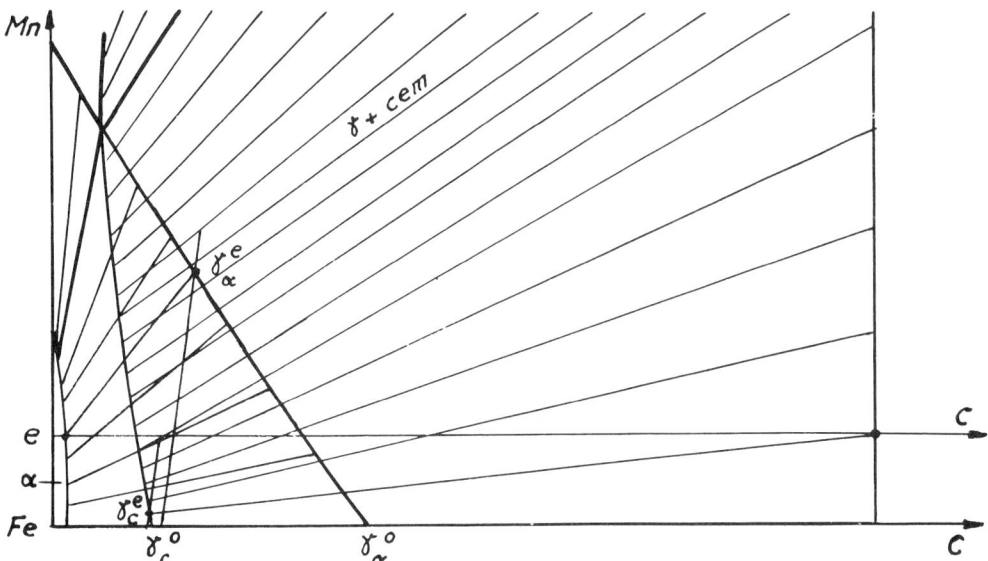

Fig. 9 Local equilibrium construction for finding the effect of a manganese addition on the driving force for pearlite. Without manganese one obtains the difference in carbon activity between γ_α^o and γ_c^o. With a manganese addition to the level e one obtains the difference in carbon activity between γ_α^e and γ_c^e. The thin, parallel lines are isoactivty lines for carbon in austenite, going through these points. From Hillert (1953).

A NEW TYPE OF IMPINGEMENT

For transformations in alloyed steels there are two types of impingement. One is similar to impingement in binary alloys but the other is different because it occurs with a gradual shift of the tie-line representing the local equilibrium at the phase interface. The first type is illustrated by Fig. 10. With the initial alloy composition represented by the cross in Fig. 10a there will be a rapid reaction and no-partition ferrite of a composition represented by a circle will form. That point defines the operating tie-line, the other end-point of which gives the austenite at the top of the spike. The spike is very thin and falls in a region small enough to have an almost constant carbon activity. Points within the spike are thus represented by points on the isoactivity line going through the high end-point on the tie-line. In front of the spike the alloy composition is the same as the initial one but the carbon content will be higher. That is why carbon diffuses away from the growing ferrite and towards the interior of the austenite. All points on that diffusion path fall on the horizontal line but with decreasing carbon contents until the cross is reached. All points in the austenite are thus described by points along the thick line on Fig. 10a.

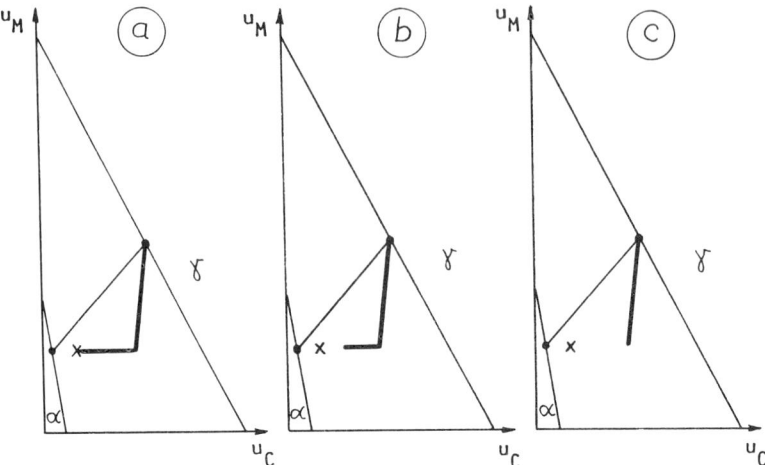

Fig. 10 Fe-C-M phase diagram illustrating the rapid growth of α from γ, controlled by carbon diffusion. The initial austenite is represented with a cross, the local equilibrium at the interface is represented by a tie-line and the variation of composition in γ is represented by the thick line. Its vertical part holds for the spike and its horizontal part for the main portion of the austenite. (a) illustrates all situations before impingement has occurred. (b) illustrates that no austenite has the initial carbon content when impingement has started. (c) illustrates the end of the carbon controlled reaction. The main portion of the austenite has a uniform composition represented by the lowest point on the thick line.

The start of impingement may be defined as the time when there is no longer any austenite represented by the cross. The carbon increase has now reached the interior of the austenite. The thick line in Fig. 10b illustrates such a case. The rate of reaction is still controlled by carbon diffusion but the driving force is gradually diminishing. Finally, the driving force for carbon diffusion approaches very low values. All of the austenite is now represented by points on the vertical part. See the thick line in Fig. 10c. Now the further reaction has to wait for the slow diffusion of the alloy element. By a slow reaction the amount of ferrite will continue to grow and the remaining austenite will receive more carbon. That increase goes to all the remaining austenite and the carbon activity of the system, which is now uniform, will gradually increase. The operating tie-line will thus shift to lower positions. Fig. 11a demonstrates such a case. The top of the spike has now decreased and so has the composition of the new ferrite. Diffusion of the alloy element will now take place in the parent austenite as well as in the precipitated ferrite. It seems that this stage should not be regarded as a type of impingement because the pile-ups of carbon have already impinged upon each other and the pile-ups of the alloy element in the parent austenite have not yet impinged. When that happens there will no longer be any austenite on the horizontal line through the initial composition, the cross. During this stage of impingement there will be a further shift of the tie-line and the horizontal line representing the remaining austenite grows shorter and shorter as illustrated in Fig. 11b. The system finally approaches the equilibrium tie-line which goes through the cross representing the initial composition, Fig. 11c.

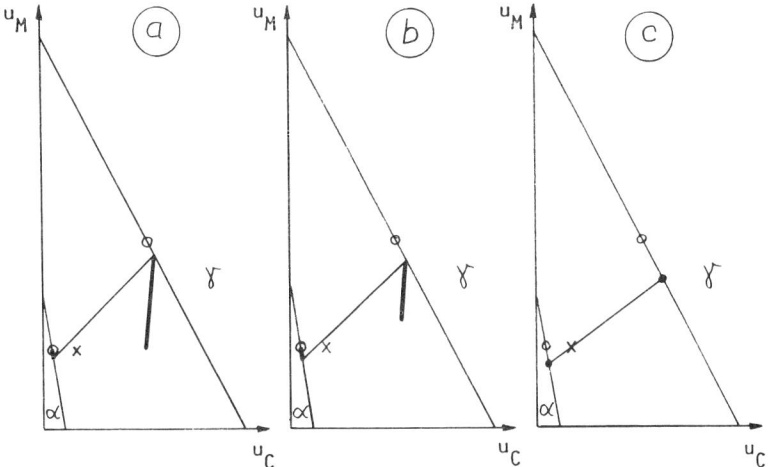

Fig. 11 Fe-C-M phase diagram illustrates the slow growth of α from γ, controlled by diffusion of the alloy element. The operating tie-line is gradually moving downwards with time. (a) illustrates a stage where the spike is growing thicker. (b) illustrates a stage where neighboring spikes have impinged and all the reamining austenite has a higher alloy content than initially. (c) illustrates the final state of equilibrium. The tie-line now goes through the cross representing the initial composition of the austenite, i.e. the average composition of the steel.

In order to evaluate the growth rate during the last two stages it is necessary first to determine the carbon activity from the amount of new phase formed and the corresponding change in carbon content of the remaining austenite. Then the operating tie-line can be identfied and it yields the differences in alloy activity which drives the further reaction. In the general case it is not possible to find an analytical solution to these two stages. Instead very simple models were applied when a number of practical cases were considered. In particular, the alloy diffusion was estimated as if it were a binary case.

It is interesting to note that there is a close connection between the carbon activity and the operating tie-line. It is thus convenient to test the applicability of the local equilibrium assumption by testing that connection experimentally. Fig. 12 shows a series of measurements of the carbon activity during annealing at 677°C of an austenitic stainless steel of the 18% Cr-8% Ni type with an impurity of 0.1% C (Stawström and Hillert, 1969). Chromium carbide precipitates along the grain boundaries and will be surrounded by a negative spike of chromium, i.e. a chromium depleted zone. If the carbon activity is high enough, the operating tie-line will fall so far down that the depleted zone obtains a chromium content too low for corrosion resistance. The material is thus sensitized. If the annealing continues more chromium carbide will form and there will be a gradual decrease of the carbon activity. The operating tie-line will move up and increase the chromium content in the depleted zone. Finally, the chromium concentration will rise above a critical value and the corrosion resistance will be restored in spite of the fact that the zone is thicker and represents a larger total loss of chromium. Fig. 12 gives a comparison between the calculated and measured decrease of

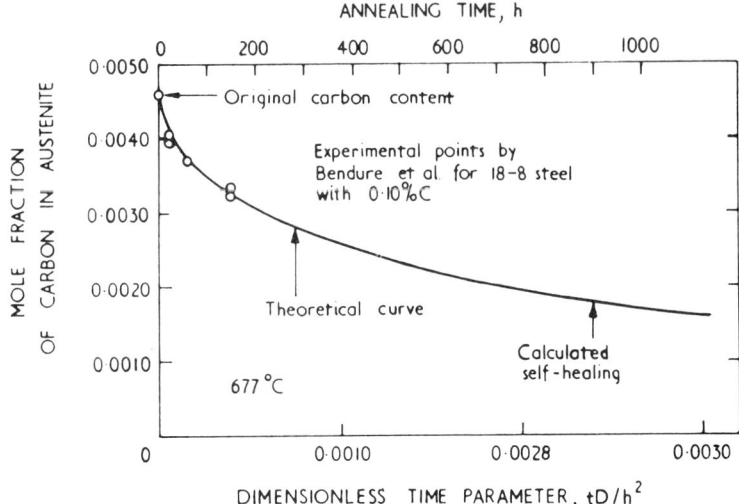

Fig. 12 Decrease of carbon activity during precipitation of chromium carbide in an 18-8 austenitic stainless steel, studied by the decrease in the carbon content of the austenite. An arrow indicates when the carbon activity has decreased to such a low value that the chromium content of austenite is increased enough to restore the corrosion resistance. From Stawström and Hillert (1969).

the carbon activity and an arrow indicates when the chromium content is expected to have risen above the critical level and restored the corrosion resistence (so-called self-healing).

The resulting carbon content of the austenite is of primary importance for the austenitization of a high-carbon steel before hardening. Fig. 13 shows how the carbon content may increase with time due to three rate controlling reactions, diffusion of carbon in austenite, diffusion of chromium in cementite and diffusion of chromium in austenite (Hillert, Nilsson and Törndahl, 1971). The agreement between experiments and calculations is reasonably good except for the first, very rapid reaction.

NUMERICAL SOLUTION OF DIFFUSION

The more or less ambitious analytical solutions for diffusion controlled transformations which have been discussed so far have limited applicability and are approximate. In this connection it may be further emphasized that the analytical solutions presented by Kirkaldy are valid only for constant values of D_{12} and D_{21} whereas eqs. 16 and 17 show that they are approximately proportional to the carbon and alloy content, respectively. Kirkaldy (Kirkaldy, Lane and Mason, 1963) has attempted an accurate solution for this case and obtained an integral-differential equation which had to be solved numerically by iteration. It may thus seem like a natural approach to try to solve the whole problem of diffusion by numerical integration of Fick's law. This seems even more natural today when thermodynamic properties are becoming available through databases on computers. One such attempt has been made by Ågren in Stockholm. He started from the most basic equations, eqs. 7 and 8, with $L_{12}=L_{21}=0$ (Ågren, 1982a, 1982b). By assuming that the diffusion takes place by exchange of carbon and vacancies on the interstitial sublattice and

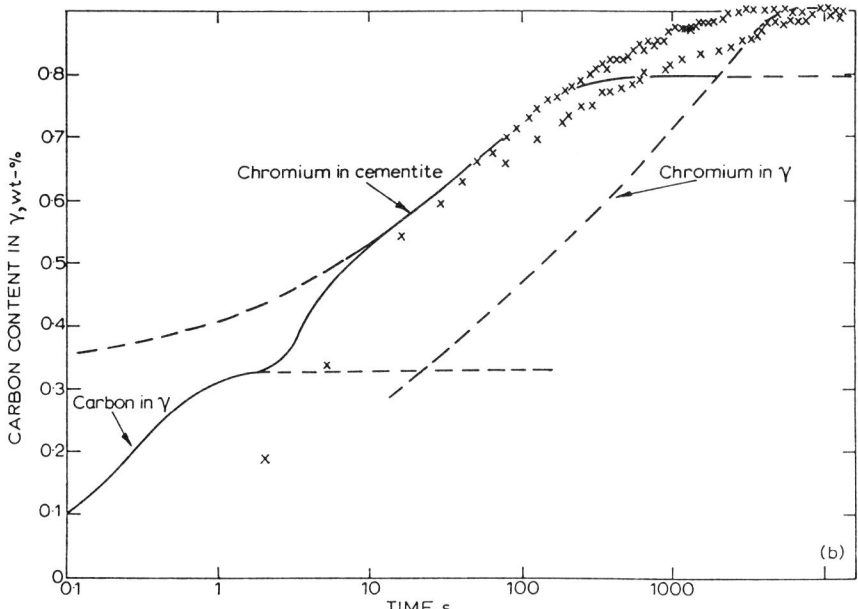

Fig. 13 Comparison between experimental values for the dissolution of cementite in austenite in a chromium steel at 910°C and calculated values for three stages of the reaction. From Hillert, Nilsson and Törndahl (1971).

of the alloy element and iron on the main sublattice he obtained

$$J_1 = -L_{11} \frac{\partial \mu_1}{\partial z} = -u_1(1-u_1)M_{1Va} \frac{\partial \mu_1}{\partial z} \qquad (22)$$

$$J_2 = -L_{22} \frac{\partial(\mu_2-\mu_3)}{\partial z} = -u_2 u_3 M_{23} \frac{\partial(\mu_2-\mu_3)}{\partial z} \qquad (23)$$

M_{1Va} and M_{23} are here mobilities and Ågren assumed that they could be approximated as independent of the alloy content and the carbon content, respectively. Their values could thus be obtained from binary information:

$$J_1 V_o = -D_{11}^{bin} \frac{\partial u_1}{\partial z} \qquad (24)$$

$$J_2 V_o = -D_{22}^{bin} \frac{\partial u_2}{\partial z} \qquad (25)$$

$$M_{1Va} = D_{11}^{bin}/u_1(1-u_1)V_o (\frac{\partial \mu_1}{\partial u_1})^{bin} \qquad (26)$$

$$M_{22} = D_{22}^{bin}/u_2 u_3 V_o (\frac{\partial(\mu_2-\mu_3)}{\partial u_2})^{bin} \qquad (27)$$

Before presenting some results obtained with this technique it is worth comparing the assumption of constant mobilities with the previous assumption of constant activity diffusion coefficient, D_C^a. For the ternary case at a constant u_M value Ågren obtained

$$D_{11}^{tern} = M_{1Va} u_1 (1-u_1) \left(\frac{\partial \mu_1}{\partial u_1}\right)^{tern} \cong M_{1Va} RT \cong D_{11}^{bin} \tag{28}$$

because

$$\frac{\partial \mu_1}{\partial u_1} \cong \frac{RT}{u_1} \tag{29}$$

for dilute solutions, Ågren thus suggests that calculations based upon isoactivity lines for carbon should be made by evaluating the difference in carbon content between the isoactivity lines at the actual level of alloy content and multiplying with the binary diffusion constant D_C^{bin}.

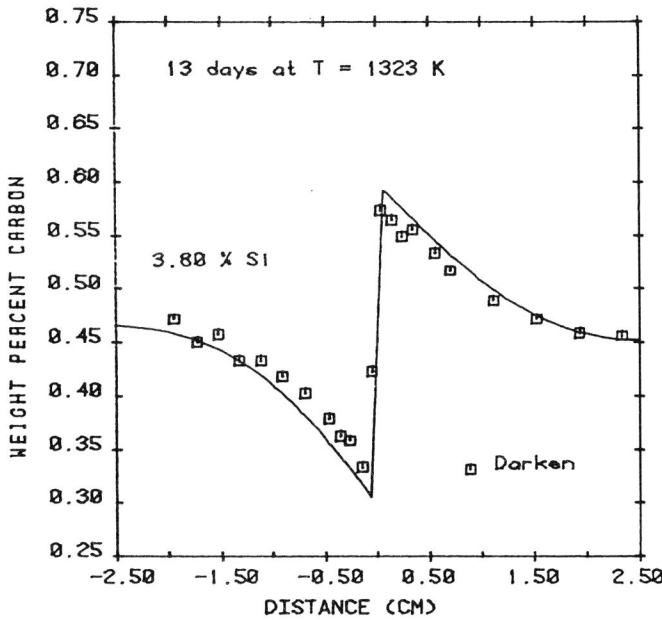

Fig. 14 Computer simulation of Darken's up-hill diffusion experiment. Compare Fig. 1. From Ågren (1982c).

Fig. 14 shows the result of Ågren's simulation of one of Darken's up-hill diffusion experiments (Ågren, 1982c). The agreement is very satisfactory considering the fact that no parameters have been adjusted in order to fit the data. In particular, the thermodynamic data come from a database which is derived from several sources of information on the Fe-C-Si system. Naturally, some parameter in that database should be modified if one would trust Darken's information more than the other information. In this connection it may be mentioned that the steep but somewhat gradual variation in the center of the system according to Kirkaldy's calculation, Fig. 1, was probably due to his choice of a relatively high value for the diffusion coefficient of silicon.

Fig. 15 shows Ågren's simulation of the case of dissolution of cementite in austenite (Ågren and Vassilev, 1984) discussed in connection with Fig. 13. The agreement is satisfactory except for the first, very rapid stage. It is particularly interesting to see that the program could handle the difficult transition between rapid reaction by carbon diffusion and slow reaction by chromium diffusion.

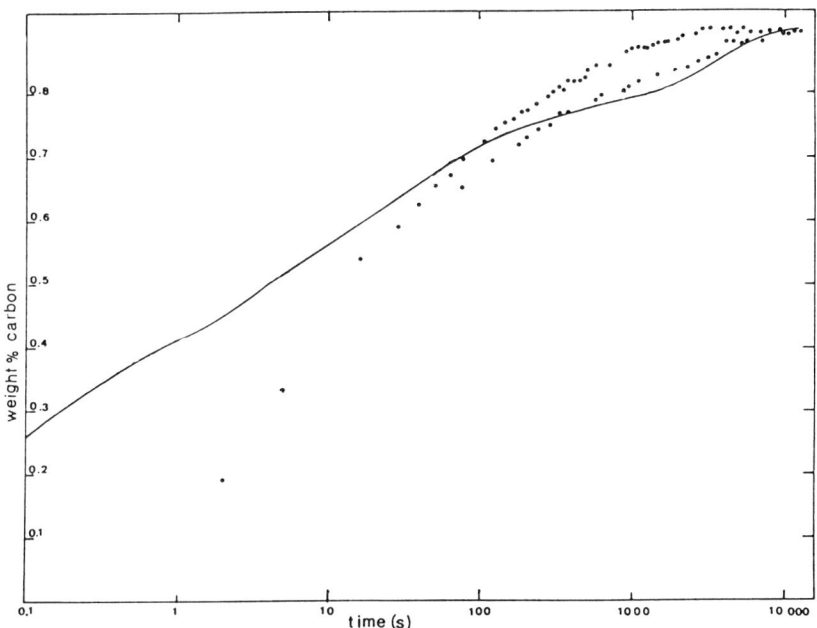

Fig. 15 Computer simulation of the dissolution of cementite in austenite in a chromium steel at 910°C (full line). All the dashed lines are from Fig. 13. From Ågren and Vassilev (1984).

The program is also capable of simulating diffusion controlled transformations during a gradual change of temperature. As an example, Fig. 16 shows how the average carbon content of ferrite increases with temperature due to the gradual dissolution of cementite during continuous heating (Ushioda and others, 1986). The calculations for manganese were performed both under the assumption of a homogeneous manganese distribution in the initial state and a distribution corresponding to equilibrium at 700°C.

Finally, Fig. 17 shows a case where the reaction is partly reversed during the last stage. This case concerns the formation of austenite in so-called intercritical annealing of a DP steel (Ågren, 1982b).

CONCLUDING REMARKS

The general impression from a considerable number of calculations with analytical or numerical technique is that the local equilibrium assumption holds fairly well even at relatively low temperatures where one could expect difficulties due to the limited mobility of the alloy elements. A reasonable way of testing whether local equilibrium can be expected would be to evaluate the width of the spike from D_M/v and compare with atomic dimensions. Such a test was made for the effect of manganese on the formation of pearlite (Hillert, 1982). The surprising result was that most of the experimental data gave a spike much thinner than atomic dimensions and

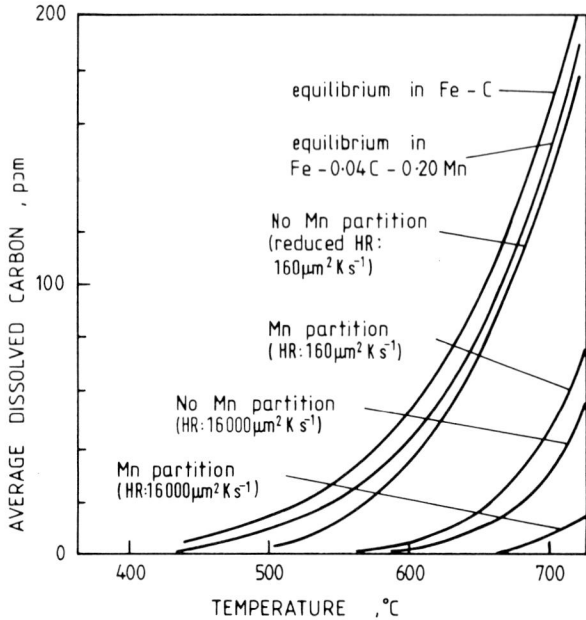

Fig. 16 Computer simulation of the dissolution of cementite in ferrite during continuous heating (HR is the normalized heating rate) in a steel with 0.04% C and 0.2% Mn. Two initial conditions were considered, one with a homogeneous distribution of manganese and the other with a partition of manganese between cementite and ferrite according to equilibrium conditions at 700°C. From Ushioda and others (1986).

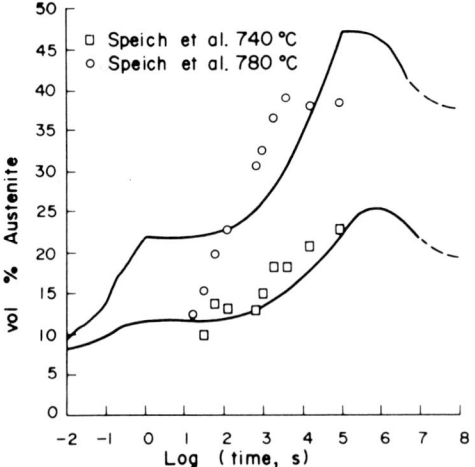

Fig. 17 Computer simulation of the formation of austenite in a manganese DP steel during annealing at 740 or 780°C. From Ågren (1982b).

still it seemed possible to explain the kinetic results from the local equilibrium assumption. This may be explained by processes going on inside the moving phase interface, leading to effects related to the so-called solute drag effect. Anyway, it seems that it should be possible to apply the local equilibrium assumption in order to explain and predict a large number of transformations in alloyed steels.

REFERENCES

Darken, L.S. (1942), Trans AIME 150, 157.
Darken, L.S. (1949), Trans AIME 180, 430.
Darken, L.S. (1951), Atom Movements, ASM, Cleveland, 1-25.
Darken, L.S. and R.M. Fisher (1962), in Decomposition of Austenite by Diffusional Processes (Eds. Zackay and Aaronson) AIME 249-288.
Gilmour, J.B., G.R. Purdy and J.S. Kirkaldy (1972), Met. Trans. 3, 3213.
Hillert, M. (1952), Jernkont. Ann 136, 25.
Hillert, M. (1953), Paraequilibrium (Internal Report) Swedish Inst. Metal Res.
Hillert, M. and A. Hultgren (1953), Jernkont. Ann 137, 217.
Hillert, M. and R.D. Sharp (1953), Jernkont. Ann 137, 785.
Hillert, M. (1954), Acta Met. 2, 11.
Hillert, M. (1955), Acta Met. 3, 34.
Hillert, M. (1957), Jernkont. Ann 141, 67.
Hillert, M. (1971), Suppl. Trans Iron Steel Inst. Japan 11, 1153.
Hillert, M., K. Nilsson and L.-E. Törndahl (1971), J. Iron Steel Inst. 209, 49.
Hillert, M. (1982), in Solid State Phase Transformations (Eds. H.I. Aaronson, D.E. Laughlin, R.F. Sekerka and C.M. Wayman) AIME, 789-806.
Hultgren, A. (1920), Metallographic Study on Tungsten Steels, John Wiley and Sons, New York, 30
Hultgren, A. (1947), Trans ASM 39, 915.
Hultgren, A. and E. Hägglund (1948), Jernkont. Ann 132, 1.
Hultgren, A. (1951), Jernkont. Ann 135, 403.
Hultgren, A. (1953), Kgl. Vetenskapsakad. Handl. Ser. 4, Bd, 4, No. 3.
Kirkaldy, J.S. (1957), Can. J. Phys. 35, 435.
Kirkaldy, J.S. (1958a) Can. J. Phys. 36, 899.
Kirkaldy, J.S. (1958b) Can. J. Phys. 36, 907.
Kirkaldy, J.S. (1958c) Can. J. Phys. 36, 917.
Kirkaldy, J.S. (1959), Can. J. Phys. 37, 30.
Kirkaldy, J.S. and G.R. Purdy (1962), Can. J. Phys. 40, 208.
Kirkaldy, J.S., D. Weichert and Zia-Ul-Haq (1963), Can. J. Phys. 41, 2166.
Kirkaldy, J.S., J.E. Lane and G.R. Mason (1963), Can. J. Phys. 41, 2174.
Kirkaldy, J.S. and J.E. Lane (1966), Can. J. Phys. 44, 2059.
Kirkaldy, J.S. and G.R. Purdy (1969), Can. J. Phys. 47, 865.
Lane, J.E. and J.S. Kirkaldy (1964), Can. J. Phys. 42, 1643.
Mehl, R.F. and H. Seltz (1942), Trans AIME 150, 169.
Onsager. L. (1931), Phys. Rev. 37, 405.
Onsager, L. (1932), Phys. Rev. 38, 2265.
Onsager, L. (1945-6), Ann. N.Y. Acad. Sci. 46, 241.
Puls, M.P. and J.S. Kirkaldy (1972), Met. Trans. 3, 2777.
Purdy, G.R., D.H. Weichert and J.S. Kirkaldy (1964), Trans AIME 230, 1025.
Sharma, R.C., G.R. Purdy and J.S. Kirkaldy (1979), Met. Trans 10A, 1129.
Stawström, C. and M. Hillert (1969), J. Iron Steel Inst. 207, 77.
Ushioda, K., W.B. Hurchinson, J. Ågren and U. von Schlippenbach (1986), Materials Science Techn. 2, 807.
Wagner, C. (1952), Thermodynamics of Alloys, Addison-Wesley, Cambridge.
Ågren. J. (1982a), J. Phys. Chem. solids 43, 421.
Ågren. J. (1982b), Acta Met. 30, 841.
Ågren, J. (1982c), Scand. J. Metall. 11, 3.
Ågren, J. and G.P. Vassilev (1984), Materials Sci. Eng. 64, 95.

Influence of the Chemistry and Structure of Austenite:Ferrite Boundaries upon Growth Kinetics and Composition of Grain Boundary Ferrite Allotriomorphs in Fe-C-X Alloys

H. I. AARONSON*, M. ENOMOTO** AND W. T. REYNOLDS, JR.*

*Department of Metallurgical Engineering and Materials Science,
Carnegie Mellon University, Pittsburgh, PA 15213, USA
**National Research Institute for Metals, Tsukuba Laboratories, 1-2-1, Sengen,
Sakura-Mura, Niihari-Gun, Ibaraki 305, Japan

ABSTRACT

A review is presented of the growth kinetics of grain boundary ferrite allotriomorphs in Fe-C and Fe-C-X alloys, of the models which have been developed to explain these kinetics and of the deficiences which repeated comparisons with experiment have brought to light in these models. These models, principally that of "local equilibrium" and of paraequilibrium, also encounter problems in explaining experimental data on the composition of ferrite allotriomorphs as determined earlier by electron probe analysis and more recently by STEM. Addition of the rejector plate mechanism to the "local equilibrium" model is useful in explaining growth kinetics and composition of partitioned ferrite in Fe-C-Mn and Fe-C-Ni alloys, but the rejector plate component seems on firmer ground than that of "local equilibrium". Effects of carbide precipitation at austenite:ferrite boundaries and, especially, the partially coherent structure which allotriomorphs now appear to have instead of the incoherent structure originally predicted can explain some of the discrepancies. However, the paraequilibrium model, as modified by a solute drag-like effect, seems presently to be the most promising avenue of approach to the growth kinetics and composition of allotriomorphs formed at lower temperatures. At higher temperature, though, the comparisons with experiment so far reported suggest the need for further theoretical developments.

KEYWORDS

Grain boundary ferrite allotriomorphs, local equilibrium model, paraequilibrium model, rejector plate mechanism, growth kinetics, alloying element partition, partial coherency, disordered boundaries, carbide precipitation at $\alpha{:}\gamma$ boundaries, solute drag-like effect, ledge mechanism.

INTRODUCTION

For many years, the proeutectoid ferrite reaction in steel has been the phase transformation in solid metals most thoroughly investigated by physical metallurgists and their allies in related professions. Despite the immense publicity and research attention which have been lavished upon the so-called "engineered materials" in more recent times, ferrite formation remains, as it has been for more than a century, the solid state reaction of greatest world-wide technological importance. The claim is frequently made by those who support diversion of research support to non-metallic materials, rapid solidification processing and the like that our understanding of phase transformations and other aspects of steel is now so far advanced that little scientific or industrial profit remains to be garnered by continued research upon them. And indeed, in Prof. Hillert's splendid contribution to this symposium (Hillert, 1988) a highly optimistic view is taken of our present understanding of the ferrite reaction in steel. While the present authors agree that substantial progress has been made in this area, they also consider that the summary of evidence presented here suggests that many years of difficult but exciting experimental and theoretical research lie ahead of us before the mechanisms and kinetics of this reaction can be considered to be well understood. Considerably more time is likely to elapse before multi-component commercial steels have been thoroughly analyzed.

For the twin purposes of simplifying this contribution and concentrating it upon the most thoroughly documented aspect of the proeutectoid ferrite reaction, we shall restrict our considerations to the particular morphology of proeutectoid ferrite known as grain boundary allotriomorphs. This morphology, illustrated in Fig. 1, nucleates at and grows preferentially, and more or less smoothly (as observed with low-power optical microscopy) along austenite grain boundaries, probably mainly of the high-angle type (Aaronson, 1956; Aaron and Aaronson, 1971). For many years, allotriomorphs were considered, upon the basis of indirect evidence, to have a predominantly disordered or incoherent interfacial structure, more so than any other ferrite morphology (Smith, 1953; Aaronson, 1962). This circumstance made allotriomorphs the preferred subject for growth kinetics measurements, since the substantial body of theory which has been developed for ferrite growth kinetics in Fe-C-X alloys is based on the assumption that the structure of austenite:ferrite boundaries does not affect their migration. In the penultimate section of this paper, however, this assumption will be challenged, primarily though not solely upon the basis of direct experimental evidence recently secured on grain boundary allotriomorphs in a non-ferrous alloy system. The modifications in the theory of growth kinetics needed in order to accommodate these new considerations will then be discussed.

Theory of Local Equilibrium and of Paraequilibrium Growth Kinetics

The local equilibrium model of ferrite growth kinetics in Fe-C-X alloys was evolved independently by Kirkaldy and his students (Kirkaldy, 1958; Purdy, Weichert and Kirkaldy, 1964) and by Hillert (Hillert, 1953, 1955, 1969) and subsequently further elaborated by Coates (1972, 1973a, 1973b). The thermodynamic basis for this model was stated by Gilmour, Purdy and Kirkaldy (1972):

$$\overline{G}_{i\alpha}^{\alpha\gamma} = \overline{G}_{i\gamma}^{\gamma\alpha} \qquad i = 1,2,3 \tag{1}$$

where $\overline{G}_{i\alpha}^{\alpha\gamma}$ and $\overline{G}_{i\gamma}^{\gamma\alpha}$ are the partial molar free energies of element i in ferrite at the $\alpha/(\alpha+\gamma)$ and in austenite at the $\gamma/(\alpha+\gamma)$ phase boundaries and i = 1 designates Fe, i = 2 refers to C and i = 3 indicates X. The kinetic basis for this model derives from the need for compatible fluxes of C and X in austenite attending the migration of α:γ boundaries. However, since the diffusivity of C in austenite in the temperature region of interest is as much as 10^7-10^9 greater than that of X, the concentration gradient of X in austenite must be greater than that of C by a similar factor. Particularly at the early stages of growth, this would require a diffusion distance of X in austenite less than a lattice parameter. Since such short diffusion path lengths are physically meaningless, another solution must be sought. Fig. 2 (DeHoff, 1983) describes schematically the very ingenious one found by Hillert and Kirkaldy. Isothermal sections are shown at successively lower temperatures through the Fe-rich corner of an Fe-C-X phase diagram in which X is an austenite-stabilizer, such as Mn or Ni. At T_1 (Fig. 2a) the alloy, of composition A, is wholly austenitic. Reduction of the temperature to T_2 (Fig. 2b) places the alloy within the $\alpha+\gamma$ region. The relevant tieline is shown as a diagonal connecting points on the $\alpha/(\alpha+\gamma)$ and $\gamma/(\alpha+\gamma)$ phase boundaries. However, this tieline does not pass through composition A. Instead, A lies upon an iso-activity line for carbon in austenite. Hence only X can diffuse; the problem of the much higher diffusivity of carbon than of X is eliminated by removing the driving force for carbon diffusion. The assumption is made that local equilibrium is at all times maintained. Thus the terminii of the tieline represent the compositions of the austenite and ferrite phases at disordered α:γ boundaries. The difference between the X concentration in austenite at an α:γ boundary and that at A represents the driving force for X diffusion. Reducing the reaction temperature to T_3 is seen in Fig. 2c to increase the driving force for X diffusion but to leave the situation with respect to carbon diffusion unchanged. However, once the reaction temperature is reduced to the point where the alloy composition falls on the dashed line connecting the highest X concentration in ferrite (in the Fe-X system) with the carbon concentration in austenite at the $\gamma/(\alpha+\gamma)$ phase boundary of the Fe-C (actually, Fe - Fe_3C) system, the bulk X concentration in the alloy becomes identical to the concentration of X in ferrite. Hence the supersaturation driving X diffusion in austenite becomes unity and the diffusion distance of X into austenite can become vanishingly small, thereby fully compensating for the much smaller diffusivity of X than of C. With further reductions in reaction temperature, the bulk composition of the alloy leaves the region marked I in Fig. 2 and enters region II. In the latter, point A lies along the line connecting the I:II boundary and the ferrite terminus of the tieline. This line is almost parallel to the carbon concentration axis; it is one of a constant ratio of X to Fe. In region II, the carbon diffusion gradient is again between the phase boundary and bulk carbon concentrations. However, since the bulk concentration no longer lies on the iso-C-activity line, a finite diffusion gradient is now established. Because the X diffusion distance in austenite is now very short, growth kinetics are controlled by the long-range diffusion of C in austenite.

An alternative to the foregoing "local equilibrium" model, applicable at larger undercoolings, was originally suggested in qualitative form by Hultgren (1947) and later described quantitatively by Hillert (1952, 1953) and by Gilmour, Purdy and Kirkaldy (1972). On this model, the ratio of the atomic concentration of X to that of Fe in ferrite is identical to that in austenite. Only the carbon atoms partition between the two phases. The C concentration in austenite is again much higher, but both carbon concentrations are adjusted so that, given the (usually) non-equilibrium concentrations of X in austenite and in ferrite, the partial molar free energies of C are identical in the two phases. Gilmour, Purdy and Kirkaldy (1972) have formulated the equations for paraequilibrium as:

$$\overline{G}_{2\ \alpha}^{\alpha\gamma} = \overline{G}_{2\ \alpha}^{\alpha\gamma} \tag{2}$$

$$\overline{G}_{3\ \gamma}^{\gamma\alpha} - \overline{G}_{3\ \alpha}^{\alpha\gamma} = -(G_{1\ \gamma}^{\gamma\alpha} - G_{1\ \alpha}^{\alpha\gamma})(x_1/x_3) \tag{3}$$

where x_1 and x_3 represent the atom fractions of Fe and of X in the bulk alloy. As illustrated in Fig. 3 (DeHoff, 1983), the paraequilibrium phase boundaries of the $\alpha+\gamma$ region lie within those for complete or "orthoequilibrium" (Hillert, 1953). Again assuming that $\alpha:\gamma$ boundaries are disordered, paraequilibrium growth is controlled by the diffusion of carbon in austenite.

Fig. 1. Optical micrograph of typical grain boundary ferrite allotriomorph; 0.29% C, 0.76% Mn steel, austenitized 1800 s. at 1300°C, reacted 7 s. at 725°C.

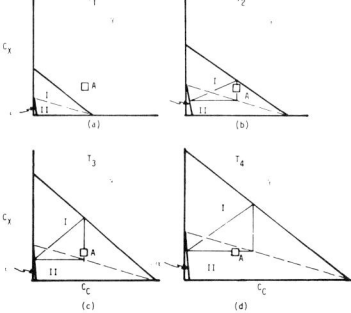

Fig. 2. Schematics of the Fe-rich corner of an Fe-C-X system at four temperatures, where $T_1>T_2>T_3>T_4$, showing how an alloy of composition A enters Region I and then passes into Region II. DeHoff (1983).

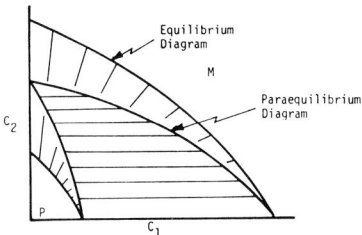

Fig. 3. "Equilibrium Diagram" = $\alpha+\gamma$ region under condition of equality of partial molar free energies of each species in the two phases; "paraequilibrium diagram" shows $\alpha+\gamma$ region when only carbon partial molar free energies are equilibrated and X/Fe ratio is the same in both phases. DeHoff (1983).

These two models, "local equilibrium" and paraequilibrium, have provided the basis for virtually all analyses which have been made of experimental measurements on the growth kinetics of ferrite allotriomorphs. We shall now ascertain how these analyses have fared when compared with experimental measurements of grain boundary allotriomorph growth kinetics in Fe-C and Fe-C-X alloys.

COMPARISONS OF THEORY AND EXPERIMENT

Growth Kinetics

Experimental Considerations. A reasonable assessment can be made of the growth kinetics of grain boundary allotriomorphs in polycrystalline austenite if the following conditions are fulfilled (Bradley and Aaronson, 1977; Bradley, Rigsbee and Aaronson, 1977). The grain boundaries should be oriented perpendicular to the plane of polish. The length of the longest and the thickness of the thickest allotriomorph are measured (at room temperature) as a function of isothermal reaction time from a sufficiently large sample of allotriomorphs at each of a number of reaction times, performed over the widest usable range of such times. Measurements are confined to allotriomorphs nucleated at grain faces, and the reaction times short enough so that extensive overlap of the carbon diffusion fields from allotriomorphs located at other austenite grain boundaries is avoided. (Use of a large austenite grain size extends the useful range of reaction times under given conditions of alloy composition and reaction temperature.) Conspicuously faceted allotriomorphs as well as those which have grown predominantly into one of the two bounding matrix grains are not measured. Obviously the alloys studied must have sufficient hardenability so that transformation takes place only at the intended isothermal reaction temperature. In studies of this type performed upon high-purity Fe-C alloys, whose hardenability, even when specimens are only 0.025 cm. thick, is just sufficient for this purpose, it was found that plots of the logarithm of the half-thickness vs. the logarithm of the isothermal reaction time have a slope of ca. one-half when transformation is conducted isothermally, but different--often larger--slopes when it is not, e.g., when specimens of excessive thickness are employed. Although such plots should actually be made as a function of the growth time rather than of the reaction time, it is found experimentally that in Fe-C alloys (Bradley, Rigsbee and Aaronson, 1977) and in most Fe-C-X alloys (Bradley and Aaronson, 1981) the difference between the parabolic rate constants obtained from the two types of plot is small enough to be disregarded. Measurements of lengthening kinetics have been found to be less useful and are now of diminishing importance. Earlier overlap with the diffusion fields of adjacent allotriomorphs is one obvious problem with these measurements. Another is that it is difficult to identify when adjacent allotriomorphs have physically impinged (this is known as "hard impingement", whereas diffusion field overlap is usually termed "soft impingement") (Robertson, 1931). Hence one may, in effect, be incorporating allotriomorph coalescence into measurements of lengthening kinetics.

It should finally be mentioned that in substitutional alloys, or in interstitial alloys containing a substitutional component, diffusion along matrix grain boundaries and the broad faces of allotriomorphs can markedly expedite both lengthening and thickening (Aaron and Aaronson, 1968; Brailsford and Aaron, 1969), as well as the kinetics of dissolution (Pasparakis, Coates and Brown, 1973; Pasparakis and Brown, 1973). This phenomenon, known during growth as the "collector plate mechanism" when the allotriomorphs are solute-rich and the "rejector plate mechanism" when they are solute-poor relative to the matrix phase, is particularly important when the matrix has an fcc crystal structure, becoming operative at temperatures less than ca. $0.9\ T_m$ (Goldman, Aaronson and Aaron, 1970). Presumably a similar situation obtains when the matrix has an hcp structure, though this point has yet to be tested experimentally. The relatively high ratio of boundary to volume diffusivity in close packed structures gives the collector plate mechanism a kinetic advantage relative to solute transport by volume diffusion directly to or from the allotriomorph. As would be anticipated from the lower packing density of bcc crystals, the collector plate mechanism is distinctly less prominent in alloys with a bcc matrix phase (Menon and Aaronson, 1986). In a later section, the rejector plate mechanism will be shown to play a significant role in growth kinetics at higher temperatures in some Fe-C-X systems.

Growth Kinetics in Fe-C Alloys. Figures 4-6 show the variation of the ratio of the parabolic rate constant measured experimentally to that calculated for an oblate ellipsoid with the experimentally measured aspect ratio in three Fe-C alloys (Bradley, Rigsbee and Aaronson, 1977). The calculated rate constant was obtained from the Atkinson analysis (Atkinson, 1969; Atkinson and co-workers, 1973) of this problem, wherein the variation of the diffusivity of carbon in austenite with carbon concentration is taken fully into account[1]. Note that this ratio is much less than unity primarily in the lowest carbon alloy and at the highest reaction temperature studied, and that it increases with decreasing reaction temperature, tending toward but not reaching unity. These deviations from unity have been attributed to faceting of the grain boundary allotriomorphs, with the facets requiring displacement by the usually slower acting ledge mechanism (Aaronson, 1962; Bradley, Rigsbee and Aaronson, 1977). The mechanistic aspects of this problem are considered further in a later section on interphase boundary structure effects upon allotriomorph growth kinetics. For the present, we note that these data on Fe-C alloys were used to correct empirically counterpart data on Fe-C-X alloys for the faceting effect by assuming that it is the same at a given carbon concentration and undercooling. While this correction can hardly be very accurate, it does appear to represent a useful improvement of the Fe-C-X data--and is in any event a minor change except at low carbon concentrations and undercoolings, as illustrated in Fig. 7.

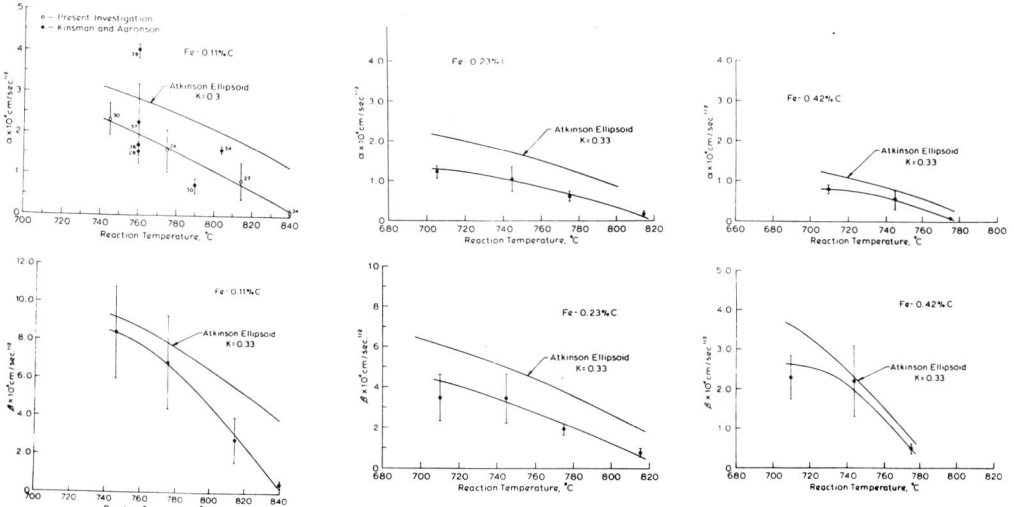

Fig. 4. Ratio of the experimentally measured to the Atkinson-analysis calculated parabolic rate constants for thickening (α) and lengthening (β) vs. reaction temperature in an Fe-0.11% C alloy. Bradley, Rigsbee and Aaronson (1977).

Fig. 5. Same as Fig. 4, for an Fe-0.23% C alloy.

Fig. 6. Same as Fig. 4, for a Fe-0.42% C alloy.

---

[1] The only "arbitrary constant" utilized in calculating the parabolic rate constant is the aspect ratio of the allotriomorphs. This has been found to be c. 1/3, independent of reaction time, temperature and carbon concentration (Bradley, Rigsbee and Aaronson, 1977), and usually (but not always (Abe, Aaronson and Shiflet, 1985)) of alloying element identity and concentration (Bradley and Aaronson, 1981).

Purdy and Kirkaldy (1963) have evaluated the parabolic rate constant for the migration kinetics of presumably disordered α:γ boundaries in a high-purity Fe-0.57% C alloy by means of a diffusion couple technique. Decarburization of the ends of a bar of this alloy in the austenite region, followed by slow cooling into the $\alpha+\gamma$ range, resulted in the formation of an essentially planar polycrystalline layer of ferrite at each end of the bar. Subsequent heat treatment homogenized the carbon concentration in the austenite remaining untransformed. Finally, heating the autogenous diffusion couple thus formed to a higher temperature in the $\alpha+\gamma$ region caused the austenite to grow back into the ferrite layers. Measurement of growth distance vs. time (after periodically quenching to room temperature) permitted ready evaluation of the parabolic rate constant, since the measured distance between α:γ boundaries at opposite ends of the couple represents the true distance as long as the plane of polish is parallel to the diffusion direction. This method is excellent for investigations confined to the $\alpha+\gamma$ region. However, if applied to sub-eutectoid reaction temperatures, which are often the region of greatest interest, the probability that ferrite will nucleate at austenite grain boundaries within the remaining austenite, i.e., ahead of the advancing or retreating layers of ferrite, is quite high, particularly at larger undercoolings. When such nucleation becomes relatively frequent, the accuracy of this method is quickly destroyed by the interference thus created with the diffusion fields of the planar α:γ boundaries. It is perhaps for this reason that, despite its clarity, this method seems not to have been further utilized. It should be noted, though, that Purdy and Kirkaldy suggested that crystallographic effects upon the migration kinetics of the ferrite layer ought to be minor because the ferrite nucleated at the surfaces of the bar and then grew through several austenite grains prior to measurement. This seems a reasonable conclusion, but nonetheless must later be reconsidered.

Data on grain boundary allotriomorph growth kinetics in Fe-C alloys have also been gathered by the at-temperature technique of thermionic electron emission microscopy (THEEM) (Kinsman and Aaronson, 1967, 1973; Atkinson and co-workers, 1973). Despite the exceptional precision of these data, however, a combination of stereological factors and, especially, local variations in growth kinetics attributed to interfacial structure effects-- again to be discussed in a later section--led to so much scatter in the values of the parabolic thickening rate constant that data from a given alloy are not easily compared either with those from other alloys or with theory.

In Fe-C-X Alloys. Much of the early data on the thickening kinetics of grain boundary allotriomorphs in Fe-C-X alloys were also secured with THEEM (Kinsman and Aaronson, 1967, 1973). Although these data were useful in indicating trends, their internal scatter again gave rise to difficulties in respect of quantitative comparisons. Application of the room temperature measurement technique previously described for Fe-C alloy, however, produced data on Fe-C-Mn, Fe-C-Si, Fe-C-Ni and Fe-C-Cr alloys which permitted detailed testing of the local equilibrium and paraequilibrium hypotheses (Bradley and Aaronson, 1981). The ratios of the experimental parabolic rate constant for thickening, corrected for faceting as previously described, to that calculated from the local equilibrium, pile-up (Hillert, 1969) and paraequilibrium analyses, are plotted as a function of isothermal reaction temperature in Figs. 8-10. Although Fig. 8 shows that while the "local equilibrium" analysis yields reasonably good predictions for the Fe-C-Si and Fe-C-Cr alloys, this model clearly underestimates thickening kinetics drastically at smaller undercoolings in one of the Fe-C-Ni alloys and at all reaction temperatures studied in both a higher Ni and in an Fe-C-Mn alloy. These errors arose because the model predicted that transformation was occurring in Region I (Fig. 2), where X partition between austenite and ferrite is required, whereas experimentally this did not take place. The Hillert (1969) "pile-up" model, tested in Fig. 9, is a differently formulated version of the "local equilibrium" model. In Region II, this model predicted parabolic rate constants usually within a factor of two of the Coates version of the local equilibrium model. In Region I, however, the Hillert model persistently failed because the effective carbon concentration in austenite at α:γ boundaries appeared to exceed the bulk carbon concentration. This is understandable, since matching the flux of X at the α:γ boundaries with that of C requires such a very small flux of C away from these boundaries that a sufficiently precise calculation of the concentration difference needed to produce it is unlikely at the present state of either the theory or the data on carbon activity in austenite.

Fig. 10 shows levels of agreement between faceting-corrected-experimental and the paraequilibrium-calculated values of the parabolic rate constant which are quite good for both Fe-C-Ni alloys studied. Following a recent correction of a typographical error in a Hillert-Staffanson (Uhrenius, 1978) constant for Fe-C-Si (Enomoto, 1988), good agreement is also obtained for this alloy as well. However, appreciable discrepancies are found between the calculated and measured values of the parabolic rate constant in the Fe-C-Mn and Fe-C-Cr alloys.

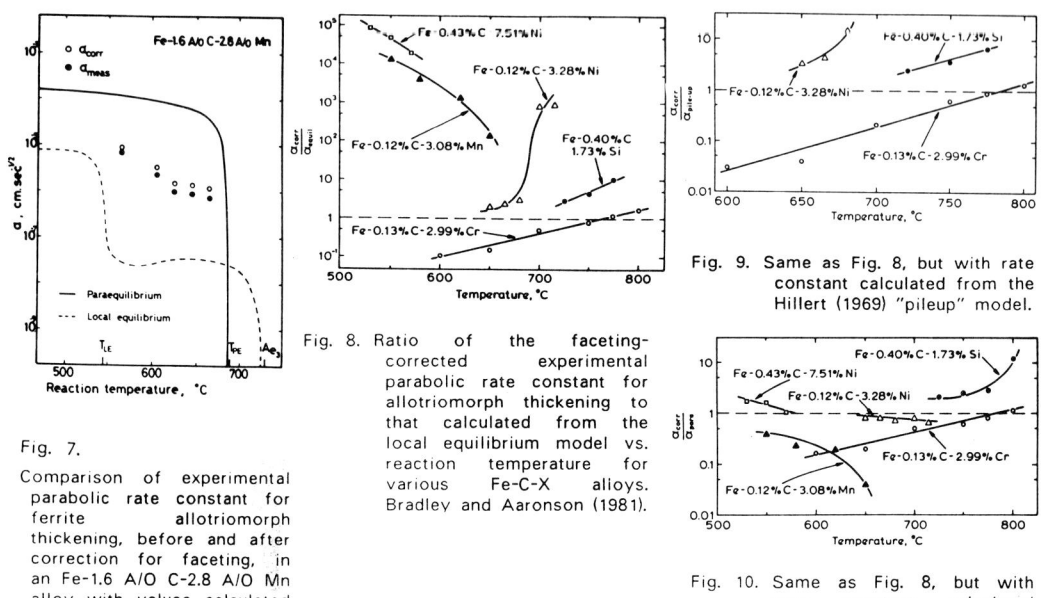

Fig. 7. Comparison of experimental parabolic rate constant for ferrite allotriomorph thickening, before and after correction for faceting, in an Fe-1.6 A/O C-2.8 A/O Mn alloy with values calculated from the local equilibrium and the paraequilibrium models. Enomoto and Aaronson (1987).

Fig. 8. Ratio of the faceting-corrected experimental parabolic rate constant for allotriomorph thickening to that calculated from the local equilibrium model vs. reaction temperature for various Fe-C-X alloys. Bradley and Aaronson (1981).

Fig. 9. Same as Fig. 8, but with rate constant calculated from the Hillert (1969) "pileup" model.

Fig. 10. Same as Fig. 8, but with rate constant calculated from the paraequilibrium model.

Inasmuch as the a_{corr}/a_{calc} ratios fall within the range 0.1 - 10 for all three models at some reaction temperatures in at least three of the alloys studied, another means is required for weighing the relative merits of these models. Both the local equilibrium and the pile-up models require that local equilibrium be established at $\alpha{:}\gamma$ boundaries. Unless the tieline across the $\alpha+\gamma$ region is nearly parallel to the carbon concentration axis (see Fig. 2), achievement of local equilibrium requires that appreciable partition of substitutional solute, X, occur between austenite and ferrite to a depth of at least one lattice parameter within austenite. The maximum penetration distance of X into austenite is readily calculated once the parabolic rate constant has been measured and the longest reaction time used to evaluate this constant has been noted (Kinsman and Aaronson, 1973; Bradley and Aaronson, 1981). Alloying element diffusivities in austenite can be estimated from the compilation of Fridberg, Torndahl and Hillert (1969). Fig 11 shows that only at the highest reaction temperatures used in the Fe-C-Si and the Fe-C-Cr alloys could sufficient volume diffusion of X have occurred to have permitted the achievement of local equilibrium.

Even the paraequilibrium calculations, however, display appreciable differences with respect to the faceting-corrected experimental data on thickening kinetics. A number of possible explanations for these discrepancies have been considered. The three principal ones, the effects of carbide precipitation at $\alpha{:}\gamma$ boundaries, the structure of $\alpha{:}\gamma$ boundaries and the solute drag-like effect on these boundaries, are examined in the last three sections of this review.

Bulk Composition

On the local equilibrium hypothesis, partition of X between austenite and ferrite is normally to be expected at all temperatures in region I of Fig. 2. The lowest temperature in this region for a given alloy will be denoted as T_{LE}. The paraequilibrium hypothesis requires the absence of X partition below the paraequilibrium $\gamma/(\alpha+\gamma)$ phase boundary, whose temperature at a given alloy composition will be termed T_{PE}. The earliest detailed experimental study of X partition between austenite and proeutectoid ferrite is that of Bowman (1946), who reported, on the basis of lattice parameter measurements, that no partition of Mo could be observed at 650° and 705°C in fully reacted specimens of Fe-C-Mo alloys.

During the 1960's, a series of electron probe microanalysis studies of X partition between austenite and ferrite was reported in Fe-C-X alloys. It was first shown that no bulk partition of Cr occurs at temperatures from 800° down to at least 500°C (Aaronson, 1962a). Since the Ae3 temperature of the Fe-C-Cr alloy used is c. 825°C, these results limit the region in which Cr partition could have occurred to a narrow range just below Ae3. These results also demonstrated, contrary to a suggestion by Hultgren (1947), that bulk partition of Cr is absent above as well as below the bay in the TTT-curve for initiation of transformation. Next, Purdy, Weichert and Kirkaldy (1964) examined the growth of ferrite in an elegant diffusion couple investigation in Fe-C-Mn alloys. Electron probe studies revealed the absence of Mn partition in an Fe-0.21%C-1.52% Mn diffusion couple reacted at 725°C and the presence of Mn partition in an Fe-0.28%C-3.16% Mn couple reacted at 742°C. The partitioned region in austenite was too narrow to be detected with electron probe analysis, whose resolution was estimated at 1.5μ. The concept of the "envelope of zero partition", i.e., the boundary between regions I and II in Fig. 2, was introduced in this paper, but thermodynamic analysis of Fe-C-X phase diagrams had not yet advanced to the point where the location of this boundary, as a function of temperature, could be calculated with reasonable certainty.

Two years later, the results of an extensive electron probe study of X partition between ferrite allotriomorphs and austenite during the early stages of transformation were reported in a number of Fe-C-X alloys wherein X was successively Si, Mn, Ni, Co, Mo, Al, Cr, Cu and Pt (Aaronson and Domian, 1966). For all but Fe-C-Pt, two levels of carbon concentration, usually c. 0.1 W/O (0.5 A/O) and 0.4 W/O (2 A/O) were employed. In most of the alloys, c. 3 A/O X was utilized, but hardenability or phase diagram considerations limited the proportions of Co and Mo to considerably lower levels. In each alloy, a considerable effort was made to examine partition behavior from as close as practicable to the experimentally determined Ae3 temperature down to the highest temperature at which bainitic carbides were visible in ferrite as observed with optical microscopy. However, partition could be detected only in the two Fe-C-Mn, the two Fe-C-Ni and the one Fe-C-Pt alloys employed. The partition data are summaried graphically in Fig. 12.

The absence of partition in the other ternary alloys studied was attributed to the paraequilibrium Ae3 falling too close to the equilibrium Ae3 to permit sufficient undercooling so that partitioned ferrite could form with detectable kinetics in (or just below) the temperature interval between the two phase boundaries (Aaronson and Domian, 1966). Since diffusion of X rather than of carbon should control growth with partition, once a little undercooling below the paraequilibrium Ae3 had been supplied it was expected that unpartitioned growth of ferrite would occur so much more rapidly than partitioned growth, despite the smaller driving force available for the former process, that any partitioned transformation product would be quickly overwhelmed. Although a simple analysis of Fe-C-X solution thermodynamics provided support for this explanation (Aaronson, Domian and Pound, 1966), it would appear appropriate to repeat these calculations with the much more accurate thermodynamic "apparatus" now available. In the considerations on the Mn and Ni alloys next to be recounted, some interesting discrepancies appear between the T_{PE} calculated and that experimentally observed which may well carry over into the ternary systems not exhibiting partition within the temperature regions examined.

Reynolds, Enomoto and Aaronson (1984) calculated T_{PE}, T_{LE} and the ratio of the atom fraction of X in ferrite to that in austenite predicted by the local equilibrium model (as refined by Coates (1972, 1973a, 1973b)) as a function of temperature in the Fe-C-Mn and Fe-C-Ni alloys examined in the electron probe study. These calculations have since been further refined, with results which are plotted in Fig. 12 atop the (1966) experimental data. In the Ni-containing alloys, the errors in prediction of T_{LE} are quite small; while there are significant inaccuracies in predicting the ratio of the Ni concentration in ferrite to that in the parent (compositionally undisturbed) austenite, the calculated variation of partition ratio with temperature closely parallels the experimentally measured one for both alloys. In the Fe-C-Mn alloys, however, a quite different situation prevails. T_{LE} is predicted to lie outside the temperature range studied, and the predicted partition ratios are in gross disagreement with those secured experimentally[2]. Similarly, while the calculated T_{PE} calculated falls above that experimentally determined, this discrepancy is much larger for the Mn than for the Ni alloys. Unfortunately, the Hillert-Staffansson parameters needed to make counterpart calculations for the Fe-C-Pt alloy studied are unavailable, thus preventing this set of experimental data from being utilized. On the basis of the comparisons which have been made, however, the compositional predictions of the local equilibrium theory are little more encouraging than its kinetic predictions. Similarly, the failures of the paraequilibrium model to predict more accurately the experimentally determined highest temperature at which paraequilibrium develops require explanation. Both will be further examined in the next section.

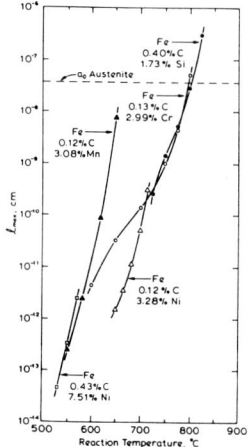

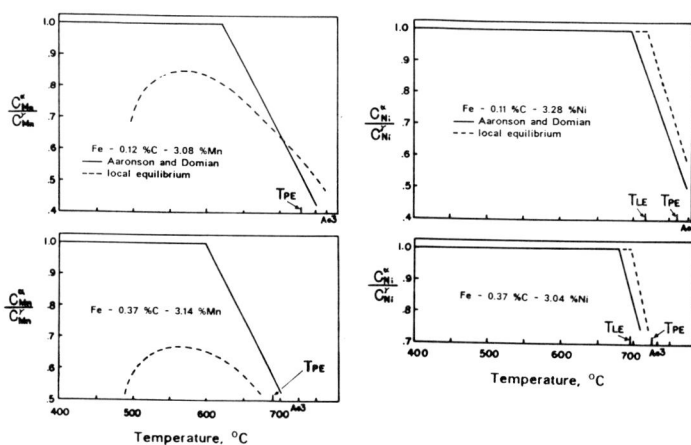

Fig. 11. Maximum X penetration distance into austenite during longest time used to determine parabolic rate constant for thickening vs. reaction temperature in Fe-C-X alloys studied in Figs. 8-10. Bradley and Aaronson (1981).

Fig. 12. Ratio of the X concentration in ferrite to that in the parent austenite, determined via electron probe microanalysis, vs. reaction temperature in the indicated Fe-C-X alloys. Aaronson and Domian (1966). Calculated curves determined from the local equilibrium model, using Hillert-Staffansson (1970) phase boundaries.

---

[2] It should be noted that the main difference between the present and previous calculations is that the Hillert-Staffansson (1970) analysis was used to compute phase boundary compositions reported here while the Central Atoms Model (Lupis and Elliott, 1967) was previously employed.

LOCAL EQUILIBRIUM CUM REJECTOR PLATE MECHANISM

During a comprehensive electron probe study of X partition between ferrite and austenite (Aaronson and Domian, 1966), it was observed that at reaction temperatures where partition of Mn or Ni did take place, diffusion of these elements also occurred away from the austenite grain boundary along which the allotriomorph being studied had been growing on either side of the allotriomorph. A generation later, this finding inspired an attempt to explain some of the discrepancies between the ferrite allotriomorph growth kinetics and compositions predicted by "local equilibrium" theory and those experimentally observed by incorporating the "rejector plate" mechanism (Fig. 13) into "local equilibrium" analysis.

In the initial study of this type (Enomoto and Aaronson, 1987), the electron probe observations were first confirmed with the much higher spatial resolution technique of STEM. Figure 14 illustrates the buildup of Mn along an austenite grain boundary in front of a lengthening allotriomorph and the diffusion of Mn normal to the austenite grain boundary. The Coates (1973) version of the local equilibrium equations was modified to fit the different circumstances of the rejector plate mechanism. Now, the fluxes of carbon and X may be regarded as (geometrically) independent. Hence the local equilibrium equation for carbon diffusion can be used without the ternary interaction term, and is simply:

$$\frac{x_C^\gamma - x_C^\infty}{x_C^\gamma - x_C^\alpha} = \left[(\pi)^{1/2}/2\right] \eta e^{\eta^2/4} \text{erfc}(\eta/2) \tag{4}$$

where $\eta = a/D_C^{V~1/2}$, a = parabolic rate constant, D_C^V = volume diffusivity of carbon in austenite, x_C^γ and x_C^α = atom fractions of carbon in austenite and in ferrite at the α:γ boundary, and x_C^∞ = carbon concentration in the bulk alloy. For X diffusion along the austenite grain boundary away from a growing allotriomorph, an approximate solution was developed based upon Whipple's (1954) solution to a similar problem. Applying mass balance as in the original treatment of the collector plate mechanism (Aaron and Aaronson, 1968) and modeling an allotriomorph as an ellipsoid of revolution, the following relationship for Mn transport was obtained:

$$\frac{x_2^\gamma - x_2^\alpha}{x_2^\gamma - x_2^\infty} = \frac{4(D_2^V)^{1/2}}{(\pi)^{1/2} a \eta^2} I(\eta^2/4) \tag{5}$$

where the subscript "2" changes the various x's to atom fractions of the alloying element, D_2^V = volume diffusivity of X in austenite and:

$$I(\eta^2/4) = -\frac{e^{\eta^2/4}}{E_i(-\eta^2/4)} - \frac{\eta^2}{4} \tag{5a}$$

where:

$$E_i(-\xi) = -\int_\xi^\infty e^{-\xi}/\xi \, d\xi, \text{ where } \xi = -\eta^2/4 \tag{5b}$$

If $(D_2 t)^{1/2}$ is too small to permit local equilibrium in austenite at the α:γ boundary to be achieved, x_2^∞ may be more appropriate than x_2^γ. Figure 7 compares the experimental parabolic rate constants for allotriomorph thickening with those calculated from the usual (Coates) "local equilibrium" and from the paraequilibrium models for an Fe-1.6 A/O C-2.8 A/O Mn alloy. On a logarithmic scale, the measured rate constants fall about midway between the two sets of calculated constants. With experimental data points now omitted, Fig. 16 demonstrates that incorporation of the rejector plate mechanism much improves agreement between the calculated and measured rate constants. (Curve (a) in this Figure utilizes Eq. (5) as is, whereas curve (b) was calculated by replacing x_2^γ with $x_2^{\gamma 0}$ in the numerator of the l.h.s. of this equation.) However, Fig. 16 indicates that the accounting for the STEM measurements of the average composition of ferrite allotriomorphs in this Fe-C-Mn alloy, though improved, is still not very good when rejector plate mechanism is added to the "local equilibrium" model. Mirroring the better results obtained for the "local equilibrium" model in Fe-C-Ni alloys, addition of the rejector plate mechanism yielded almost exact agreement with the electron probe partition data (Fig. 17).

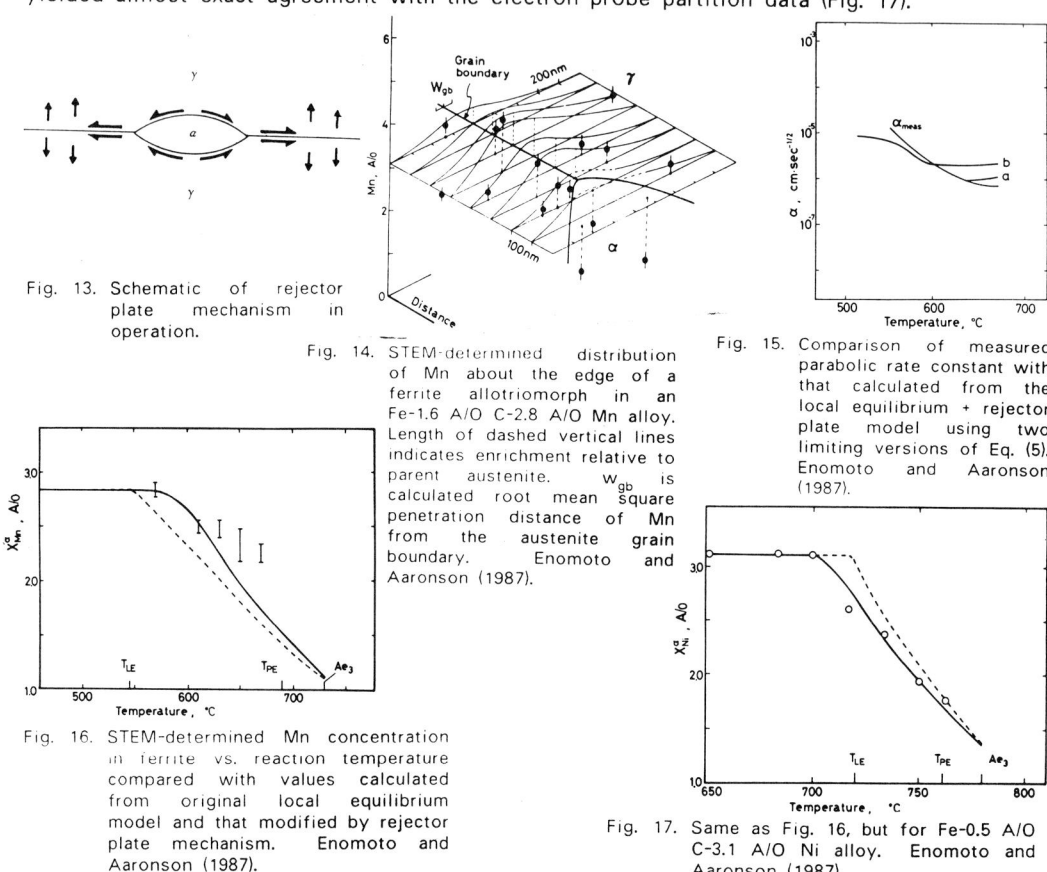

Fig. 13. Schematic of rejector plate mechanism in operation.

Fig. 14. STEM-determined distribution of Mn about the edge of a ferrite allotriomorph in an Fe-1.6 A/O C-2.8 A/O Mn alloy. Length of dashed vertical lines indicates enrichment relative to parent austenite. w_{gb} is calculated root mean square penetration distance of Mn from the austenite grain boundary. Enomoto and Aaronson (1987).

Fig. 15. Comparison of measured parabolic rate constant with that calculated from the local equilibrium + rejector plate model using two limiting versions of Eq. (5). Enomoto and Aaronson (1987).

Fig. 16. STEM-determined Mn concentration in ferrite vs. reaction temperature compared with values calculated from original local equilibrium model and that modified by rejector plate mechanism. Enomoto and Aaronson (1987).

Fig. 17. Same as Fig. 16, but for Fe-0.5 A/O C-3.1 A/O Ni alloy. Enomoto and Aaronson (1987).

Despite these improvements, however, and the clear electron probe and STEM evidence that the rejector plate mechanism is operative, Fig. 18 shows that even at 670°C, the highest reaction temperature used, where measurable Mn partition did occur between austenite and ferrite, the maximum penetration distance of Mn into austenite via volume diffusion at the longest reaction time used for the STEM measurements, 8×10^4 sec., is just greater than one lattice parameter. At 610°C, the lowest temperature at which partition was found, the maximum distance which Mn could have penetrated into austenite is seen in this Figure to be much less than a lattice parameter. Previous investigations have shown that the maximum penetration distance in the absence of bulk partition is normally also much smaller than one austenite lattice parameter (e.g. Fig. 11). This repeated finding, now extended into the region wherein bulk partition is observed, provides the most important empirical evidence that the "local equilibrium" hypothesis cannot be considered to represent accurately the situation with respect to alloying element compositions in contact with $\alpha:\gamma$ boundaries, even when the "rejector plate" mechanism is incorporated.

Further evidence in support of this point was developed in a follow-on investigation by Enomoto (1988). Fe-0.13%C-4.44%Mn and Fe-0.14% C-6.95% Ni alloys were employed. Since these X concentrations exceed the solubilities of X in ferrite, partition must occur at all temperatures, i.e., the alloys must always lie within region I of Fig. 2, if the local equilibrium hypothesis is correct. However, Fig. 19 shows that below a characteristic temperature in each alloy, partition of X between austenite and ferrite is again absent. The maximum penetration depth of X into austenite in both alloys (for allotriomorphs 1μ thick) at the highest temperature studied in each (where partition did take place) is much less than a lattice parameter in the Mn alloy and just barely above it in the Ni alloy. These two results are consistent with the absence of local equilibrium. However, the root mean square penetration distance of X into austenite from a stationary austenite grain boundary is significantly greater than a lattice parameter (see Fig. 19), indicating that the "rejector plate" mechanism can still assist growth even in the absence of "local equilibrium".

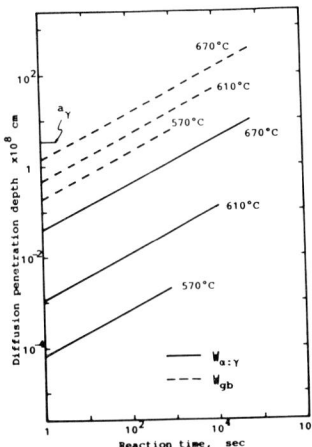
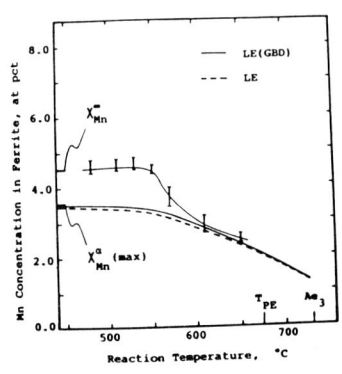

Fig. 18. Maximum Mn penetration depth ahead of moving $a''\gamma$ boundaries vs. reaction time at three reaction temperatures (dashed lines) and root mean square penetration distance of Mn from stationary austenite grain boundary under the same conditions in an Fe-1.6 A/O C-2.8 A/O Mn alloy. Enomoto and Aaronson (1987).

Fig. 19. STEM-determined and calculated Mn concentrations in ferrite vs. reaction temperature in Fe-0.6 A/O C-4.5 A/O Mn vs. reaction temperature. Calculated curves determined from original local equilibrium (LE) and local equilibrium + rejector plate mechanism (LE+GBD) models. Enomoto (1988).

ROLE OF CARBIDE PRECIPITATION AT AUSTENITE:FERRITE BOUNDARIES

This section, and the two following, deal with discrepancies between the growth kinetics calculated from the paraequilibrium model and those measured experimentally in temperature regimes where bulk partition of X between austenite and ferrite is not found at reaction times short enough so that the carbon diffusion fields of allotriomorphs on opposite sides of austenite grains do not significantly overlap. (Gilmour, Purdy and Kirkaldy (1972) have investigated the $(Dt)^{1/2}$ regime where extensive carbon diffusion field overlap does occur.) Pinning of α:γ boundaries by carbide precipitation at them is a frequently suggested possibility for decreasing the migration kinetics of these boundaries, e.g. Purdy (1978). However, a TEM examination of the alloys on which the growth kinetics data were obtained within the time-temperature regions used to secure these data (Shiflet, Aaronson and Bradley, 1981) disclosed that no carbide precipitation took place at α:γ boundaries in the Fe-C-Mn and Fe-C-Cr alloys (in which the measured growth kinetics were slower than those predicted by the paraequilibrium model) and in the low-Ni Fe-C-Ni alloy (where paraequilibrium behavior was accurately followed). Carbide precipitation did take place at higher temperatures in the higher Ni, Fe-C-Ni alloy and in the Fe-C-Si alloy.

However, in the Si alloy, such precipitation took place only occasionally at α:γ boundaries, and then only at a temperature below the region in which growth kinetics measurements were made. Similarly, carbides rarely formed at α:γ boundaries in the higher Ni Fe-C-Ni alloy.

Analysis of the nucleation kinetics of carbides at α:γ boundaries vs. the growth kinetics of those boundaries has indicated that carbide precipitation is likely only at the terraces of partially coherent interfaces, not on their risers, because the risers, though of higher interfacial energy, should usually overrun embryos prior to achievement of critical nucleus size (Aaronson and co-workers, 1978). Early TEM observations yielded just this result (Davenport and Honeycombe, 1971). Later studies, though, claimed that precipitation of carbides can also take place on disordered α:γ boundaries (Ricks and Howell, 1983). These boundaries were so classified when the ferrite allotriomorph and the austenite grain forming them were shown not to bear a Kurdjumow-Sachs orientation relationship with respect to one another (by making use of thin films of retained austenite between adjacent martensite plates). As will be shown in a later section, however, such evidence is no longer sufficient to identify a disordered α:γ boundary.

An indirect disproof of the ability of carbides to precipitate on disordered α:γ boundaries has been provided by Obara, Shiflet and Aaronson (1983). An Fe-C-Mo alloy, partially transformed in the $\alpha+\gamma$ region, was quenched to room temperature, severely cold rolled, recrystallized at a somewhat higher temperature in the $\alpha+\gamma$ region and finally isothermally reacted at a temperature just below that of the eutectoid range. Interphase boundary carbides and fibrous carbides, which formed at the reaction temperatures used in conventionally heat treated specimens of this alloy, were now replaced by isolated carbides. (In both types of heat treatment cycle, the carbides were Mo_2C.) Extensive faceting of the ferrite remaining after the recrystallization anneal in the $\alpha+\gamma$ region suggests that the α:γ boundaries were not predominantly disordered but instead were still partially coherent, at least in the faceted areas. However, it seems probable that the partially coherent areas had a considerably higher density of growth ledges than would otherwise have been present; hence the time available for a carbide to nucleate at a partially coherent terrace on an α:γ boundary prior to being overrun by an advancing growth ledge would be proportionately reduced. Whether the recrystallized α:γ interfaces are predominantly disordered or partially coherent with closely spaced ledges, however, their inability to sustain high nucleation rates of Mo_2C is consistent with the earlier view that interphase boundary carbide precipitation is favored at immobile, partially coherent terraces upon which growth ledges are widely spaced (Davenport and Honeycombe, 1971; Aaronson and co-workers, 1978).

Hence the analogy of interphase boundary pinning by precipitates to the pinning of grain boundaries by precipitates or inclusions seems inappropriate. However, carbide precipitation on terraces can affect the migration kinetics of the risers of ledges by altering the carbon concentration in austenite in the vicinity of the risers--in principle, as far as that corresponding to the extrapolated γ/(carbide+γ) phase boundary. Bradley, Shiflet and Aaronson (1983) have investigated this possibility by calculation for Fe-C-X alloys containing 0.1 W/O (0.5 A/O) or 0.4 W/O (2 A/O)C and 3 A/O of Si, Mn, Ni or Mo. When the alloy composition and reaction temperature lie between the extrapolated Ae3 (γ/($\alpha+\gamma$)) and Acm (γ/(carbide+γ)) phase boundaries, carbide precipitation at α:γ boundaries increases the driving force for the ledgewise growth of ferrite. When alloy composition and reaction temperature intersect outside this region, the reverse situation obtains. These effects are most pronounced at higher reaction temperatures, as would be anticipated from the geometry of the phase diagrams and their metastable equilibrium extrapolations. A decrease of more than an order of magnitude is thus predicted for Fe-0.5 A/O C-3 A/O Mn near 650°C and an order of magnitude increase is expected for Fe-2 A/O C-3 A/O Mo at c. 770°C. It should be emphasized, though that these changes are based on the upper limiting assumption that a carbide precipitated on a terrace adjacent to the riser of a ledge is able to change the carbon concentration in contact with the riser from the extrapolated Ae3 to the extrapolated Acm. The actual change in the average carbon concentration in the austenite in contact with a riser is surely less and hence the influence of such carbide precipitation upon ferrite growth kinetics must be smaller.

ROLE OF INTERPHASE BOUNDARY STRUCTURE

The interphase boundary structure of grain boundary allotriomorphs has long been considered to be predominantly of the disordered type (Smith, 1953; Aaronson, 1962b). However, an undercurrent of concern about the validity of this assumption--or at best, deduction--has been present in the literature on allotriomorph growth kinetics for many years. Thus, Purdy and Kirkaldy (1963) emphasized the probably disordered nature of the boundaries they studied in their Fe-C diffusion couple by noting that the individual crystals comprising the layers of ferrite formed at opposite ends of the couple during controlled decarburization grew through many austenite grains. Hence low energy orientation relationships appear unlikely to have survived. However, their micrograph of the couple suggest that faceting may still have occurred. As previously noted, the slower than anticipated growth kinetics of ferrite were taken to signify the presence of partially coherent facets along the broad faces of the allotriomorphs (Bradley, Rigsbee and Aaronson, 1977). Analysis of the nucleation kinetics of proeutectoid ferrite allotriomorphs at austenite grain faces in Fe-C (Lange, Enomoto and Aaronson, 1988) and Fe-C-X (Enomoto and Aaronson, 1986) alloys, of proeutectoid alpha allotriomorphs at beta grain faces in Ti-Cr and Ti-Co alloys (Menon and Aaronson, 1986) and of α_m during the $\beta \rightarrow \alpha_m$ massive transformation at beta grain faces in Ti-Ag and Ti-Au alloys (Plichta and co-workers, 1980), using classical heterogeneous nucleation theory, provided convincing evidence that incoherent nuclei, formed by abutting spherical caps with disordered interfaces, are usually not viable at moderate undercoolings. A coherent pillbox model of the critical nuclei in these transformations, while not at all rigorous, did account sufficiently well for the nucleation kinetics measurements to indicate that critical nuclei must be as coherent as possible with both matrix grains forming grain faces. Linear facets on both sides of grain boundary allotriomorphs formed during precipitation (Aaronson, 1956, 1962) and massive transformations (Plichta and Aaronson, 1980), as well as secondary sideplates developing into both matrix grains (Hillert, 1962), provide useful indirect confirmation of this deduction, with the coherent boundaries originally present being replaced by partially coherent ones during the (usually) early stages of growth (van der Merwe, 1963a, b; Aaronson and Russell, 1983).

Direct observation of the interphase boundary structure of grain boundary ferrite allotriomorphs with TEM performed at room temperature is prevented by transformation of the remaining (90 - 99+%) proportion of the austenite to martensite during quenching to room temperature. The hot-stage TEM technique effectively employed by Purdy (1978, 1987) to study other austenite decomposition problems, however, should be capable of surmounting this difficulty, though the concommitant reduction in resolution and changes in ferrite morphology should be at least somewhat troublesome. Recently, Furuhara, Dalley and Aaronson (1988) have taken advantage of the more rapid decline in the M_s temperature with composition in some hypoeutectoid Ti-X alloy systems to examine the interfacial structure of hcp proeutectoid alpha allotriomorphs at bcc beta grain faces in a Ti-7.15 W/O Cr alloy. The untransformed beta matrix is readily retained at room temperature, even in thin foils, in this alloy. They have found that alpha allotriomorphs exhibit a Burgers orientation relationship, more or less exactly, with respect to one beta grain and a usually irrational, non-Burgers relationship with respect to the other. However, arrays of linear defects are visible (with sufficient tilting of the thin foils) everywhere on both interfaces of the allotriomorphs. The defects so far observed appear to be predominantly ledges, often closely spaced; extensive experience in many alloy systems (Aaronson, 1974) permits the deduction to be made, with good certainty, that the terraces of these ledges contain misfit dislocations, and thus that these boundaries are partially coherent. Enomoto (1987) has shown that at relatively late reaction times a planar ledged interphase boundary will exhibit parabolic kinetics, as does a planar disordered boundary (Zener, 1949), though of course with a usually different parabolic rate constant. Hence the parabolic kinetics exhibited by both proeutectoid ferrite (Bradley, Rigsbee and Aaronson, 1977) and proeutectoid alpha (Menon and Aaronson, 1986) allotriomorphs are also consistent with a ledge mechanism of growth. However, further development of the Enomoto analysis will be needed to take into account the contributions to growth of the multiple sets of ledges observed on alpha allotriomorphs.

ROLE OF THE SOLUTE DRAG-LIKE EFFECT

Some of the more important failures of the paraequilibrium model to explain the growth kinetics of ferrite allotriomorphs shown in Fig. 10 cannot be explained by either carbide precipitation or, insofar as we can deduce, by interfacial structure effects. These deficiencies are currently ascribed mainly to a solute drag-like effect (SDLE) (Kinsman and Aaronson, 1967; Aaronson, 1969; Reynolds, Enomoto and Aaronson, 1984; Boswell and co-workers, 1986). Unlike the solute drag effect upon grain boundary motion during grain growth or recrystallization (Cahn, 1962), diffusion of X to $\alpha:\gamma$ boundaries cannot occur because ferrite growth kinetics are too rapid relative to $(D_X^\gamma t)^{1/2}$. Instead, if X exhibits an appreciable size misfit with respect to Fe atoms in austenite, there will be a tendency for X atoms in the austenite matrix overrun by an $\alpha:\gamma$ boundary to remain within the boundary. This retention is facilitated by the similarity of the boundary diffusivities of X and the volume diffusivity of C in austenite (Fridberg, Torndahl and Hillert, 1969). The thermodynamic basis of the SDLE is that the partial molar free energies of X and Fe in ferrite are not equal to their counterparts in austenite under paraequilibrium conditions (Gilmour, Purdy and Kirkaldy, 1972) (Eq. 3). Hence X atoms in $\alpha:\gamma$ boundaries are not in equilibrium with either phase (Reynolds, Enomoto and Aaronson, 1984). Thus, if X decreases the activity of carbon in austenite, the carbon activity in the bulk austenite in contact with the boundary will be reduced and the driving force for ferrite growth will be diminished. Conversely, if X increases the activity of C in austenite, growth kinetics should be accelerated (as in Fe-C-Si alloys--see again Fig. 10), leading to an inverse SDLE[3].

When X moderately reduces C activity in austenite, e.g., Mn, a reduction in growth kinetics is the main effect of the SDLE. However, when the reduction is larger, a number of other phenomena develop. These include passage of growth kinetics through a minimum at a temperature near that of the bay, diminution or even disappearance of Widmanstatten morphologies as the bay temperature is approached, extreme degeneration of Widmanstatten morphologies when they reappear below the bay (Boswell and co-workers, 1986) and extensive changes in the nature of carbide precipitation in association with ferrite at temperatures below that of the bay (Tsubakino and Aaronson, 1987). The incomplete transformation phenomenon also appears to derive from the SDLE (Aaronson, 1969).

Field ion microscope/atom probe data have recently provided some evidence for Mo segregation at $\alpha:\gamma$ boundaries (Stark and Smith, 1988; Reynolds, Aaronson and Brenner, 1988). This segregation appears to be of non-equilibrium extent, as predicted by the foregoing concept of the SDLE (Reynolds, Aaronson and Brenner, 1988). A quantitative theory of the SDLE is badly needed but is proving very difficult to develop. The problem of acquiring an accurate thermodynamic description of the mobile areas of $\alpha:\gamma$ boundaries, long recognized by Hillert (1969), is an important obstacle to the achievement of this goal.

ACKNOWLEDGEMENTS

Appreciation is expressed to the Office of Naval Research and the U. S. Air Force Office

---

[3] The marked decrease in bainite plate lengthening kinetics by Ni (Rao and Winchell, 1967) is said to contradict these views (Hehemann, Kinsman and Aaronson, 1972), since Ni also increases the activity of carbon in austenite, though less effectively than does Si (Krikaldy, Thomson and Baganis, 1978). However, Fig. 10 indicates that the paraequilibrium model accounts satisfactorily for the influence of Ni upon allotriomorph thickening kinetics. This apparent contradiction can be resolved, it has been proposed, if Ni increases the inter-ledge spacing on the edges of ferrite/bainite plates (Reynolds and Aaronson, 1985)--a suggestion which should be testable with TEM if sufficient austenite can be retained.

of Scientific Research for support present, to the Army Research Office, Electric Power Research Institute, and American Iron and Steel Institute for support past and to the Ford Motor Company for support considerably further past.

REFERENCES

Aaron, H. B., and H. I. Aaronson (1968). *Acta Metall.*, 16, 789.
Aaron, H. B., and H. I. Aaronson (1971). *Met. Trans.*, 2, 23.
Aaronson, H. I. (1956). In *The Mechanism of Phase Transformations in Metals*. Inst. of Metals, London, p. 47.
Aaronson, H. I. (1962a). *Trans. TMS-AIME*, 224, 870.
Aaronson, H. I. (1962b). In *Decomposition of Austenite by Diffusional Processes*. Interscience, NY, p. 387.
Aaronson, H. I. (1969). In *The Mechanism of Phase Transformation in Crystalline Solids*. Inst. of Metals, London, p. 270.
Aaronson, H. I. (1974). *Jnl. of Microscopy*, 102, 275.
Aaronson, H. I., and H. A. Domian (1966). *Trans. TMS-AIME*, 236, 781.
Aaronson, H. I., and K. C. Russell (1983). In *Proceedings of an International Conference on Solid-Solid Phase Transformations*. TMS-AIME, Warrendale, PA, p. 371.
Aaronson, H. I., H. A. Domian and G. M. Pound (1966). *Trans. TMS-AIME*, 236, 768.
Aaronson, H. I., M. R. Plichta, G. W. Franti and K. C. Russell (1978). *Met. Trans.*, 9A, 363.
Atkinson, C. (1969). *Trans. TMS-AIME*, 245, 801.
Atkinson, C., K. R. Kinsman, H. B. Aaron and H. I. Aaronson (1973). *Met. Trans.*, 4, 783.
Boswell, P. G., K. R. Kinsman, G. J. Shiflet and H. I. Aaronson (1986). In *Mechanical Properties and Phase Transformations in Engineering Materials--Earl R. Parker Symposium on Structure Property Relationships*. TMS-AIME, Warrendale, PA, p. 445.
Bowman, F. E. (1946). *Trans. ASM*, 36, 61.
Bradley, J. R., and H. I. Aaronson (1977). *Met. Trans.*, 8A, 317.
Bradley, J. R., and H. I. Aaronson (1981). *Met. Trans.*, 12A, 1729.
Bradley, J. R., J. M. Rigsbee and H. I. Aaronson (1977). *Met. Trans.*, 8A, 323.
Bradley, J. R., G. J. Shiflet and H. I. Aaronson (1983). In *Proceedings of an International Conference on Solid-Solid Phase Transformations*. TMS-AIME, Warrendale, PA, p. 819.
Brailsford, A. D., and H. B. Aaron (1969). *Jnl. App. Phys.*, 40, 1702.
Cahn, J. W. (1962). *Acta Metall.*, 10, 789.
Coates, D. E. (1972). *Met. Trans.*, 3, 1203.
Coates, D. E. (1973a). *Met. Trans.*, 4, 1077.
Coates, D. E. (1973b). *Met. Trans.*, 4, 2313.
Davenport, A. T., and R. W. K. Honeycombe (1971). *Proc. Roy. Soc. London*, 322, 191.
DeHoff, R. T. (1983). In *Proceedings of an International Conference on Solid-Solid Phase Transformations*. TMS-AIME, Warrendale, PA, p. 503.
Enomoto, M. (1987). *Acta Metall.*, 35, 935.
Enomoto, M. (1988). In *Phase Transformations '87*. Inst. of Metals, London, in press.
Enomoto, M., and H. I. Aaronson (1986). *Met. Trans.*, 17A, 1095.
Enomoto, M., and H. I. Aaronson (1987). *Met. Trans.*, 18A, 1547.
Fridberg, J., L. Torndahl and M. Hillert (1969). *Jern. Ann.*, 153, 263.
Furunara, T., A. Dalley and H. I. Aaronson (1988). Submitted to *Scripta Met.*
Gilmour, J. E., G. R. Purdy and J. S. Kirkaldy (1972). *Met. Trans.*, 3, 3213.
Goldman, J., H. I. Aaronson and H. B. Aaron (1970). *Met. Trans.*, 1, 1805.
Hillert, M. (1952). *Jern. Ann.*, 136, 91.
Hillert, M. (1953). Internal Report, Swedish Inst. for Metals Research, Stockholm, Sweden.
Hillert, M. (1955). *Acta Metall.*, 3, 34.
Hillert, M. (1969). In *The Mechanism of Phase Transformations in Crystalline Solids*. Inst. of Metals, London, p. 231.
Hillert, M. (1988). Proceedings of this symposium.
Hillert, M. and L.-I. Staffanson (1970). *Acta Chem. Scand.*, 24, 3618.
Hultgren, A. (1947). *Trans. ASM*, 39, 915.
Kinsman, K. R., and H. I. Aaronson (1967). In *Transformation and Hardenability in Steels*. Climax Molybdenum Co., Ann Arbor, MI, p. 39.
Kinsman, K. R. and H. I. Aaronson (1973). *Met. Trans.*, 4, 959.

Kirkaldy, J. S. (1958). *Canadian Jour. of Phys.*, 36, 907.
Lange, W. F. III, M. Enomoto and H. I. Aaronson (1988). *Met. Trans.*, 19A, 427.
Menon, E. S. K. and H. I. Aaronson (1986). *Met. Trans.* 17A, 1703.
Obara, T., G. J. Shiflet and H. I. Aaronson (1983). *Met. Trans.*, 14A, 1159.
Pasparakis, A., and L. C. Brown (1973). *Acta Metall.*, 21, 1259.
Pasparakis, A., D. E. Coates and L. C. Brown, (1973). *Acta Metall.*, 21, 991.
Plichta, M. R. and H. I. Aaronson (1980). *Acta Metall.*, 28, 1041.
Plichta, M. R., J. H. Perepezko, H. I. Aaronson and W. F. Lange III (1950). *Acta Metall.*, 28, 1031.
Purdy, G. R. (1978). *Acta Metall.*, 26, 477.
Purdy, G. R. (1987). *Scripta Met.*, 21, 1035.
Purdy, G. R. and J. S. Kirkaldy (1963). *Trans. TMS-AIME*, 227, 1255.
Purdy, G. R., D. H.Weichert and J. S. Kirkaldy (1964). *Trans. TMS-AIME*, 230, 1025.
Reynolds, W. T. Jr., M. Enomoto and H. I. Aaronson (1984). In *Phase Transformations in Ferrous Alloys*. TMS-AIME, Warrendale, PA, p. 155.
Reynolds, W. T. Jr., H. I. Aaronson and S. S. Brenner (1988). Submitted to *Scripta Met.*
Ricks, R. A., and P. R. Howell (1983). *Acta Metall.*, 31, 853.
Robertson, J. M. (1931). *Carnegie Scholarship Memoirs*, Iron and Steel Inst., 20, 1.
Shiflet, G. J., H. I. Aaronson and J. R. Bradley (1981). *Met. Trans.*, 12A, 1743.
Smith, C. S. (1953). *Trans. ASM*, 45, 533.
Stark, I., and G. D. W. Smith (1988). In *Phase Transformations '87*. Inst. of Metals, London, in press.
Tsubakino, H. and H. I. Aaronson (1987). *Met. Trans.*, 18A, 2047.
Uhrenius, B. (1978). In *Hardenability Concepts with Applications to Steel*. TMS-AIME, Warrendale, PA, p. 28.
Van der Merwe, J. H. (1963a). *Jour. Appl. Phys.*, 34, 117.
Van der Merwe, J. H. (1963b). *Jour. Appl. Phys.*, 34, 123.
Whipple, R. T. P. (1954). *Phil. Mag.*, 45, 1225.
Zener, C. (1949). *Jour. Appl. Phys.*, 20, 950.

Thermodynamics of Fe-Ti-C and Fe-Nb-C Austenites and Nonstoichiometric Titanium and Niobium Carbides

K. BALASUBRAMANIAN AND J. S. KIRKALDY

Department of Materials Science and Engineering
McMaster University, Hamilton, Ontario, Canada

ABSTRACT

The thermodynamics of Fe-Ti-C and Fe-Nb-C austenites and that of nonstoichiometric titanium and niobium carbides have been experimentally investigated using a dynamic gas equilibration technique in the temperature range 1273K-1473K. Methane-hydrogen gas mixtures have been used for fixing carbon potentials and the carbon contents have been determined as dynamic weight changes via a sensitive Cahn microbalance. The effects of titanium-carbon and niobium - carbon interactions in austenite have been observed (i) as a miniumum in the carbide solubility curve, (ii) as increases in carbon content due to Ti/Nb additions at constant carbon activity and (iii) as the variation of solubility limit of the carbide with carbon content at high carbon concentrations. The results on the isoactivity measurements in the ternary Fe-Ti-C and Fe-Nb-C austente have been analyzed using the modified Wagner formalism. The ternary interaction parameter ε_C^M (M=Ti,Nb) has been quantitatively related to the solubility minimum and the relative increase in carbon content at constant carbon activity. The variation of the solubility limit of the carbide with carbon content has been described using an additional term to the classical solubility relationships. This additional term is related to the self- and cross-interaction of carbon. Using the solubility relations the dissolution free energy of bcc Ti in fcc Fe has also been determined. The results on the activity - composition relationship in the binary nonstoichiometric titanium and niobium carbide phases have been analyzed using the sublattice - subregular model suggested by Hillert and Staffansson and the interaction parameters in the model determined. A correlational relationship between the ternary interaction parameter and the free energy of formation of carbides has been established.

KEYWORDS

Thermodynamics; Fe-Ti-C; Fe-Nb-C; austenite; interaction parameters; solubility; binary carbides.

INTRODUCTION

The thermodynamics of iron rich ternary Fe-M-C alloys (M=Mn,Ni,Cu,Si, Cr,Mo etc.) have been extensively investigated in the past two decades. The experimental data have also been analyzed in terms of thermodynamic models and considerable progress has been made in utilising the ternary information for predicting higher order systems. The development of microalloyed steels containing titanium, niobium and vanadium has been one of the important endeavours in the design of steels. These alloy additions form fine carbide, nitride, and carbonitride precipitates during hot deformation of austenite, therby influencing its recrystallization and growth kinetics. The thermodynamics of microalloyed austenite and the precipitate phases is one of the many important factors that are involved in the complex interplay of deformation and precipitation in these steels. The solubility of carbides, nitrides, and carbonitrides of Ti, Nb, V etc. in alloyed austenite has been the focus of many investigations in the past two decades. The mutual stability is generally understood in terms of solubility limits of binary stoichiometric compounds of these elements in austente devoid of solute interactions. The carbides and nitrides of Ti and Nb are nonstoichiometric and hence their composition can vary when pecipitated in steels. As group IV and group V transition metals ae very strong carbide and nitride formers, their interaction with carbon and nitrogen in austenite is very pronounced. All the earlier investigations were performed at low carbon levels (below 0.2 Wt%C) and hence are not capable of predicting the solubility at higher levels where effects due to solute interactions become significant. The present work was undertaken to gain insight into the nature and the magnitude of these interactions in microalloyed austenite.

EXPERIMENTAL PROCEDURE

A dynamic gas equilibration technique together with a Cahn 1000 microbalance has been used for obtaining the activity – composition relationships in Fe-Ti-C and Fe-Nb-C austenites and binary nonstoichiometric carbides of Ti and Nb. The details regarding the design and construction of the experimental apparatus, materials and sample preparation are discussed elsewhere(2). The experiments essentially involve equilibration of ten Fe-Ti or Fe-Nb samples ranging from 0.005Wt% to 1.0Wt% in composition with a gas mixture, the carbon potential of which is fixed. Carbon contents ranging from 0.2Wt%C to 2.0Wt%C were accessed. The equilibration experiments were carried out in two stages. Intially all the ten alloys along with a pure Fe sample were simultaneously equilibrated to various carbon levels. The carbon contents were determined by weighing the sample before and after equilibration. These 'simultaneous' runs were performed to determine the austenite – carbide phase boundary (solubility limit) as a discontinuity in the variation of the carbon content as a function of Nb or Ti content for a given carbon potential. Once the solubility limit is known, five compositions, four lying in the single phase region and the fifth lying in the two phase region but close to the solubility limit were chosen for 'individual' equilibration. These individual equilibration experiments were performed for accurately determining the increase in the carbon content due to addition of Ti or Nb via dynamic weight change measurements. The =carbon potentials used in the 'simultaneous' runs' were repeated for these five compositions. The equilibration was accessed from both higher and lower potential sides. The activity – composition relationships in binary Ti an Nb carbides have been determined by equilibrating thin foils (10-15 microns thick) of Ti or Nb to gas mixtures of various carbon potentials. In most cases a single specimen is sufficient for accessing many carbon levels. However, as the carbides are ceramic the specimen tends to disintegrate after few runs and new specimens have to be introduced as and when required.

EXPERIMENTAL RESULTS

For the sake of brevity only typical experimental results obtained from isoactivity measurements in Fe-Ti-C system are shown in Figs. 1 and 2. Three important features are borne out in Fig. 1, viz., (i) the solubility limit obtained as the discontinuity in the isoactivity curves shows a minimum between 0.35Wt%C and 0.5Wt%C in the Fe-Ti-C system; (ii) the increase in carbon content over and above the binary level is very significant at higher carbon levels; and (iii) there is a substantial increase in the solubility limit at high carbon levels. These three features arise out of strong interactions between titanium and carbon in austenite. Fig. 2 depicts the typical variation of the composition of titanium carbide with carbon potential (proportional to the gas ratio). It can be seen from this illustration that (i) large changes in gas ratio are necessary

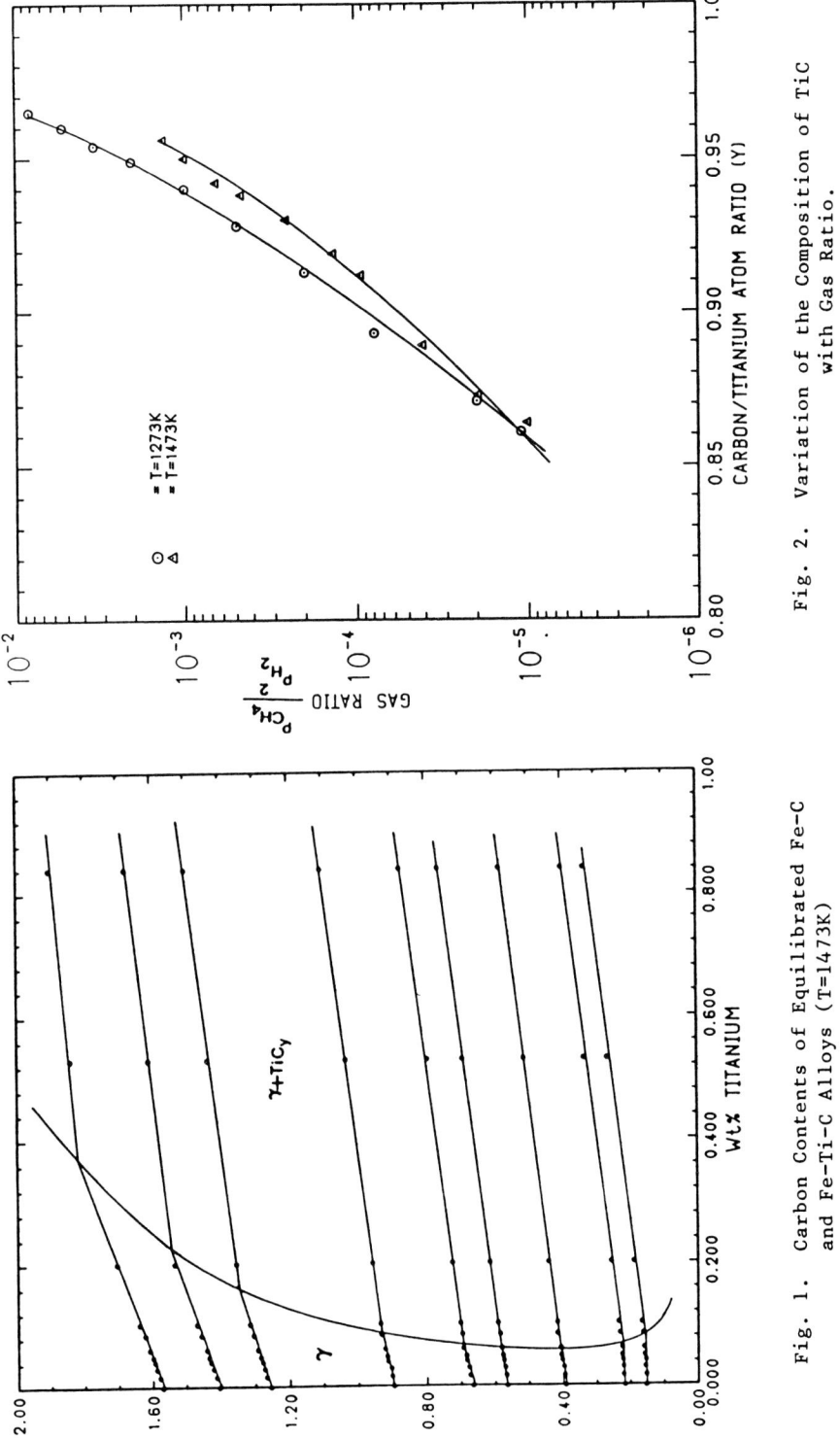

Fig. 1. Carbon Contents of Equilibrated Fe-C and Fe-Ti-C Alloys (T=1473K)

Fig. 2. Variation of the Composition of TiC with Gas Ratio.

for obtaining small changes in composition; (ii) carbon contents below y=0.85 are not accessible via gas equilibration methods using methane-hydrogen mixtures; and (iii) the variation of composition with temperature at constant gas ratio is very small in the temperature range 1273K-1473K. Similar results are obtained for Fe-Nb-C system.

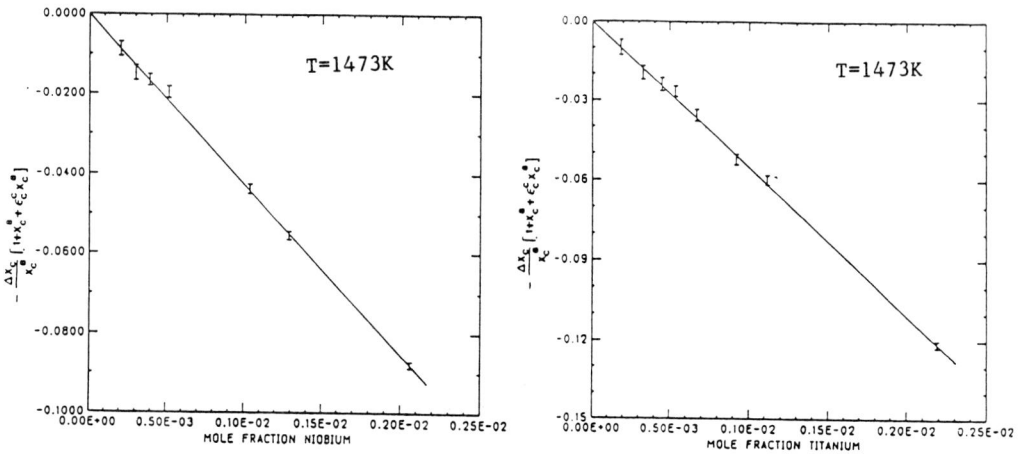

Fig. 3. Determination of C-Nb Interaction Parameter

Fig. 4. Determination of C-Ti Interaction Parameter

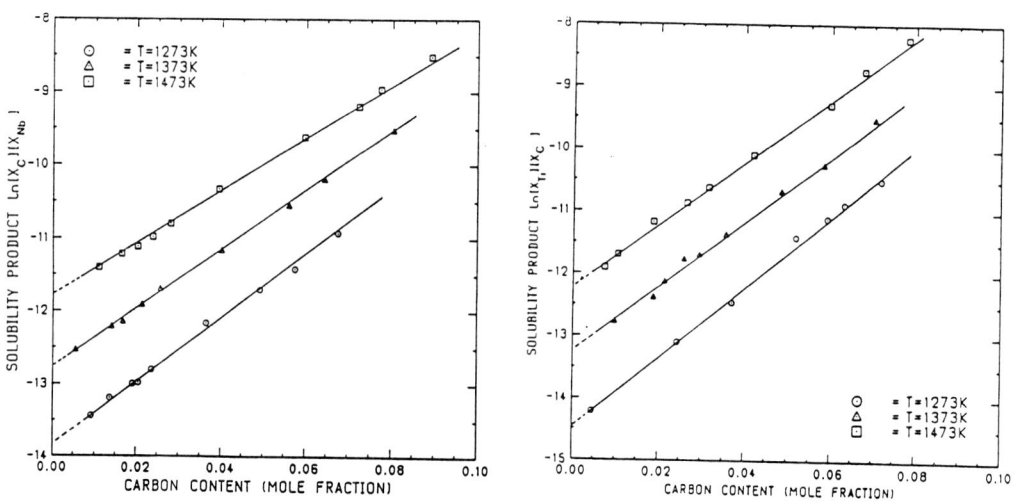

Fig. 5. NbC in Fe-Nb-C Austenite

Fig. 6 TiC in Fe-Ti-C Austenite

Variation of the Solubility Limit of of Carbides in Austenites

ANALYSIS AND DISCUSSION

Modified Wagner Formalism for Dilute Ternary Austenites

The thermodynamics of dilute ternary austenites can be satisfactorily described using the modified Wagner formalism(37,4,28,8). According to this formalism the variation of chemical potentials of solutes in a ternary Fe-M-X austenite (M = Ti,Nb etc., X=C,N) can be written as

$$G_M = RT\ln a_M = RT\ln X_M + RT\left[\ln{}^0\gamma_M + \varepsilon_M^M X_M + \varepsilon_M^X X_X\right]$$
$$- RT/2\left[\varepsilon_M^M X_M^2 + \varepsilon_X^X X_X^2 + 2\varepsilon_X^M X_M X_X\right] \quad (1)$$

$$G_X = RT\ln a_X = RT\ln X_X + RT\left[\ln{}^0\gamma_X + \varepsilon_X^X X_X + \varepsilon_X^M X_M\right]$$
$$- RT/2\left[\varepsilon_M^M X_M^2 + \varepsilon_X^X X_X^2 + 2\varepsilon_X^M X_M X_X\right] \quad (2)$$

and

$$RT\ln\gamma_{Fe} = - RT/2\left[\varepsilon_M^M X_M^2 + \varepsilon_X^X X_X^2 + 2\varepsilon_X^M X_M X_X\right] \quad (3)$$

where X_i and ${}^0\gamma_i$ refer to the concentration (mole fraction) and Henry's law coefficients of solute i respectively. The ε_i^j's and ε_j^i's are the self- and cross- interaction parameters of the solutes. Pure metal and pure nonmetal are taken as standard states. Equations (1) and (2) contain the extra term $RT\ln\gamma_{Fe}$ which is a modification to the classical Wagner expressions. This term is necessary for overcoming two problems inherent in the classical expressions(3), namely, (i) inconsistency, i.e., the reciprocity relation $\varepsilon_i^j = \varepsilon_j^i$ arising from the mathematical imperative

$$\frac{\partial}{\partial n_i}\left(\frac{\partial G}{\partial n_j}\right) = \frac{\partial}{\partial n_j}\left(\frac{\partial G}{\partial n_i}\right) \quad (4)$$

is satisfied only at finite concentrations and hence in general the classical expressions for the partial molar free energies are not consistent with equation (4); (ii) the Gibbs-Duhem relation, the other imperative, is also not satisfied at finite concentration of the solutes. Darken(4) proposed a 'quadratic' formalism which reduced to Wagner's expressions at infinite dilution but which is thermodynamically consistent at finite concentrations. Pelton and Bale(28) following the lines of Darken proposed an extra term ($RT\ln\gamma_{Fe}$) to the classical expressions to overcome the inconsistency. These authors also extended this 'modified' formalism to encompass higher order interactions while preserving the interaction parameter notation due to Wagner. A major advantage of this modification is that the existing compilations of the interaction parameters can be used directly used in the modified formalism. Hillert(8) showed the correspondence between the modified formalism and the regular solution expressions. This correspondence is useful in obtaining the Wagner interaction parameters from the regular solution parameters existing in various data bases.

The Henry's law coefficient and the self interaction parameter can be obtained from the thermodynamic analysis the binary Fe-Ti and Fe-Nb and Fe-C systems. The thermodynamics of the binary Fe-C system has been well investigated. The Henry's law coeffcient and the self interaction parameter for carbon in binary Fe-C as given by Chipman(3) will be used for analysis in this study. The experimental information regarding the solute interactions in fcc Fe-Ti and Fe-Nb phases are lacking. However, due to the success of techniques like the computer coupling of the thermochemistry and the phase diagram these parameters can be determined to a high degree of accuracy. Kaufman(14) and Murray(23) analyzed the $(\alpha+\gamma)$ phase relationships in the Fe-Ti, Fe-Nb system using the analytical expressions from a subregular solution model. The interaction parameters in the Wagner formalism extracted (1) from the parameters given by these authors is listed in Table 1. The ternary titanium - carbon and niobium - carbon interaction parameters will be determined from the results obtained in this study. The dissolution free energy of bcc Ti and bcc Nb in fcc iron will also be determined from the carbide solubility results.

TABLE 1

Wagner Interaction Parameters for fcc Phase in Fe-Ti, Fe-Nb Systems

System	$RT\ln\gamma_M^0$	$RT\epsilon_M^M$	Reference
Fe-Ti	- 33372 + 3.76T	91035	Kaufman (14)
	- 42506 + 3.76T	131035	Murray (23)
Fe-Nb	9000 + 3.56T	0	Kaufman (14)

(All values in Joules/Mole of Metal)

TERNARY INTERACTIONS IN DILUTE TERNARY AUSTENITES

The parameter defining the carbon-transition metal solute (Ti,Nb) interactions is in principle determinable from three sources, namely, (i) solubility minimum, (ii) the increase in the amount of dissolved carbon in austenite at constant carbon activity due to the addition of Nb or Ti, and (iii) the variation of solubility product with carbon content at higher carbon concentrations. These three manifestations can be quantitatively related to the ternary interaction parameter(1). In this section these quantitative relationships are considered. The details involved in the derivation are given in reference (1). These relationships are obtained under following assumptions, viz., (a) the thermodynamics of the dilute ternary austenites can be satisfactorily described using the modified Wagner formalism, i.e., equations (1) and (2) shall be used to describe the partial molar free energies of the solutes in the austenites; (b) the analysis and the attendant relationships are applicable at carbon contents greater than 0.1 Wt%C and are not applicable at very high metal contents and therefore the effects due to the self interaction of the transition metal solutes and the nonstoichiometry of the carbide can be ignored. The Fe-Ti-C and Fe-Nb-C austenites adequately satisfy these requirements and therefore the following relationships can be applied to obtain the ternary interaction parameter and the solubility limit of the carbide at high carbon concentrations.

Solubility Minimum and Ternary Interactions

The ternary interaction parameter is related to the carbon content $(X_c)_{opt}$ at solubility minimum $dX_M/dX_C = 0$ as given below

$$\epsilon_M^C = - \frac{1}{(X_c)_{opt}} \left[1 + \epsilon_C^C (X_c)_{opt} \right] \tag{5}$$

In the solubility minimum occurs between 0.4Wt%C and 0.5Wt%C in the Fe-Nb-C system and between 0.35Wt%C and 0.45Wt%C in the Fe-Ti-C system. From equation (5) the C-Ti and the C-Nb interaction parameters can be determined to be around -60 and -50, respectively. It is difficult to determine the temperature dependence of the interaction parameters from as the minimum is not sharply defined.

Increase in Carbon Content and Ternary Interactions

The increase in carbon content ΔX_c due to Ti or Nb additions at constant carbon activty is directly related to the ternary interaction parameter as given below:

$$-\varepsilon_C^M X_M = \frac{\Delta X_C}{X_C^B} \left[1 + X_C^B + \varepsilon_C^C X_C^B \right] \qquad (6)$$

where X_C and ε_C^C refer to the carbon content in the binary Fe-C and the binary self interaction parameter for carbon. The above expression is ideally suited for the determination of the ternary interaction parameter as the ε_C^M (or rather $-\varepsilon_C^M$) (i) is obatined as the sum of two positive terms of similar magnitude and (ii) the terms involved in the brackets X_C and ε_C^C are accurately known and (iii) the error in the determination of the ternary interaction parameter is directly proportional to the error in the determination of the increase in carbon contents and therfore higher weights can be give for the data obtained at higher carbon concentrations. Figures 3 and 4 give the variation of right hand side of equation (6) with X_{Nb} and X_{Ti} at 1473K and the interaction parameter is obtained as the slope. The results obtained at 1273K ,1373K and 1473K obtained in this study for Fe-Ti-C and Fe-Nb-C systems have been analyzed using equation (6) and the following interaction parameters obtained.

$$\varepsilon_C^{Nb} = - \frac{(65350 \pm 4000)}{T} + (2.5 \pm 2) \qquad (7)$$

$$\varepsilon_C^{Ti} = - \frac{(85400 \pm 5000)}{T} + 2.5 \qquad (8)$$

From these results it can be seen that the variation of the interaction parameter (when expressed as $RT\varepsilon_C^M$) with temperature is very small and therefore the entropic part of the interaction parameter should not be very significant. The free energy of formation of carbides which is a good measure of the interaction between carbon and the transition metals, as will be shown in a latter section correlates well with the ternary interaction parameters ($RT \varepsilon_C^M$, M=transition metal) in austenite. The small change in interaction parameter with temperature is in agreement with the low entropy of formation observed for these carbides.

The Carbide Solubility and the Ternary Interactions

In general the mass action law approach is not feasible for the analysis of the austenite - nonstoichiometric carbide equilibria. However, if the composition of the carbide can be considered as being close to stoichiometry (a condition realised at high carbon levels), then one can use the mass action law to obtain simpler expressions for the solubility of the carbide. Within the assumptions listed in the earlier paragraph the increase solubility of the carbide in austenite can be obtained as

$$\ln[X_M X_C] = \ln K - [\varepsilon_C^C + \varepsilon_M^C] X_C \qquad (9)$$

where K refers to the classical solubility product derived from the free energy of formation of the stoichiometric carbide from austenite. The second term on the right hand side is the correction for the increased solubility of the carbide arising due to ternary interactions at higher carbon contents. The solubility limit obtained from the change in slope of the isoactivity curves can be fitted to the above equation in order (i) to cross-check the value of the ternary interaction parameter and (ii) to determine the leading term defining the classical solubility product. Since the free energy of formation of stoichiometric carbide (from bcc Ti, Nb and graphite) is known from calorimetric measurements (Hultgren(11)) and since the dissolution free energy of graphite if fcc Fe is known from the the thermodynamics of the binary Fe-C system(3), the dissolution free energy of bcc Ti and bcc Nb in fcc Fe can be obtained for the leading term in equation (9). Thus a graph of $\ln[X_M \cdot X_C]$ vs X_C as shown in Figures 5 and 6 is a straight line with lnK as intercept and the slope is directly related to the interaction parameters as given in equation (9)

Dissolution Free Energy of bcc Ti and Nb In fcc Fe

The dissolution free energy (Henry's law coefficient) as mentioned in the previous paragraph has been determined from the solubility limits obtained in this study. A few other investigations (see Table 2) on the solubility of NbC in Fe-Nb-C and TiC in Fe-Ti-C austenite were confined to low carbon levels (less than 0.2 Wt%C) and these results were expressed in the form of the classical solubility product relationship. The solubility limits given in these investigations and the dissolution free energy of bcc Ti and bcc Nb obtained from them are given in Table 2 along with the results from this study It can be seen from Table 2 that wide range of values are obtained for niobium while there is fair agreement in the case of titanium.

TABLE 2

Dissolution Free Energy of bcc Ti and Nb from Solubility Studies

$$\overset{0}{G}_{NbC} - \overset{0}{G}^{bcc}_{Nb} - \overset{0}{G}_{Gr} = -137.65 + 1.78\text{E-}03\ T\ \text{KJ/Mole of Nb} \quad (\text{Ref.}(11))$$

$$\overset{0}{G}_{TiC} - \overset{0}{G}^{bcc}_{Ti} - \overset{0}{G}_{Gr} = -188.3 + 14.34\text{E-}03\ T\ \text{KJ/Mole of Ti} \quad (\text{Ref.}(11))$$

$$RT\ln\overset{0}{\gamma}_C = 44.18 - 17.7\text{E-}03\ T\ \text{KJ/Mole of C} \quad (\text{Ref. }(3))$$

Solubility Product log[M][X] = A+B/T			$RT\ln\overset{0}{\gamma}_M$ (Joules) (M = Nb or Ti)			
A	B	1273K	1373K	1473K	Reference	
Fe-Nb-C						
2.90	- 7500	2557	5749	8940	de Kazinczy	(5)
3.18	- 7700	- 438	2218	4873	Mori	(21)
3.42	- 7900	- 2458	- 262	1933	Narita	(26)
3.04	- 7290	- 4875	- 1952	971	Meyer	(20)
3.70	- 9100	13693	15352	17013	Smith	(31)
4.37	- 9290	1000	1337	1754	Johansen	(13)
3.31	- 7970	1563	3970	6376	Koyama	(16)
1.18	- 4880	- 5684	800	7285	Ohtani	(27)
3.89	- 8030	-11425	-10130	- 8832	Balasubramanian	(2)
Fe-Nb-N						
3.79	- 10150	- 7604	- 6766	- 5707	Mori	(22)
4.04	- 10230	-12166	-11807	-11226	Smith	(32)
2.80	- 8500	-15067	-12333	- 9378	Narita	(24)
4.2	- 10000	-11840	-10960	-10070	Balasubramanian	(2)
Fe-Ti-C						
5.33	- 10475	-41423	-42180	-42936	Narita	(25)
2.75	- 7000	-45073	-40890	-36707	Irvine	(12)
4.03	- 8720	-43340	-41607	-39874	Shiraiwa	(30)
4.47	- 9420	-42696	-41289	-39883	Ohtani	(27)
4.20	- 8970	-40661	-39874	-38881	Balasubramanian	(2)
Fe-Ti						
Thermodynamic Analysis		-28585	-28210	-27834	Kaufman	(14)
		-37720	-37344	-36968	Murray	(23)

A few other features can be inferred from the results illustrated in Table 2. Regarding values for Nb, both the enthalpic and entropic components exhibit a wide range of values and secondly the values determined from the NbN solubility are more consistent than those determined from the NbC solubility. In fact the Fe-Nb-N system satisfies all the assumptions made in the derivation of the classical solubility relationship. Most of the studies were done close to and at one atmosphere nitrogen and hence the precipitate is close to stoichiometry, while only studies done at very high carbon levels like the present one satisfy this requirement in the Fe-Nb-C system. The contributions arising from solute interactions are minimal due to very low concentrations of N in austenite and hence the solubility limits in Fe-Nb-N austenite reflect the dissolution characteristics of N and Nb to a greater degree than the corresponding carbon austenite. Therefore the values obtained from NbN studies should more reliable. An average of the values obtained from the consistent NbN solubility results(2,24,32) is recommended as the free energy of dissolution of bcc niobium in fcc iron, viz.,

$$RT\ln{}^0\gamma_{Nb} = -25500 + 10.9\,T \text{ Joules/Mol Nb} \quad (10)$$

In the case of Ti there is a general agreement in the values obtained from various solubility studies barring some minor discrepancies. The value determined from Narita's(25) data becomes more negative with temperature (positive entropy of dissolution) while all other data sets show the opposite trend. The value taken from Irvine's(12) data shows a very strong dependence on temperature compared to other three data sets. The following average of the values obtained from the results of other three data sets is recommended as the free energy of dissolution of bcc Ti in fcc iron.

$$RT\ln{}^0\gamma_{Ti} = -60590 + 14.08T \text{ Joules/Mol} \quad (11)$$

In Table 2 the values obtained via thermodynamic analysis of bcc-fcc phase equilibria in Fe-Ti system as given by Murray(23) is in fair agreement with those obtained from solubility studies. Thus both Ti and Nb when dissolved in fcc iron exhibit strong negative deviations from ideality.

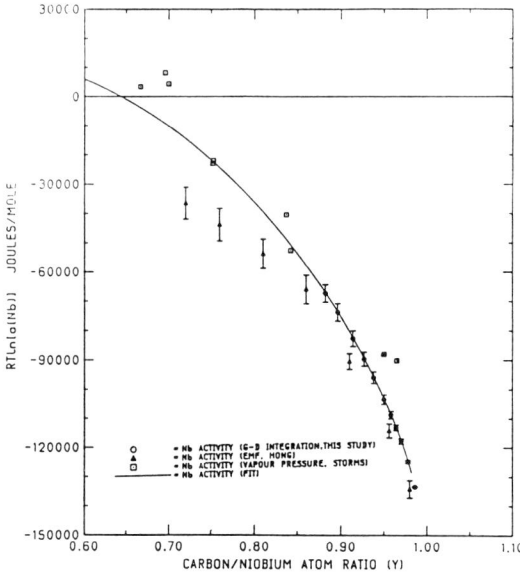

Fig. 7. Variation of the Partial Molar Free Energy of Nb in NbCy (T=1273K).

ANALYSIS OF NbC_y and TiC_y PHASES

In this section the sublattice - subregular model suggested by Hillert and Staffansson(9) has been used to describe the thermodynamics of NbC_y and TiC_y phases. The molar free energy of NbC_y or TiC_y phase existing in the NaCl structure as given by the sublattice - subregular model is

$$G_m = y\,^0G_{MX} + (1-y)\,^0G_{MV_a} + RT\left[y\ln y + (1-y)\ln(1-y)\right]$$
$$+ y(1-y)\left[L_0 + L_1[y - (1-y)]\right] \quad (12)$$

where y refers to the fraction of nonmetal atoms in the interstitial sublattice (which is same as y in the formula MX_y), $^0G_{MX}$ is the free energy of the stoichiometric MX and L_0 and L_1 are the interaction parameters describing the excess free energy of the system. $^0G_{MV_a}$ (Va=vacancy) refers to the free energy of the system when the interstitial lattice is completely vacant, i.e., the free energy of niobium in the pure fcc state. The partial molar expressions are obtained as

$$\bar{G}_X = (^0G_{MX} - \,^0G_M^{fcc}) + RT\ln\left(\frac{y}{1-y}\right) + \bar{EG}_X \quad (13)$$

and

$$\bar{G}_M = \,^0G_M^{fcc} + RT\ln(1-y) + \bar{EG}_M \quad (14)$$

where $\bar{EG}_X$ and $\bar{EG}_M$ refer to the partial molar excess free energies which are given below.

$$\bar{EG}_X = -L_0(2y-1) - L_1(6y^2+6y-1) \quad (15)$$

and

$$\bar{EG}_M = y^2\left[(L_0 - L_1) + 2L_1(2y-1)\right] \quad (16)$$

It is to be noted that the functions multiplying the interaction parameters in equation (15) are the first and second order Legendre polynomials in y and are orthogonal in the range zero to one. In equation (16) the function $\bar{EG}_M/(y^2)$ is also expressed in terms of Legendre polynomials. The constant term (L_0-L_1) becomes the coefficient of the zeroth order polynomial, $P_0 = 1$, while $2L_1$ is the coefficient of the first order polynomial $P_1 = (2y-1)$. Orthogonal polynomials remove the ill-conditioning associated with curve-fitting procedures.

Analysis of NbC_y Phase

The thermodynamics of the cubic NbC_y has not been adequately investigated by experiment. Most of the studies are confined to the calorimetric determination of integral prperties and heat capacities of essentially stoichiometric phases. There have been only two experimental investigations pertaining to the variation of partial molar free energies, namely, the high temperature (2300K-2500K) vapour pressure measurements by Storms et al.(34) and the EMF study by Hong et al.(10) in the temperature range 1100K-1350K. The carbon activity-composition measurements made in this study at 1273K and 1473K complement the niobium activities determined in the other two studies. The NbC_y phase is stable up to 4300K and hence it is necessary to include the only available high temperature vapour pressure data in the evaluation of the parameters. The results from all three data sets have been used in optimising the parameters in the sublattice model. The parameters in the sublattice model (equations (15,16)) were evaluated at three temperatures, 1273K, 1600K and 2400K using selected data from all of the investigations. A linear temperature dependence of these parameters was then obtained from the evaluations at these three temperatures. For 1273K data from all the investigations except those corresponding to y=0.72, 0.76 from the EMF study were used, the reason for their deletion being the uncertainty over the partial molar entropies(27). At 1600K, there is good agreement between the two niobium data sets and hence all the data points were used. At 2400K, only the vapour pressure data was used as most of EMF measurements could not be satisfactorily extrapolated from 1200K to 2400K. Figures 7 and 8 show the variation of partial molar free energy of niobium as a function of composition at 1273K and 2400K.

Analysis of TiC_y Phase

There have been a few low temperature (1050-1473K) studies (see Table 3) and one high temperature (1900K) vapour pressure measurement by Storms(33) on the partial molar free energies of titanium and carbon in the TiC_y. Table 3 outlines the summary of these experimental investigations on the partial molar free energies of binary titanium carbide phase. Even though data on both the titanium and carbon activity are available from these five investigations there is considerable discrepancy in the activity-composition relationships obtained from all of these investigations. Hence critical assessment of data using thermodynamic models is necessary. The thermodynamics of this phase has been analyzed by Teyssandier et al.(35), Kaufman and Agren(15) and Uhrenius(36). Teyssandier et al. used the substitutional regular solution model with a Redlich-Kister expansion for the excess free energy while Uhrenius and Agren used the sublattice subregular model suggested by Hillert and Staffansson(9) for analyzing the TiC_y phase. Of all the

TABLE 3

Summary of Experimental Investigations on the Partial Molar Free Energies in Binary Titanium Carbide

Property Measured	Composition Range	Temperature Range	Expt. Technique	Reference	
a_c	$0.70 \leq y \leq 0.89$	1173-1373K	Gas Eqn.	Alekseev	(1)
a_c	$0.67 \leq y < 1.0$	1273-1473K	Gas Eqn.	Grieveson	(7)
a_{Ti}	$0.69 \leq y \leq 1.0$	1045-1135K	EMF	Malkin	(19)
a_{Ti}	$0.71 \leq y \leq 1.0$	853K	EMF	Koyama	(17)
a_{Ti}	$0.54 \leq y \leq 0.98$	1900K	Vap. Pr	Storms	(33)

TABLE 4

Interaction Parameters for Ti and Nb Carbides

Parameter	NbC_y Phase	TiC_y Phase
$^0G_M^{fcc} - {^0G_M^{bcc}}$	$8995 + 3.56\,T$	$-1004 + 3.77\,T$
L_0	$-42625 + 4.42\,T$	$-67365 + 1.50\,T$
L_1	$-114525 + 33.58\,T$	$-162340 + 46.15\,T$
$G_{MC} - {^0G_M^{fcc}} - {^0G_{gr}}$	$-146650 - 1.76\,T$	$-187325 + 10.58\,T$

(All values in Joules/mole of metal)

available experimental results, only the vapour pressure measurements by Storms(33) and the low temperature EMF measurements by Koyama(17) have been found to be consistent with the phase diagram in the assessments by the above-mentioned authors. As there was insufficient data at high temperatures, Kaufman and Agren used the high and low carbon phase boundaries of the TiC_y phase together with Storms data at 1900K to evaluate the parameters. Figure 9 shows the variation of the partial molar free energy of carbon in TiC_y determined at 1473K in this study along with predictions by Teyssandier et al. and Kaufman and Agren. The data obtained in this study is in fair agreement with the predictions of Kaufman and Agren at compositions above y = 0.9 but the difference increases as carbon content is decreased. Teyssandier et al. used the titanium activity data given by Storms at 1900K and Koyama at 853K to obtain the parameters in their substitutional regular solution model. Their predictions are higher than those given by Kaufman and Agren and the discrepancy is larger at lower carbon contents. As Kaufman and Agren used the low carbon phase boundary in addition to Storms titanium activity data and the high carbon phase boundary for evaluating the parameters their predictions should be more reliable at low carbon levels. As the accuracy of the data obtained at carbon levels below y=0.9 are limited and as the data at higher carbon levels are resonably close to the predictions of the latter authors, their parameters in the sublattice model prove sufficient for the description of the data obtained in this study. The parameters given by Kaufman and Agren are given in Table 4 along with the evaluated parameters for the NbC_y phase.

Correlational Relationships and the Ternary Interaction Parameters

It is fairly well established(6) that there exists a regular pattern in the crystal structures of carbides and nitrides as one traverses along a period. The tendency to form closed packed structures as interstitial solutes are added increases as one moves away from iron to the left of the periodic table. The group IV, V elements which exist in bcc and hcp lattices on addition of interstitital solutes stabilize the fcc phase in the form of a carbide or nitride. To the right of iron we see Ni, Co, Cu, etc., which already exist in fcc form and the carbides and nitrides of these elements are mostly unstable. Near iron we see the transition from bcc metals forming fcc carbides to fcc metals forming unstable carbides and nitrides. Thus Cr, Mn and to some extent iron itself form complex carbides which exist in orthorhombic, cubic and hexagonal structures. The periodic pattern in structure is accompanied by the anticipated variation of the free energies of formation of carbides and nitrides, large negative values for group IV and V compounds to positive values for the unstable carbides in group VII and VIII. Since iron lies near the transition, the solutes to the far left form very stable carbides and nitrides (TiC, NbC etc.) and the solutes near iron form complex carbides with iron dissolved $((Fe,Cr)_3C$ etc) in them.

The free energy of formation of a binary compound is one of the indicators of the degree of attraction/repulsion between the two components that make up the compound. The more negative the free energy of formation higher is the degree of attraction. When the two elements that form the compound are dissolved in a solvent like Fe, depending on the nature and the magnitude of interaction of the individual solutes with the solvent the one to one correspondence between the free energy of formation and the degree of solute interaction will be affected. Figure 10 illustrates the correlation between the ternary interaction parameters and the free energies of formation of stoichiometric carbides from pure elements as well as from austenite. The free energy values have been taken from compilations of Hultgren et al.(11) and Kubaschewski(18). The carbon interaction parameters for Ti, Nb and V have been taken from the results of Ohtani et al.(27) and that obtained in this study(2). The values for carbon with Ni, Mn, Cr and Mo have been taken from the compilation by Sharma et al.(29). The free energies of formation from austenite have been determined using the dissolution free energy values obtained from the analysis of binary Fe-M systems(reference 2). From Figure 10 it can be seen that there is a linear variation of the carbon-metal interaction with the free energy of formation from pure elements. This implies that the interaction of the solutes in fcc iron is very similar to the interaction between these elements in the carbide. This striking linear correlation is somewhat lost if we instead consider the free energy of formation from austenite. The dissolution free energies of transition metal solutes in the fcc form in fcc Fe vary fairly linearly, but the lattice stability of solutes (free energy difference between the fcc and bcc forms) varies in a nonlinear way with group number. Therefore the combination of these energies which is used in the calculation of free energy of formation from austenite varies nonlinearly. However, two groups of solutes can be identified, those that form cubic carbides (group IV and V) and others that form hexagonal, orthorhombic carbides (Cr,Mn,W, Mo etc.) and a fairly good correlation exists within each group.

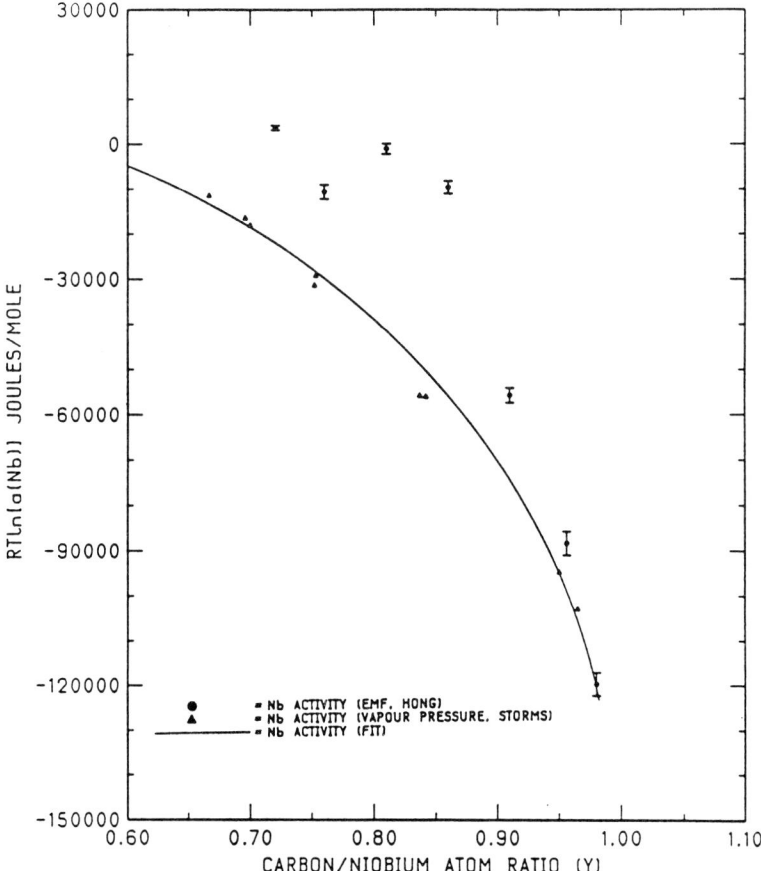

Fig. 8. Variation of the Partial Molar Free Energy of Nb in NbCy (T=2400K).

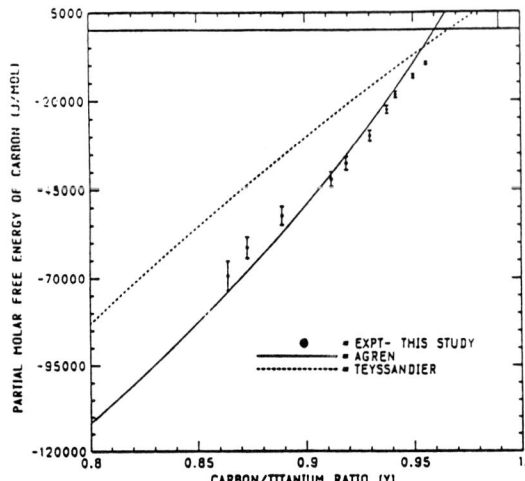

Fig. 9. Variation of the Partial Molar Free Energy of C in TiCy (T=1473K).

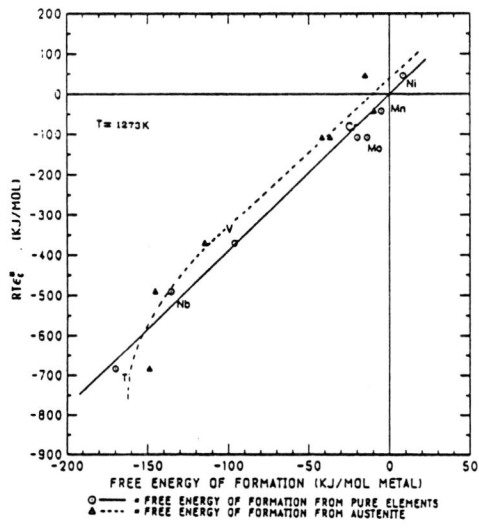

Fig. 10. Correlational Relationship: Carbon-Metal Interaction Parameter vs Free Energy of Formation of Carbides.

REFERENCES

1. Alekseev, V. I., Panov, A. S., Fiveiskii, Ye. V., and Shvartsman, L. A. (1967-68). Proc. Thermodynamics of Nuclear Materials 1967, IAEA, Vienna, Published in 1968, pp 435-447.
2. Balasubramanian, K. (1988). Ph.D Thesis, McMaster University, Canada.
3. Chipman, J. (1972). Metall. Trans. 3, pp 55-64.
4. Darken, L. S. (1967). TMS-AIME, 239 pp 90-96.

5. de Kazinsky, F., Axnas, A., and Pachleiter, P. (1963). Jernkont. Ann., 147 p 408.
6. Goldschmidt, H. J. (1967). Interstitial Alloys, Sections 2.5 and 4.1, Plenum Press, Butterworths, London, pp 44-46 and 88-101.
7. Grieveson, P. (1967). Proc. Brit. Ceramic Soc., 8, pp 137-153.
8. Hillert, M. (1986). Metall. Trans. A 17, pp 1878-1879.
9. Hillert, M., and Staffansson, L. I. (1970). Acta Chem. Scand., 24, pp 3618-3626.
10. Hong, Y. R., Kumar, R. V., Balasubramanian(*), K., and Kay(*), D. A. R. (1988). to be submitted to Metall. Trans. B (* - Department of Materials Science and Engineering, McMaster University, Canada).
11. Hultgren, R., Desai, P., Hawkins, D. T., Gleiser, M., and Kelly, K. K. (1973). Selected Values of the Thermodynamic Properties of Binary Alloys, American Society of Metals p 520.
12. Irvine, K. J., Pickering, F.B., and Gladman, T., (1967). J. Iron and Steel Inst., 205 p 171.
13. Johansen, T. H., Christensen, N., and Augland, B. (1967). Trans. TMS-AIME 239, p 1651.
14. Kaufman, L., and Nesor, H. (1978). CALPHAD, 2, pp 56-63 and p 117.
15. Kaufman, L., Agren, J., (1982). Technical Report : F 49620-80-C-0020, Air Force Office of Scientific Research (AFSC), Bolling Air Force Base, D.C. 20332.
16. Koyama, S. (1972). Japan Inst. Metals, Journal, 52, p 1090.
17. Koyama, K., and Hashimoto, Y. (1973). Nippon Kinzoku Gakkai-si 37 pp 406-411.
18. Kubaschewski, O., Evans, E. LL., and Alcock, C. B. (1967). Metallurgical Thermochemistry, Fourth Edition, Pergamon Press, pp 421-429.
19. Malkin, V. I., and Pokidshev, V. V. (1971). Russ. J. Phys. Chem., 45 pp 1159-1161.
20. Meyer, L. (1966). Dissertation, Clausthal Berg., Hulten., and Z. Metallkunde, 58, pp 334-339.
21. Mori, T. (1964). Tetsu-to-Hagane, 50, p 911.
22. Mori, T., Tokizane, M., Yamaguchi, K. (1968). Tetsu-to-Hagane, 54, p 763.
23. Murray, J. L., (1981). Bulletin of Alloy Phase Diagrams, 2 p 320.
24. Narita, K. (1956). J. Chem. Soc. Japan, 77 p 1536.
25. Narita, K. (1959). J. Chem. Soc. Japan, 80 p 266.
26. Narita, K, and Koyama, S. (1966). Jap. Inst. Metals, Journal, 52 p 292.
27. Ohtani, H., Nishizawa, T., Tanaka, T., and Hasebe, M. (1985). Proc. Japan-Canada Seminar on Secondary Steelmaking, Tokyo, Japan., Publ. The Canadian Steel Industry Research Association (CSIRA) and The Iron and Steel Institute of Japan (ISIJ), p J-7-1.
    Also: Nishizawa, T., Ohtani, H., and Hasebe, M. (1985). Presentation, CALPHAD XIV, Boston, Mass.
28. Pelton, A. D., and Bale, C. D. (1986). Metall. Trans. A 17, pp 1211-1215.
29. Sharma, R. C., Lakshmanan, V. K., and Kirkaldy, J. S. (1984). Metall. Trans A 15, p 545.
30. Shiraiwa, T., Fujino, N., and Murayama, J., (1970). Trans. Iron and Steel Inst. Japan, 10 p 406.
31. Smith, R. P. (1962). Trans. TMS-AIME, 224, p 190.
32. Smith, R. P. (1966). Trans. TMS-AIME, 236, p 220.
33. Storms, E. K. (1967). Refractory Carbides, Academic Press, New York and London, pp 16-17.
34. Storms, E. K., Calkin, B., and Yencha, A. (1969). High Temp. Sci. 1, p 430.
35. Teyssandier, F., Ducarrior, M., and Bernard, C. (1984). CALPHAD, 8 p 233.
36. Uhrenius, B. (1984). CALPHAD 8, p 101.
37. Wagner, C. (1952) Thermodynamics of Alloys, Addison Wesley, Reading, Mass., p 51.

Phase Equilibrium Calculations in Multicomponent Systems

A.D. PELTON*, W. T. THOMPSON**, C. W. BALE* and N. G. ERIKSSON*

CRCT – Centre de Recherche en Calcul Thermochimique
*Ecole Polytechnique de Montreal
P.O. Box 6079, Station A, Montréal, Québec, Canada, H3C 3A7
**Royal Military College, Kingston, Ontario, Canada, K7K 5L0

ABSTRACT

Extensive computer databases are being prepared for the thermodynamic properties of a large variety of multicomponent solutions. Evaluated data for binary and ternary phases are stored in the form of parameters of model equations. The same models are then used to calculate properties of multicomponent phases. The modified interaction parameter formalism, which we have developed to be thermodynamically correct beyond infinite dilution, is used to model dilute solutions such as steel. We have developed an extended quasichemical model which provides very good representations of properties of multicomponent silicate melts and other ordered solutions. We have extended the sublattice model for use in multicomponent salt solutions. The Pitzer equations are used for concentrated aqueous solutions. Other models are employed where appropriate. The solution databases, along with a large database for the properties of pure compounds, are accessible by general interactive on-line programs which calculate multicomponent phase equilibria. The equilibria can be calculated under a variety of constraints such as mass constraints, chemical potential constraints, total enthalpy constraints, etc. Users of the program can also employ their own private data in conjunction with data from the public databases.

INTRODUCTION

With the advent of electronic computing, there has been a renaissnce of applied chemical thermodynamics. Numerical methods of free energy minimization have been developed for the computation of phase equilibria in multicomponent systems. Several publicly-accessible database computing systems are now available in which such programs are coupled with extensive thermodynamic data banks.

For multicomponent solution phases, evaluated data for binary and sometimes ternary sub-systems are stored in the form of parameters of model equations. The same models are then used to estimate the properties of the multicomponent solutions from the binary and ternary parameters.

We shall begin with two examples of phase equilibrium calculations using the multicomponent solution databases of the F*A*C*T (Facility for the Analysis of Chemical Thermodynamics) computing system which we have developed. Following this, we shall outline some of the solution models which we have proposed.

Calculation of Slag/Metal Equilibrium

In Fig. 1 is shown the input/output for a computer calculation of changes in metal and slag compositions during steelmaking in the BOF. In the first line of Fig. 1 the user indicates that the hot metal contains 94.0g Fe, 1.1g Si, 0.8g Mn and 4.1g C. In addition, a flux consisting of 6.25g CaO and 1.25g CaO.MgO is added and 8.0g of O_2 are blown in. In subsequent lines of input (not shown on Fig. 1) the user specifies a total pressure of 1.0 atm and a temperature of 1873 K. All required thermodynamic data are then automatically retrieved from the various databases and the total amount and composition of each phase at equilibrium is calculated as shown in Fig. 1. The calculated compositions of the metal and slag phases agree closely with measured values (Lankford, 1985).

Other phases such as solid CaO and Ca_2SiO_4 were considered in the calculation, but it can be seen from Fig. 1 that they do not form under the given conditions. However, from their activities which are given in the line of output following each solid, it can be seen that CaO, with an activity of 0.985 is very close to precipitating.

For the metal and slag phases, the modified interaction parameter model and the modified quasichemical model respectively were used. These are discussed below.

This is an example of a mass-contrained equilibrium calculation since the total mass of each component is fixed along with an equilibrium T and P. The program can also calculate equilibria under other constraints such as fixed chemical potential of a species, constant volume, adiabatic reaction from a specified initial state, etc.

Calculation of Gas/Molten Salt Equilibrium

In Fig. 2 is shown the input/output for a computer calculation of equilibria occurring in a molten carbonate fuel cell. In the line of input, a molten carbonate electrolyte containing 0.1 mol each of Li_2CO_3 and K_2CO_3 as well as 0.02 mol of Na_2CO_3 is placed in contact with a gas phase containing 7.5 mol CO_2, 59.4 mol H_2, 23.3 mol H_2O and 0.3 mol H_2S. In a subsequent line of input (not shown on Fig. 2), a temperature of 973 K and a pressure of 10.0 atm are specified. All data are then automatically retrieved from the databases and the equilibrium amount and composition of each phase are calculated. As can be seen in Fig. 2, substantial amounts of hydroxide dissolve in the molten salt phase along with small amounts of sulphate. Electrolyte losses to the gas phase occur through formation of $KOH.H_2O$ and $LiOH.H_2O$ (Pelton, 1981). No solid phases precipitate under the given conditions.

The sublattice model discussed below was used for the molten salt phase.

SOLUTION MODELS

Modified Interaction Parameter Model

The interaction parameter formalism first proposed by Wagner (1962) is frequently used for representing the thermodynamic properties of dilute solutions, particularly for iron-based alloys. In the usual formulation for a system of N solutes (numbered 1, 2, ... N) and a "solvent", $\ln \gamma_i / \gamma_i^\circ$ for each solute is expanded as:

$$\ln \gamma_i / \gamma_i^\circ = \epsilon_{i1} X_1 + \epsilon_{i2} X_2 + \epsilon_{i3} X_3 + \ldots + \epsilon_{iN} X_N \quad [i = 1, \ldots, N] \quad [1]$$

where γ_i is the activity coefficient, γ_i° is the value at infinite dilution, X_i is the mole fraction and the ϵ_{ij} are "first order interaction parameters".

It was first shown by Darken (1967) that the formalism is limited to infinite dilution and is thermodynamically inconsistent at finite concentrations. Nevertheless, the formalism is often used to compute activity coefficients at solute concentrations of several atomic percent. Depending upon the system, at these concentrations the thermodynamic inconsistency can result in significant errors.

The thermodynamic inconsistency is easily seen (Pelton, 1986a) by noting that eq [1] does not obey the necessary relationship:

$$(\partial \ln \gamma_j / \partial n_i) = (\partial \ln \gamma_i / \partial n_j) \quad [2]$$

The Gibbs-Duhem equation is also not respected.

We have shown (Pelton, 1986a) however, that a very simple modification can be made to the formalism in order to render it thermodynamically consistent at finite concentrations. Eq. [1] is modified to:

$$\ln \gamma_i / \gamma_i^\circ = \ln \gamma_{solvent} + \epsilon_{i1} X_1 + \epsilon_{i2} X_2 + \epsilon_{i3} X_3 + \ldots + \epsilon_{iN} X_N \quad [3]$$

where:
$$\ln \gamma_{solvent} = -\frac{1}{2} \sum_{j=1}^{N} \sum_{k=1}^{N} \epsilon_{jk} X_j X_k \quad [4]$$

Eq. [4] for $\ln \gamma_{solvent}$ contains only quadratic terms which may be ignored in very dilute solutions. That is, at infinite dilution the modified formalism reduces exactly to the standard formalism. Furthermore, the notation of the standard formalism is preserved as are the numerical values of the ϵ_{ij} parameters. Hence, the existing extensive compilations of interaction parameters can be used directly in the modified formalism.

The modified interaction parameter formalism has been extended (Pelton, 1986a) to include higher-order terms (quadratic, cubic etc.) for systems of any number of components.

Sublattice Model

A two-lattice model, one for cations and one for anions, is appropriate for molten salt solutions. The sublattice model for molten salts was first proposed by Temkin (1945) and extensively developed by Blander (1963, 1964), Førland (1964), Reiss and others (1962-1976), Saboungi and others (1975, 1980). A general extension of the sublattice model to systems of any number of components has recently been proposed by one of the authors (Pelton, 1983).

Let cations be denoted by A, B, C, ... and anions by X, Y, Z, ... Let q_i be the absolute charge on an ion. Cationic and anionic site fractions are denoted by X_A, X_B, X_C, ... and X_X, X_Y, X_Z, ... while equivalent site fractions are denoted by Y_A, Y_B, Y_C, ... and Y_X, Y_Y, Y_Z, ... For example, Y_A is defined as:

$$Y_A = q_A n_A / (q_A n_A + q_B n_B + q_C n_C + \ldots) \qquad [5]$$

The integral Gibbs energy per equivalent of solution is given by:

$$g = \sum_c \sum_a Y_c Y_a\, g^\circ_{c/a} + \Delta g(\text{ideal}) + g^E \qquad [6]$$

where $g^\circ_{c/a}$ is the standard Gibbs energy per equivalent of the neutral salt consisting of cations c and anions a and the summation is over all cations c and anions a. g^E is the excess Gibbs energy per equivalent. The ideal mixing term is given by assuming that the cations and anions mix randomly on their respective sites regardless of charge:

$$\Delta g(\text{ideal})/RT = (\Sigma q_c X_c)^{-1} (\Sigma X_c \ln X_c) + (\Sigma q_a X_a)^{-1} (\Sigma X_a \ln X_a) \qquad [7]$$

In a binary common-anion system A,B/X containing one anion and two cations, the excess Gibbs energy per equivalent is expressed as a polynomial in the equivalent cationic fractions:

$$g^E = \sum_{i \geq 1} \sum_{j \geq 1} \phi^{AB/X}_{ij} Y_A^i Y_B^j \qquad [8]$$

where the $\phi^{AB/X}$ are constant binary coefficients obtained by fitting experimental data.

For example, in the system K,Li/CO$_3$ a least-squares optimization of available phase diagram and thermodynamic data (Sangster, 1988) gives, for the liquid solution:

$$g^E = Y_K Y_{Li} [(-13780+8.197T)+(-5500-2.329T) Y_{Li}] \text{ J/equiv.} \qquad [9]$$

That is:

$$\phi_{11}^{KLi/CO_3} = (-13780+8.197T) \text{ and } \phi_{12}^{KLi/CO_3} = (-5500-2.329T)$$

In this example, the data considered by the least-squares optimization program were calorimetric data for liquid-liquid mixing (Andersen, 1976), and the measured phase diagram (Rolin, 1964; Janz, 1961). Gibbs energies of fusion of the components were taken from (Barin, 1973). The resultant optimized equation [9] reproduces the measured phase diagram very well as can be seen in the computer-generated diagram in Fig. 3. The measured enthalpy of mixing (Adersen, 1976) is also reproduced very closely by eq. [9].

The least-squares optimization technique also gives properties of solid compounds and solutions in the system. In the present example, the Gibbs energy of formation of the compound $KLiCO_3$ from the pure liquid carbonates was given as:

$$\Delta G°_{form} = -89654 + 78.136 \, T \quad J/mol \qquad [10]$$

The technique of least-squares optimization of binary thermodynamic and phase diagram data has been developed by the authors (Bale, 1983). By simultaneously optimizing experimental thermodynamic and phase equilibrium data, one set of self-consistent equations for the thermodynamic properties of all phases is obtained as illustrated above. The binary parameters obtained in this way are then stored in the computer database. To date we have critically evaluated and optimized the data on approximately 150 binary molten salt systems.

Model equations for estimating g^E of eq [6] for multicomponent salt solutions from the optimized parameters of the binary g^E expansions have been presented (Pelton, 1983). These are incorporated in the F*A*C*T programs, and were used in the example of Fig. 2 to calculate the properties of the 5-component liquid $Li,Na,K/CO_3,OH,SO_4$ solution from the optimized binary parameters stored in the database.

The multicomponent model has been tested against experimental data where available. For example, in Fig. 4 is shown the phase diagram of the $Li,K/OH,CO_3$ reciprocal ternary system which was calculated (Pelton, 1981) from the sublattice model solely from optimized parameters from the four binary sub-systems and from tabulated values of $g°_{c/a}$ for the pure components. Comparison with the reported (Reshetnikov, 1968) phase diagram shows excellent agreement. Along the $LiOH-K_2CO_3$ diagonal the maximum deviation is 20°C.

Sublattice models can be applied to many systems other than molten salts. Hillert (1970, 1975, 1985), Guillermet (1981), Sundman (1981) and co-workers have had much success in using the sublattice concept for a wide variety of cases. In many ceramic phases for instance, one can distinguish two or more cationic sublattices. In both metals and ionic solids the interstitial lattice may also be treated as a sublattice. For example, in solid Fe-Ni-C solutions the Fe and Ni atoms occupy the metal sublattice while carbon atoms and vacancies are distributed on the interstitial lattice. The sublattice model is also suited to treating departures from perfect ordering in systems with long-range order. In an ordered alloy A/B for example, with A and B on "cation" and "anion" sites respectively, we might consider, as lattice defects, A atoms on B sites and B atoms on A sites. The equilibrium concentration of defects is then obtained by minimizing the Gibbs energy expression given by the sublattice model.

The Modified Quasichemical Model

Simple polynomial expansions such as used in eq. [9] cannot adequately represent the excess properties of binary liquid solutions which exhibit strong structural "ordering" about a certain composition. In a binary liquid phase with ordering, the enthalpy of mixing tends to exhibit a negative peak near the composition of maximum ordering while the entropy of mixing tends to have the shape of the letter "m" with a minimum near this composition. In binary liquid-silicate solutions $MO-SiO_2$ (M = Ca, Mg, Pb, Fe, Mn, etc.) for example, such ordering is observed about a mole fraction of SiO_2, $X_{SiO_2} = 1/3$ corresponding to the orthosilicate composition M_2SiO_4. For such solutions, a model is required which accounts for ordering and which thus gives the proper characteristic shape of the ΔH and ΔS functions.

In previous publications (Pelton, 1984, 1986b; Blander, 1987) a modification of the quasichemical theory of Guggenheim (1935) was proposed which satisfies this requirement. In a binary system with components "1" and "2", the "1" and "2" particles are considered to mix substitutionally on a quasi-lattice. The relative amounts of the three types of nearest neighbour pairs (namely, 1-1, 2-2 and 1-2 pairs) are determined by the energy change $(\omega - \eta T)$ associated with the formation of two 1-2 pairs from a 1-1 and 2-2 pair according to:

$$[1-1] + [2-2] = 2[1-2] \qquad [11]$$

If this energy change is zero, then the solution is an ideal mixture. As this energy change becomes more and more negative, the formation of 1-2 pairs is favoured. As a result, the entropy and enthalpy functions have an increasing tendency to exhibit minima at the equimolar composition.

A "quasichemical" equilibrium constant can be written:

$$\frac{X_{12}^2}{X_{11} X_{22}} = 4e^{-(\omega - \eta T)/RT} \qquad [12]$$

where X_{11}, X_{22} and X_{12} are the fractions of each type of pair in solution.

From the mass balances we can write:

$$2X_1 = 2X_{11} + X_{12} \qquad [13]$$

$$2X_2 = 2X_{22} + X_{12} \qquad [14]$$

where X_1 and X_2 are the mole fractions of the components.

In order to permit the composition of maximum ordering to be chosen to correspond to that which is observed in any given system, we replace the mole fractions with "equivalent fractions", Y_1 and Y_2, defined by:

$$Y_1 = \frac{b_1 X_1}{b_1 X_1 + b_2 X_2} \qquad Y_2 = \frac{b_2 X_2}{b_1 X_1 + b_2 X_2} \qquad [15]$$

where b_1 and b_2 are numbers chosen so that $Y_1 = Y_2 = 1/2$ at the composition of maximum ordering. In systems MO-SiO_2 (M = divalent cation), this occurs at $X_{SiO_2} = 1/3$, so that $b_{SiO_2}/b_{MO} = 2$. As discussed previously (Pelton, 1984), in order that the configurational entropy be zero at the composition of maximum ordering when $(\omega - \eta T) = -\infty$, we choose $b_{MO} = 0.6887$ and $b_{SiO_2} = 1.3774$.

Eqs. [13, 14] now are replaced by:

$$2Y_1 = 2X_{11} + X_{12} \qquad [16]$$

$$2Y_2 = 2X_{22} + X_{12} \qquad [17]$$

Eqs. [12, 16, 17] may now be solved to give values of X_{11}, X_{22} and X_{12}. The partial Gibbs energies and activities are then given as:

$$\Delta G_i = RT\ln X_i + b_i RT\ln \frac{X_{ii}}{Y_i^2} + b_i \left(\frac{X_{12}}{2}\right)(1-Y_i)\frac{\partial(\omega-\eta T)}{\partial Y_i} \qquad [18]$$

In order to obtain precise representations of experimental data, it is necessary to take into account empirically the configurational dependence of the energy parameters. This is done by expanding ω and η as polynomials:

$$\omega = \omega_0 + \omega_1 Y_2 + \omega_2 Y_2^2 + \omega_3 Y_2^3 + \ldots \qquad [19]$$

$$\eta = \eta_0 + \eta_1 Y_2 + \eta_2 Y_2^2 + \eta_3 Y_2^3 + \ldots \qquad [20]$$

The coefficients ω_i and η_i are chosen empirically so as to give the best representation of all the available experimental data for a system. A computer program described elsewhere (Pelton, 1988) has been written to perform such optimizations via a least-squares technique. With this program, simultaneous optimizations of thermodynamic and phase diagram data are being carried out systematically on many oxide systems involving the components SiO_2, Al_2O_3, MgO, CaO, MnO, FeO, Na_2O, etc. In this way, a database of evaluated binary parameters is being built up. The representations which are obtained for these oxide systems with the quasichemical model are as precise as those obtained for simpler solutions with the polynomial model as discussed in the preceding section. Examples of optimizations of several binary oxide systems have been given (Pelton, 1984, 1986b; Blander 1987).

For ternary systems, the modified quasichemical equations may be extended in a straightforward manner. Complete details were given previously (Pelton, 1986b; Blander, 1987). In a ternary system we consider three pair-bond equilibria:

$$[1-1] + [2-2] = 2[1-2] \qquad [21]$$

$$[2-2] + [3-3] = 2[2-3] \qquad [22]$$

$$[3-3] + [1-1] = 2[3-1] \qquad [23]$$

The energy and entropy changes for these reactions (ω_{12}, ω_{23}, ω_{31}; η_{12}, η_{23}, η_{31}) are obtained from the three binary subsystems as illustrated in the preceding section.

Three "quasichemical equilibrium constants" as in eq. [12] may be written for the three equilibria [21-23]. In addition, there are three mass balance constraints (similar to eqs. [16, 17]). These may be solved simultaneously to give the six bond fractions X_{11}, X_{22}, X_{33}, X_{12}, X_{23}, X_{31} which can then be used to calculate all the thermodynamic properties of the ternary solution.

The MgO-CaO-SiO$_2$ System

As an example, we shall consider the MgO-CaO-SiO$_2$ ternary system. Optimized ω_{ij} and η_{ij} parameters for the MgO-SiO$_2$, CaO-SiO$_2$ and MgO-CaO systems have been determined (Pelton, 1984, 1986b; Blander, 1987). When these binary parameters are used to estimate the thermodynamic properties of the ternary liquid phase as discussed above, excellent agreement with measured activities of SiO$_2$ are obtained (Blander, 1987).

From the estimated Gibbs energy surface for the ternary liquid phase, the ternary phase diagram can also be calculated. The phase diagram computed in this way is shown in Fig. 5. For the four ternary compounds, Gibbs energies of formation as given by Helgeson (1978) were used with very slight adjustments permitted to the ΔH°_{298} terms in order to reproduce exactly the reported (Osborn, 1960) melting points of the compounds. For the compounds merwinite, monticellite, akermanite and diopside, the values of ΔH°_{298} given by Helgeson (1978) are respectively: -4566.65, -2262.71, -3878.30 and -3203.26 kJ/mol, while our adjusted values are, respectively: -4566.65, -2260.07, -3875.21 and -3199.55 kJ/mol. Limited solid solubility was assumed of Mg$_2$SiO$_4$ and Ca$_2$SiO$_4$ in each other, of CaSiO$_3$ in MgSiO$_3$, and of "Mg$_3$SiO$_5$" in Ca$_3$SiO$_5$. The limits of solid solubility were adjusted so as to reproduce the reported (Osborn, 1960) liquidus surface, but were always within the range of the reported (Levin, 1964-1987) solid solubilities.

Further details of the calculation of the ternary phase diagram will be reported elsewhere. The calculated diagram in Fig. 5 may be compared to the diagram reported by Osborn (1960). With two exceptions, the calculated ternary invariant points agree with the reported values within 20°C and 3 wt.% or better. The two exceptions are the invariants calculated at 1827°C and 1901°C which border on the phase field of Ca_3SiO_5. These are reported by Osborn (1960) at 1790°C and 1850°C respectively. The calculated compositions are also displaced about 6 wt.% from their reported (Osborn, 1960) positions.

The use of the model thus permits virtually quantitative calculation of the ternary phase diagram. Only binary ω_{ij} and η_{ij} parameters obtained from optimization of data in the three binary sub-systems were used. It is possible to include ternary correction terms in the model. This will be discussed in future publications. However, from the ternary systems which we have examined to date, it would appear that for systems consisting of SiO_2 with two basic oxides, such ternary terms, if required at all, are very small. When Al_2O_3 is a component, some small ternary terms are required. These may also be found by least-squares optimization as will be discussed in forthcoming publications.

For a ternary system, the solution of the "quasichemical equilibria" of reactions [21-23] is very similar to a complex chemical equilibrium calculation and can be performed by similar computer algorithms. The extension to calculations involving more than three components is straightforward. Only minor modifications of existing computer programs are required. Hence, the thermodynamic properties of multicomponent oxide melts and glasses can be estimated from the optimized binary (and possibly, ternary) parameters.

As illustrated above by the calculation shown in Fig. 1, the multicomponent quasichemical model has already been incorporated into the F*A*C*T on-line system.

CONCLUSIONS

Extensive computer databases are being prepared for the thermodynamic properties of a large variety of multicomponent solutions. Evaluated data for binary, and sometimes ternary, sub-systems are stored in the form of parameters of model equations. The same models are then used to estimate the properties of the multicomponent solutions from the binary and ternary parameters.

These solution databases, along with a large database for the properties of pure compounds, are accessible by the general free-energy minimization programs of the F*A*C*T on-line system thereby permitting the calculation of complex multicomponent equilibria.

For dilute solutions such as steel, we have modified the interaction parameter formalism to be thermodynamically correct at finite concentrations. For silicate melts and other ordered liquids we have modified the quasichemical model. This model can give quantitative predictions of ternary silicate phase diagrams based only upon optimized parameters for the liquid phases of the binary sub-systems. For multicomponent molten salt solutions we have extended the sublattice model. Other models such as the Pitzer model for aqueous solutions are used where appropriate.

ACKNOWLEDGEMENTS

The authors are indebted to Dr. M. Blander. Financial assistance from the Natural Sciences and Engineering Research Council of Canada and from the TIP program of the Canadian Ministry of External Affairs is gratefully acknowledged.

REFERENCES

Andersen, B.K. and O.J. Kleppa, (1976). <u>Acta Chem. Scand.</u>, A30, p. 751.

Bale, C.W. and A.D. Pelton, (1983). <u>Met. Trans.</u>, vol. 14B, pp. 77-83.

Barin, I., O. Knacke and O. Kubaschewski, (1973, 1977). <u>Thermochemical Properties of Inorganic Substances</u> (and supplement), Springer-Verlag, N.Y.

Blander, M. (1964). In M. Blander (Ed.), <u>Molten Salt Chemistry</u>, Interscience, N.Y., chap. 3.

Blander, M. and A.D. Pelton, (1987). Thermodynamic Analysis of binary liquid slags and prediction of ternary slag properties by modified quasichemical equations. <u>Geochim et Cosmochim. Acta</u>, vol. 51, pp. 85-95.

Blander, M. and S.J. Yosim, (1963). <u>J. Chem. Phys.</u>, vol. 39, p. 2610.

Darken, L.S., (1967). <u>TMS-AIME</u>, vol. 239, pp. 90-96.

Førland, T, (1964). In B.R. Sundheim (Ed.), <u>Fused Salts</u>, McGraw Hill, N.Y., chap. 2.

Guggenheim, E.A., (1935). <u>Proc. Roy. Soc.</u>, vol. A148, p. 304; R.H. Fowler and E.A. Guggenheim, (1939). <u>Statistical Thermodynamics</u>, Cambridge, pp. 350-366.

Guillermet, A.F., M. Hillert, B. Jansson and B. Sundman, (1981). <u>Metall. Trans.</u>, vol. 12B, pp. 747-54.

Helgeson, H.C., J.M. Delany, H.W. Nesbitt and D.K. Bird, (1978). Summary and critique of the thermodynamic properties of rock-forming minerals, <u>Am. J. Sci.</u>, vol. 278-A.

Hillert, M., B. Jansson, B. Sundman and J. Agren (1985). <u>Met. Trans</u>, vol. 16A, pp 261-266.

Hillert, M. and L.-I. Staffansson, (1970). <u>Acta Chem. Scand.</u>, vol. 24, pp. 3618-26.

Hillert, M. and L.-I. Staffansson, (1975). <u>Met. Trans.</u>, vol. 6B, pp 37-41.

Janz, G.J. and M. Lorenz, (1961). J. Chem. Eng. Data, vol. 6, p. 321.

Lankford Jr., W.T., N.L. Samways, R.F. Craven and H.E. McGannon (Editors) (1985). The Making, Shaping and Treating of Steel, 10th Ed., U.S. Steel Co., p. 469.

Levin, E. and co-workers, (1964-1987). Phase Diagrams for Ceramists, and supplements, American Ceramic Society, Columbus, Ohio.

Osborn, E.F. and A. Muan, (1960). Phase Equilibrium Diagrams of Oxide Systems, American Ceramic Society, Columbus, Ohio; see also Levin (1964-1987), Fig. 598.

Pelton, A.D., (1988). A Database and Sublattice Model for Molten Salts, Calphad Journal, in press; A Thermodynamic Database for Multi-component Molten Salt Solutions, Proc. Sympos. Electrochem. Soc., Hawaii, The Electrochem. Soc., in press.

Pelton, A.D. and C.W. Bale, (1986a). Met. Trans., vol. 17A, pp. 1211-1215.

Pelton, A.D., C.W. Bale and P.L. Lin, (1981). Calculation of Thermodynamic Equilibria in the Carbonate Fuel Cell - Volume I; C.W. Bale, A.D. Pelton and J. Melançon, Volume II, Report to U.S. Dept. of Energy, Contract No. DE-AC02-79ET16416, Argonne National Labs. Argonne, IL.

Pelton, A.D. and M. Blander, (1984). Computer assisted analysis of the thermodynamic properties and phase diagrams of slags, Proc. 2nd Int'l Symposium on Metall. Slags and Fluxes, TMS-AIME, Warrendale, PA, pp. 281-294.

Pelton, A.D. and M. Blander, (1986b). Thermodynamic analysis of ordered liquid solutions by a modified quasichemical approach-application to silicate slags, Met. Trans., vol. 17B, pp. 805-816.

Pelton, A.D. and M. Blander, (1988). A least squares optimization technique for the analysis of thermodynamic data in ordered liquids, Calphad Journal, vol. 12, pp. 97-108.

Reiss, H. and O.J. Kleppa, (1962). J. Phys. Chem., vol. 36, p. 144; M. Blander and S.J. Yosim, (1963). J. Chem. Phys., vol. 39, p. 2610; M.-L. Saboungi and P. Cerisier, (1974). J. Electrochem. Soc., vol. 121, p. 1258; M.-L. Saboungi and M. Blander, (1975). J. Chem. Phys., vol. 63, p. 212; M. Blander and M.-L. Saboungi, (1976). Molten Salts, The Electrochem. Soc., pp. 93-106.

Reshetnikov, N.A. and O.G. Perfil'eva, (1968). Zh. Neorg. Khim., vol. 13[6], p. 1662; Russ J. Inorg. Chem., vol. 13[6], p. 870; see also Levin (1964-1987), Fig. 4672.

Rolin, M. and J. Recapet, (1964). Bull. Soc. Chim. Fr., p. 2104.

Saboungi, M.-L., (1980). J. Chem. Phys., vol. 73, p. 5800.

Saboungi, M.-L. and M. Blander, (1975). J. Am. Ceram. Soc., vol. 58, p. 1.

Sangster, J., Y. Dessureault and A.D. Pelton, (1988). To appear in Phase Diagrams for Ceramists, vol. 7.

Sundman, B. and J. Agren, (1981). J. Phys. Chem. Solids, vol. 42, pp. 297-301.

Temkin, M., (1945). Acta Phys. Chim. USSR, vol. 20, p. 411.

Wagner, C., (1962). Thermodynamics of Alloys, Addison-Wesley, Reading, MA, p. 51.

94.0 FE + 1.1 SI + 0.8 MN + 4.1 C + 8 O2 + 6.25 CA*O + 1.25(CA*O)(MG

```
        51.658      litre  (   97.378       vol% CO
                           +   2.6019       vol% CO2
                           +   0.10689E-01  vol% Mn
                           +   0.82304E-02  vol% Fe
                           +   0.10224E-02  vol% Mg
                           +   0.90245E-05  vol% Ca
                           +   0.32093E-06  vol% SiO
                           +   0.27213E-06  vol% O
                           +   0.13275E-07  vol% O2)
                           ( 1873.0,  1.00       ,G)

  +     90.720      gram   (   99.752       wt.% Fe
                           +   0.14285      wt.% Mn
                           +   0.69876E-01  wt.% C
                           +   0.34935E-01  wt.% O2
                           +   0.10289E-05  wt.% Si)
                           ( 1873.0,  1.00       ,SOLN 2)

  +     15.223      gram   (   45.835       wt.% CaO
                           +   29.603       wt.% FeO
                           +   15.459       wt.% SiO2
                           +   5.6699       wt.% MnO
                           +   3.4329       wt.% MgO)
                           ( 1873.0,  1.00       ,SOLN 3)

                       +   0.00000E+00 gram CaO
                           ( 1873.0,  1.00       ,S1, 0.98529      )

                       +   0.00000E+00 gram (CaO)(MgO)
                           ( 1873.0,  1.00       ,S1, 0.68177      )

                       +   0.00000E+00 gram (CaO)2(SiO2)
                           ( 1873.0,  1.00       ,S2, 0.31625      )

                       +   0.00000E+00 gram (CaO)(SiO2)
                           ( 1873.0,  1.00       ,S2, 0.11287E-01)
```

FIGURE 1: Input/Output to F*A*C*T program to calculate equilibrium amounts and compositions of gas, molten metal and molten slag phases at 1873 K and 1 atm. during BOF steelmaking First line is input in grams. Note that no solid oxide precipitate.

9.8 C*O + 7.5 C*O2 + 59.4 H2 + 23.3 H2O + 0.3 H2S + 0.1 LI2C*O3 + 0.1 K2C*O3 + 0.02 NA2C*O3 =

```
   85.697      (  0.41727       H2
               +  0.37732       H2O
               +  0.85200E-01   C*H4
               +  0.67281E-01   C*O2
               +  0.49419E-01   C*O
               +  0.34802E-02   H2S
               +  0.19437E-04   C*O*S
               +  0.13786E-05   K*O*H(H2O)
               +  0.77965E-06   LI*O*H(H2O)
               +  0.75660E-06   C2H6
               +  0.34476E-06   K(O*H)
               +  0.29490E-06   C*H2O
               +  0.27314E-06   C4H8
               +  0.13952E-06   C2H4
               +  0.49531E-07   LI*O*H
                 (  973.0,  10.0     ,G)

 + 0.22375     (  0.15681E-01   LI*O*H
               +  0.31373E-02   NA*O*H
               +  0.15675E-01   K*O*H
               +  0.43876       LI2C*O3
               +  0.87783E-01   NA2C*O3
               +  0.43859       K2C*O3
               +  0.16987E-03   LI2S*O4
               +  0.33986E-04   NA2S*O4
               +  0.16980E-03   K2S*O4
                 (  973.0,  10.0    ,SOLN 2)

               +  0.00000E+00   K2C*O3
                 (  973.0,  10.0    ,S1, 0.25751    )

               +  0.00000E+00   LI2C*O3
                 (  973.0,  10.0    ,S3, 0.99619E-01)

               +  0.00000E+00   NA2C*O3
                 (  973.0,  10.0    ,S2, 0.11737E-01)
```

FIGURE 2: Input/Output to F*A*C*T program to calculate equilibrium amounts and compositions of gas and molten salt phases at 973 K and 10 atm in a molten carbonate fuel cell. First line is input in moles. Output gives total number of moles of each phase and composition in mole fraction. Note that no solid carbonates precipitate.

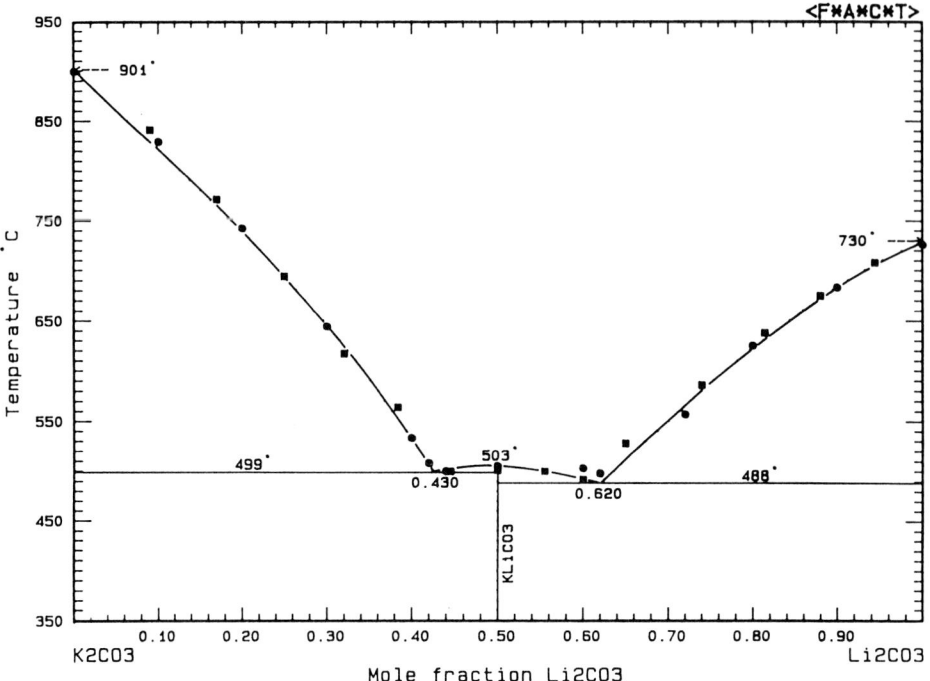

FIGURE 3: Phase diagram of the K_2CO_3-Li_2CO_3 system. Lines are calculated from the optimized thermodynamic equations. Points are experimental (Janz, 1961;, Rolin, 1964).

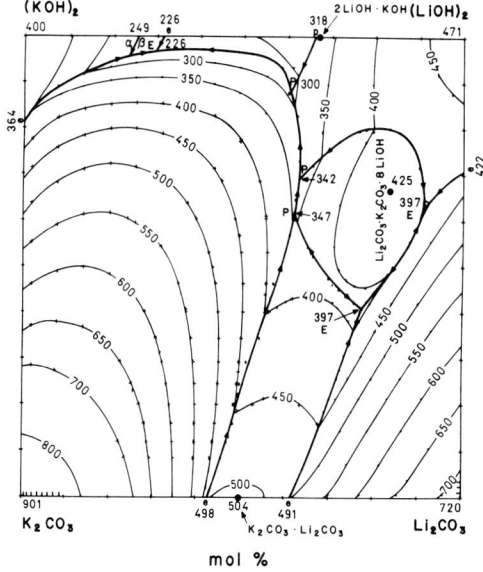

FIGURE 4: Liquidus surface of the ternary reciprocal salt system K,Li/CO_3,OH calculated by means of the sublattice model solely from the properties of the pure components and from optimized parameters of the four binary sub-systems.

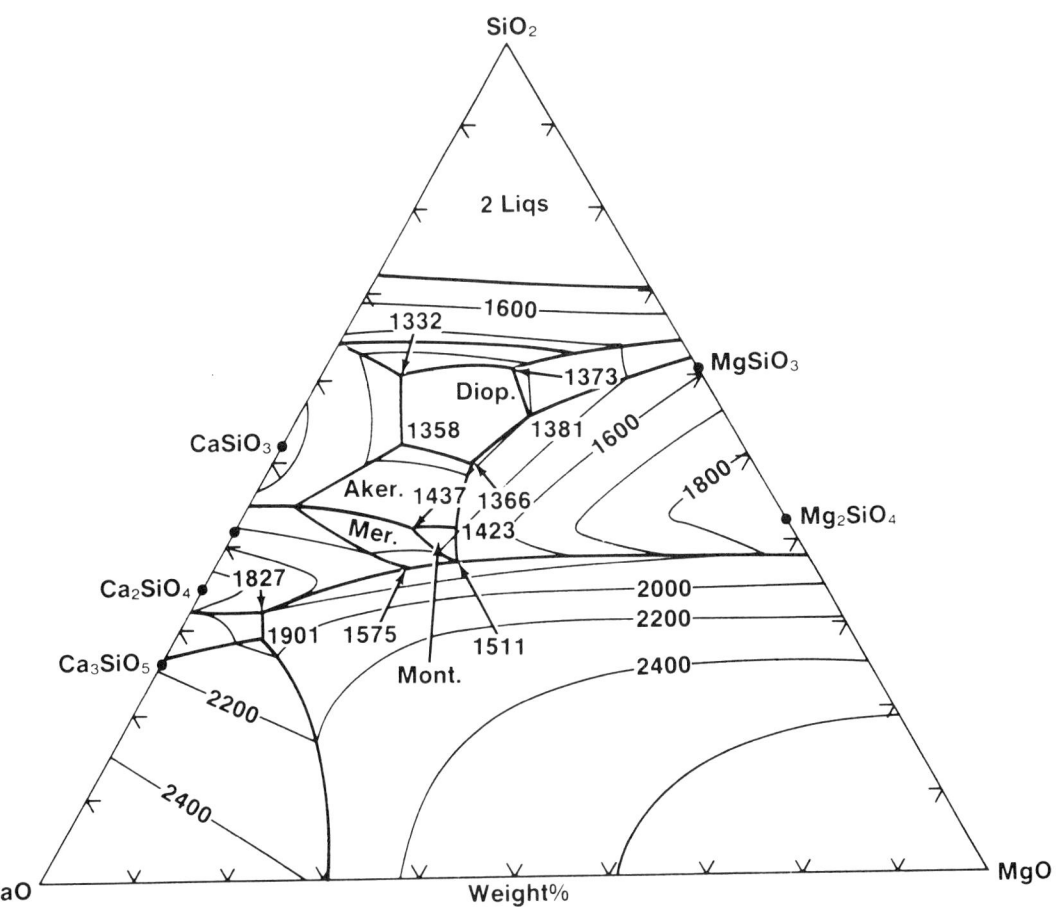

FIGURE 5: Phase diagram of the Mgo-CaO-SiO₂ system calculated from the optimized thermodynamic parameters of the liquid solutions in the three binary sub-systems. (Temperatures in °C).

Phase Transitions and The Square Root Diffusivity

J. E. MORRAL[+] AND M. S. THOMPSON[*]

[+]Department of Metallurgy and Institute of Materials Science,
University of Connecticut, U-136, 97 N. Eagleville Rd, Storrs, CT 06268
[*]United Technologies Research Center, MS24, Silver Lane, East Hartford, CT 06108

ABSTRACT

The square root diffusivity is a physical property matrix that appears in models which involve multicomponent diffusion. In the present work it is combined with multicomponent phase diagram data in a model which predicts the interface composition and growth kinetics of a plate shaped precipitate. The model assumes constant diffusivity, limited supersaturation and limited interaction with the diffusion fields of adjacent precipitates. However, any number of components can be considered and all thermodynamic and kinetic interactions between components are included in the treatment.

KEYWORDS

Multicomponent diffusion; square root diffusivity; diffusivity; phase transition; phase transformation; phase decomposition; precipitation; phase diagrams; diffusion couples.

INTRODUCTION

Diffusion limited phase transitions in multicomponent alloys can be treated as a special case of the moving boundary problem. Solutions to this problem for binary systems have been reviewed by Sekerka, Jeanfils and Heckel (1975) and have been given in a general way for multicomponent systems by Kirkaldy (1958). In the present work a solution will be derived for the thickening kinetics of a precipitate plate which is growing inside an infinite matrix. The unique feature of the solution is its simplicity, which is achieved by making use of the square root diffusivity.

As background for the current discussion, the problem of how to characterize interdiffusion in a multicomponent diffusion couple will be discussed first. This will introduce the square root diffusivity and show how it appears in solutions to Fick's Laws.

Then it will be applied to solving the multicomponent precipitation problem.

BACKGROUND

The simplest of all multicomponent diffusion problems is the single phase, constant diffusivity diffusion couple. For this reason it has been the focus of much work on multicomponent diffusion, most of which has been reviewed recently by Kirkaldy and Young (1987). Fig. 1 illustrates the problem for a system containing s solutes (i.e. an $s + 1$ component system). The concentration profile can be described by a sum of s error functions. For a system containing one solute the concentration profile looks much like the one shown in the figure. For systems with more than two components it can be more complicated and have extrema and concentrations that are outside the range of the terminal alloys, $C_j^{\mp}$ to $C_i^{\mp}$. A classification of the possibilities for ternaries has been given by Thompson and Morral (1986a).

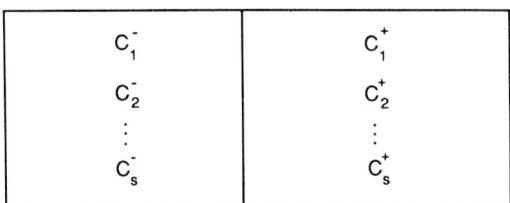

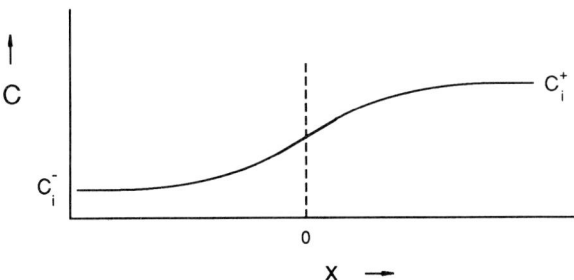

Fig. 1.   (top) An $(s + 1)$-component diffusion couple and the $2s$ composition variables which must be considered.
(below) The composition profile for one solute.

Although the equations for the concentration profiles can be lengthy and therefore cumbersome with which to work, the equation for the total solute which has crossed the initial interface, S_i, has the compact form shown in Fig. 2. Here S_i is defined to be a positive quantity when there is a net flow of solute to the right and negative when the net flow is to the left (Thompson and Morral, 1986a) and it is shown graphically as the shaded area on the figure.

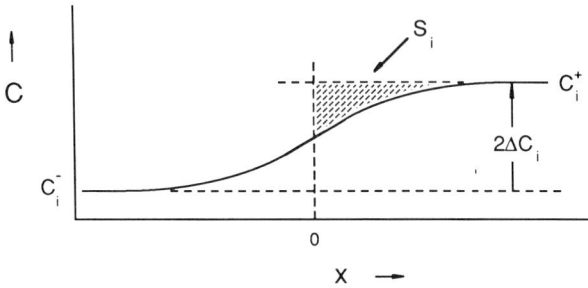

Solute crossing the x=0 plane

binary $\qquad S_1 = -\sqrt{\dfrac{4t}{\pi}}\;(r_{11}\Delta C_1)$

ternary $\qquad S_1 = -\sqrt{\dfrac{4t}{\pi}}\;(r_{11}\Delta C_1 + r_{12}\Delta C_2)$

multi-component $\qquad S_1 = -\sqrt{\dfrac{4t}{\pi}}\;(r_{11}\Delta C_1 + r_{12}\Delta C_2 + \ldots r_{1s}\Delta C_s)$

Fig. 2. The amount of solute 1, S_1, which has crossed the initial interface is illustrated by the shaded area (top) and is given analytically (below) in terms of the composition differences, C_i, and elements of the square root diffusivity, r_{ij}.

The equation for the binary system is well known (e.g. Crank, 1975) if one recognizes that r_{11} is the square root of the binary diffusivity (i.e. it is the "square root diffusivity"). The other equations have been derived more recently (Morral, 1984; Thompson and Morral, 1986a) and they are the foundation of the present treatment.

The r_{ij} coefficients in Fig. 2 are elements of the square root diffusivity matrix, the definition of which is given in Fig. 3. When the [r] matrix is squared it produces the diffusivity matrix, [D]. Also, the figure depicts matrix multiplication and the relationship between both D_{11} and r_{ij} elements and r_{11} and D_{ij} elements for a ternary system (Thompson and Morral, 1986b). By considering the expression for r_{11} it is obvious that there are significant advantages to using $[r]$ over $[D]$ in expressions for S_i.

Another important feature of $[r]$ is that all the elements of $[r]$ which are needed to calculate S_i are contained in the ith row of $[r]$. Therefore it is helpful to think of $[r]$ as consisting of s row vectors, $\vec{r}_i$, and that each vector corresponds to one of the solute components. Accordingly, S_i can be written in vector notation as:

$$S_i = (4t/\pi)^{1/2} (\vec{r}_i \cdot \vec{\Delta C}) \tag{1}$$

in which $\vec{\Delta C}$ is a vector which has coordinates that are given by half of the composition differences, ΔC_i, between the initial alloys.

$$[D] = [r][r]$$

$$\begin{bmatrix} D_{11} & D_{12} \\ D_{21} & D_{22} \end{bmatrix} = \begin{bmatrix} r_{11} & r_{12} \\ r_{21} & r_{22} \end{bmatrix} \begin{bmatrix} r_{11} & r_{12} \\ r_{21} & r_{22} \end{bmatrix}$$

$$D_{11} = r_{11}^2 + r_{12} r_{21}$$

$$r_{11} = [(D_{11} - D_{22}) + 4 D_{12} D_{21}]^{-\frac{1}{2}}$$

$$\{[\tfrac{1}{2}(D_{11} - D_{22}) + [(D_{11} - D_{22})^2 + 4 D_{12} D_{21}]^{\frac{1}{2}})]$$

$$[\tfrac{1}{2}(D_{22} - D_{11}) + ((D_{22} - D_{11}) + 4 D_{12} D_{21})^{\frac{1}{2}})^{\frac{1}{2}}$$

$$+ [(D_{22} - D_{11}) + ((D_{22} - D_{11}) + 4 D_{12} D_{21})^{\frac{1}{2}}]$$

$$[\tfrac{1}{2}(D_{22} - D_{11}) + ((D_{22} - D_{11}) + 4 D_{12} D_{21})^{\frac{1}{2}}]^{\frac{1}{2}}\}$$

Fig. 3. (first line) The definition of the square root diffusivity, $[r]$, in terms of the diffusivity, $[D]$. (following lines) The relationship between elements of $[r]$ and $[D]$ for a ternary system.

The discussion will now turn to the precipitation problem. Fig. 4 illustrates the concentration profile for one component across the interphase boundary that separates a precipitating phase, p,

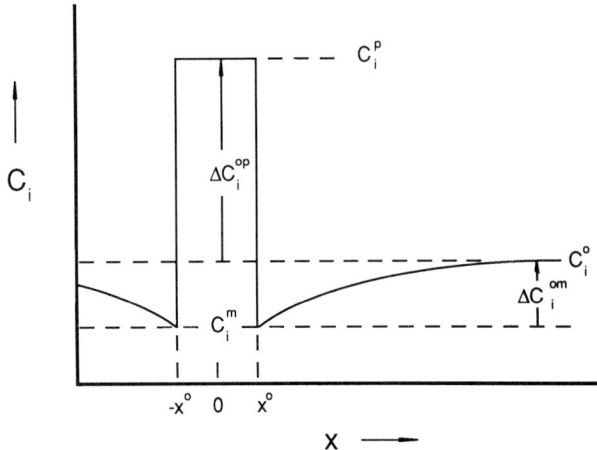

Fig. 4. Concentration profile through a growing precipitate plate with thickness $2X^o$. The concentration of i in the precipitate is C_i^p, in the orginal matrix phase is C_i^o and in the matrix at the moving interface is C_i^m.

from its matrix phase, m, in a multicomponent system. The concentration of the precipitate phase has a single value, C_i^p, while the concentration of the matrix phase varies between the interface concentration, C_i^m, and the initial concentration, C_i^o. As the figure shows, the growth of the precipitate (i.e. the movement of the boundary) is associated with diffusion of solute in the matrix. The diffusion of a particular solute can be either toward the boundary or away from it depending whether the precipitate concentration is less than or greater than the initial matrix concentration.

Previous treatments of precipitation in multicomponent systems have made different simplifying assumptions in order to obtain a solution. However, all treatments begin with a mass balance, which for this work will expressed as:

$$\Delta C_i^{op} V^p = S_i \qquad (2)$$

in which ΔC_i^{op} is the concentration difference $(C_i^p - C_i^o)$ and V^p is the precipitate volume. The quantity on the right, S_i, is equal to the solute which has left (or entered) the diffusion field of the matrix phase, while the quantity on the left is the solute added to or subtracted from the precipitate. Kirkaldy (1958) solved a differential form of equation (2) in general terms by assuming equilibrium at the interface. The solution illustrated the procedure to follow in solving multicomponent precipitation problems and showed that it was necessary to solve simultaneous equations in order to determine the interface composition and the precipitate size versus time. Somewhat later, Coates (1972) gave a detailed discussion and review of previous work on precipitation in ternary systems. He considered both the case when there is equilibrium at the interface and the case when there is "paraequilibrium" at the interface, but he assumed that the off diagonal terms in the diffusivity were negligible. Then Cooper and Gupta (1972) gave a solution for diffusion controlled crystal growth in multicomponent systems, which is similar to the present treatment both with regard to the assumptions made and the approach. The major difference is that in this work the square root diffusivity is employed.

THEORY

The present work is an extension of a recent paper entitled "The composition of a moving planar interface in a multicomponent system" (Morral and Thompson, 1988a). It is applied here to precipitation and was applied in another publication to the problem of precipitate dissolution (Morral and Thompson, 1988b). The assumptions that must be made in order to use the moving interface analysis are as follows:

1. The interface is planar.
2. There is only one precipitate in an infinite matrix.
3. The supersaturation approaches zero.
4. The diffusivity is constant.
5. The precipitate phase has a fixed composition (i.e. it is a stoichiometric compound).
6. The phases are in equilibrium at the interface.

The first two assumptions limit the treatment to the growth of a single precipitate plate which is thickening with time. The volume of a unit area of the plate is given by:

$$V^p = 2X^o \quad (3)$$

in which X^o is half of the plate thickness. Since the diffusion is one dimensional and the matrix is assumed infinite, it is possible to describe the concentration profile as a sum of error functions. The third assumption, small supersaturation, makes it possible to approximate the diffusion field in front of the moving precipitate boundary by the diffusion field in front of a stationary boundary, for example in front of $x = 0$ in the diffusion couple. The physical reason that this approximation is valid is that as the supersaturation, defined by:

$$SS = \Delta c_i^{mo}/\Delta c_i^{op} \quad (4)$$

goes to zero the boundary velocity goes to zero, also. Therefore, by assuming constant diffusivity, assumption 4, it is possible to use equation (1) for the solute which has been gained or lost by the matrix which is front of a unit area of boundary, viz:

$$S_i = (4t/\pi)^{1/2}(\vec{r}_i \cdot \vec{\Delta c}^{mo}) \quad (5)$$

in which Δc^{om} in both equations (4) and (5) refers to the concentration difference $(C_i^m - C_i^o)$.

Combining equations (2),(3) and (5) yields the following expression for the precipitate size versus time:

$$x^o = (4t/\pi)^{1/2} (\vec{r}_i / \Delta c_i^{op}) \cdot (\vec{\Delta c}^{mo}) \tag{6}$$

There are a total of $\underline{s}$ equations of this type, one for each solute, which must be satisfied in order for the precipitate concentration to be constant with time as required by the fifth assumption.

Equations (6) contain $s + 1$ unknowns. These are the $\underline{s}$ interface concentrations, C_i^m, that appear in the vector $\vec{\Delta c}^{mo}$ and the one thickness parameter x^o. The additional equation needed in order to solve for these quantities can be obtained by making the final assumption; that the phases are in equilibrium at the interface. It follows from this assumption that the interface composition must be a point on the equilibrium phase diagram solvus given by the equation:

$$Q = Q(C_i^m) \tag{7}$$

in which Q is a constant and $Q(C_i^m)$ is a function of the s interface concentrations. Solving equations (6) and (7) simultaneously gives the desired solution.

A more complete derivation of the steps leading to equation (6) as well as a broader discussion of the assumptions and the physical meaning of the solution can be found in Morral and Thompson (1988a, 1988b). However, it should be mentioned that this solution, like others before it which were derived for binaries, can be applied to a wider range of problems than may be obvious from the assumptions. For example it can be applied to an array of precipitates for short times as long as the diffusion field of one precipitate does not intersect the interface of another precipitate in the array. Also, the constant diffusivity assumption is not a serious concern as long as the supersaturation, and therefore the possible range of composition, is small.

To complete the present work, a description will be given of a graphical method for determining the interface concentrations. Once these have been determined they can be substituted into equation (6) in order to predict the precipitation kinetics. Although the graphical solution is limited to ternary systems, it offers a way of visualizing multicomponent precipitation that can be extended to higher order systems.

A GRAPHICAL CONSTRUCTION FOR THE INTERFACE COMPOSITION

The graphical construction begins by plotting the known information on a ternary phase diagram as shown in Fig. 5. Orthogonal axes are used here to facilitate the plotting of vectors. The solvus boundary, Q, has been given a general shape that is typical for ternary phase diagrams. As in the previous derivation, $\underline{m}$ and $\underline{p}$ refer to the matrix and precipitate phases, respectively. The original alloy composition, C_i^o, is plotted on the figure, as well, and it is used as the origin for plotting the "solute vectors", $\vec{r}_1/\Delta C_1^{op}$ and $\vec{r}_2/\Delta C_2^{op}$.

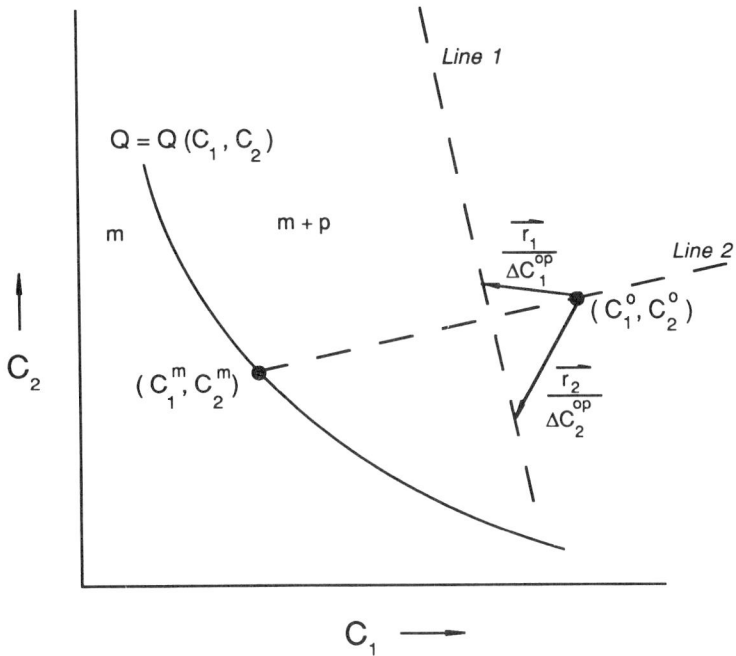

Fig. 5. A two line construction drawn on a ternary phase diagram for determining the composition of a moving interface, (C_1^m, C_2^m). All other quantities on the diagram must be known (i.e. the solvus, Q, the initial alloy composition, (C_1^o, C_2^o), and the solute vectors, $\vec{r}_i/\Delta C_1^{op}$ and $\vec{r}_i/\Delta C_1^{op}$).

According to equations (6), the dot product between each solute vector and the "composition vector", $\Delta \vec{C}^{mo}$, must be equal to each other. Accordingly, the graphical method consists of constructing the composition vector which will satisfy this condition.

The construction consists of drawing two lines. As shown in Fig. 5 the first line passes through the tips of the solute vectors. The second line passes through the initial alloy composition and is perpendicular to the first line. The composition of the interface is given by the intersection of the second line with the solvus. It follows, by definition of the dot product, that the composition vector, which extends from the initial alloy composition to the interface composition, will satisfy the requirement that its dot product with the solute vectors will be equal.

This construction can be extended to quaternaries by visualizing a plane constructed between the tips of three solute vectors. The line which passes through the initial alloy composition and is perpendicular to the plane will intersect the solvus surface at the interface concentration.

It is true in general, and it can be seen on the figure, that the composition vector will tend to lie nearest to the direction of the shortest solute vector. The effect that this has on the interface concentrations, regardless of the number of components involved, is that the concentration differences at the interface, ΔC_i^{mo}, will be larger for solutes with the shorter solute vectors. The shorter solute vectors correspond to solutes which are inherently slower moving and/or require more segregation between the phases than other solutes. The higher driving force for diffusion allows them to keep pace with the other solutes.

The multicomponent effect of how the interface concentrations adjust for the solutes present was reported before by Purdy, Weichert and Kirkaldy (1964) and later by Coates (1972). Here the effect is demonstrated again and it is seen to be related in a simple and quantitative way to the square root diffusivity, which takes into account all the thermodynamic and kinetic interactions of the solute atoms.

The extent to which multicomponent additions will alter precipitation kinetics can be determined by applying the model to individual systems for which phase diagram and square root diffusivity data are available. The model has the potential to give significantly different answers than those obtained from pseudo binary estimates in which the cross diffusion terms and phase diagram are ignored. For example in a numerical example given by Morral and Thompson (1988b) for precipitate dissolution, it was found that multicomponent predictions differed from the pseudo binary predictions by a factor of two.

CONCLUSIONS

The kinetics of a plate shaped precipitate growing in a supersaturated matrix can be described with a compact quantitative model by making use of both the square root diffusivity and the solvus of the multicomponent phase diagram for the system. By including cross terms in the square root diffusivity, the model takes into account all interactions, both chemical and kinetic, between the diffusing atoms. In addition, the model makes no restriction on the number of components which can be considered as long as the precipitate is a stoichiometric compound and the supersaturation is small.

ACKNOWLEDGMENTS

The authors are grateful for support from the University of Connecticut Research Foundation and from the National Science Foundation under grant no. DMR-8711899. Also, the help of Mr. Barry Dupen is much appreciated.

REFERENCES

Coates, D.E. (1972). Met. Trans, 3, pp. 1203-1212.

Cooper, A.R. and P.K. Gupta (1972). In L.H. Hench and W. Freiman (Ed.), Adv. in Nucleation and Crystallization of Glasses, Spec. Pub. No. 5, Amer. Ceramic Soc., pp. 131-140.

Crank, J. (1975) .The Mathematics of Diffusion, 2nd ed. Oxford University Press, London, p. 32.

Kirkaldy, J.S. (1958a). Can. J. Phys., 36, pp. 907-916.

Kirkaldy, J.S. (1958b). Can. J. Phys., 36, pp. 917-925.

Kirkaldy, J.S. and D.J. Young (1987). Diffusion in the Condensed State, The Institute of Metals, London.

Morral, J.E. (1984). Scripta MET., 18, pp. 1251-1256.

Morral, J.E. and M.S. Thompson (1988a) submitted to Scripta MET.

Morral, J.E. and M.S. Thompson (1988b) In J. Crane and H. Merchant (Ed.), Homogenization and Anealing of Aluminum and Copper Alloys, to be published by TMS, Warrendale, PA.

Thompson, M.S. and J.E. Morral (1986a). Acta metall., 34, pp. 339-346.

Thompson, M.S. and J.E. Morral (1986b). Acta metall., 34, pp. 2201-2203.

Purdy, G.R., D.H. Weichert and J. S. Kirkaldy. (1964) Trans. TMS-AIME, 230, pp 1025-1034.

Sekerka, R.F., C.L. Jeanfils and R.W. Heckel (1975). In H.I. Aaronson (Ed.), Lectures on the Theory of Phase Transformations, The Metallurgical Society, Warrendale, PA, pp. 117-169.

Molecular Dynamics Simulation of Melting

MARUTI BHANDARKAR* AND J. C. M. LI**

*Department of Chemical Engineering, University of Rochester,
Rochester, NY 14627, USA
**Department of Mechanical Engineering, University of Rochester,
Rochester, NY 14627, USA

By a rapid quenching technique, the equilibrium atomic configurations during the melting of a Lennard-Jones fcc crystal are examined at femtosecond intervals. The various processes during the "nucleation" stage are observed. A *point of no return* is found before which the system can be quenched into a perfect fcc crystal and after which it can be quenched only into a defective or amorphous structure. The melting process is further illuminated by the cavity distribution which can detect order even in an amorphous structure.

KEYWORDS

Melting; Lennard-Jones solid; molecular dynamics; fcc crystal; nucleation; cavity distribution; rapid quenching; point of no return.

INTRODUCTION

The process of transition from a crystal with definite order in the equilibrium atomic positions of all the atoms to an amorphous liquid-like structure in which there is a definite disorder can generally be called melting. Some well-known theories (*e.g.*, Lennard-Jones and Devonshire, 1939a, 1939b) were based on this order-disorder transition. While it is apparent that during melting the crystal structure must become unstable (Herzfeld and Geoppert–Mayer, 1934; Born, 1939a, 1939b), the instability of the solid phase alone should not be sufficient to predict melting (*e.g.*, Cotterill and Madsen, 1980; Ross and Wolf, 1986). Even if the crystalline instability is the cause of melting, it is desirable to know exactly how the crystal structure collapses and how the liquid structure appears. Cotterill, Kristensen and Jensen (1974), based on their earlier pseudostatic computational model (Jensen, Kristensen and Cotterill, 1973) and their subsequent molecular dynamics simulation of melting, suggested that spontaneous generation of small dislocation loops initiates the melting process. However, examination of the hot structure is difficult because the various amplitudes of atomic vibrations mask the equilibrium positions. Time-averaged positions are somewhat better but they are not equilibrium positions; and, in a rapidly varying situation in which individual atomic vibrations may degenerate into group vibrations of lower frequency, the time-averaged positions may have no more significance than instant positions. Stillinger and Weber (1982) suggested that the 'inherent structure' in a given system could be obtained from a steepest-descent mapping of the system onto a potential-energy minimum. This quenching procedure which removed the distortions in atomic configurations due to thermal vibrations, was used to study (Stillinger and Weber, 1983) the dynamics of melting and freezing. In the case of a 32-atom argon-like system, an abrupt transition from a superheated crystal to an amorphous structure occurred in about 2.7 fs, as determined by examining the potential energy in the quenched state. However, no details of the evolution of atomic arrangement during the melting process were reported in this study.

In the present molecular dynamics (MD) simulation, the equilibrium positions of individual atoms were obtained by another rapid quenching technique. In this techinque, all the kinetic energies in a hot structure were removed first. Then, all the atoms were allowed to move dynamically to their equilibrium positions. During the process some kinetic energy was developed but was repeatedly removed at prescribed intervals (after every 100 time steps or $5°C$ increase in the temperature, whichever came first). The final equlibrium configuration was almost at $0\,K$. Except for this rapid quenching technique, the simulation was otherwise similar to that of Cotterill, Kristensen and Jensen (1974).

COMPUTATION

Six layers of atoms of 18 atoms per layer were arranged in a fcc structure with the layers parallel to ⟨100⟩ planes. The system was a cube containing 108 atoms. Periodic boundary conditions were imposed in all three directions so that the cube was surrounded by 26 identical cubes. Interaction between any two atoms was assumed to be the Lennard-Jones type:

$$V(r) = 4\,\epsilon\left((\sigma/r)^{12} - (\sigma/r)^{6}\right), \tag{1}$$

where ϵ is the binding energy, r is the distance between the centers of the two atoms, and $\sigma = (0.5)^{1/6} r_0 = 0.891 r_0$ with r_0 being the equilibrium separation between two atoms. If the unit of V is ϵ and that of r is σ, Eq.(1) is a dimensionless relation with no assigned parameters. The range of interaction is taken as $2.5\,\sigma$ to save computation time.

For solid argon (Ar), $\epsilon/k = 120K$ with k being the Boltzmann constant, $\sigma = 0.34$ nm and atomic mass, $m = 39.95/N$ grams with N being the Avogadro number. The cube was $(4.76\,\sigma)^3$ so that the density was 1.688 gm/cc for Ar. It was a high density system. Simulations were carried out at constant volume. Newton's law was assumed to govern atomic motion

$$m\frac{d^2 \boldsymbol{r}_i}{dt^2} = \sum_{j\neq i} \boldsymbol{f}(r_{ij}). \tag{2}$$

The force was calculated by differentiating Eq. (1) and the direction of the force was along the center-to-center vector. The forces exerted by all the atoms in a sphere of radius $2.5\,\sigma$ were vectorially added together. The resultant force caused the acceleration of the central atom in that direction. The force in the unit of ϵ/σ is numerically the same as acceleration if the time unit is $\sigma\sqrt{m/\epsilon}$ ($= 2.16$ ps for Ar) and the distance unit is σ.

To begin the simulation, each atom was randomly assigned a certain velocity from a Gaussian distribution for each velocity component. Subsequent positions and velocities were calculated according to the Verlet (1967) scheme, namely, the acceleration at t was used to calculate the velocity change between $t - \Delta t$ and $t + \Delta t$. Similarly the velocity at t was used to calculate the displacement between $t - \Delta t$ and $t + \Delta t$. Since at $t = 0$, the acceleration was zero for all the atoms (perfect crystal), the velocity was the same between $-\Delta t$ and $+\Delta t$. Hence the velocity assignment just mentioned was considered to be at Δt. This velocity was used to calculate the displacement between $t = 0$ and $t = 2\Delta t$. Knowing the position of all the atoms at $2\Delta t$, the acceleration of each atom was calculated and used to calculate the velocity change between Δt and $3\Delta t$. The velocity at $3\Delta t$ was then used to calculate the position changes between $2\Delta t$ and $4\Delta t$. The computation continued until the average *thermal* kinetic energy (= actual kinetic energy − bulk translational energy − bulk rotational energy) became steady indicating thermal equilibrium. The time step Δt was usually 0.001 to 0.005, the higher value being appropriate initially when the total energy is low. Thermal equilibrium required a time of about 0.5. Heating and cooling was done by multiplying the velocity by a constant factor for all the atoms. Rapid quenching was carried out by eliminating all the kinetic energy, relaxing for about 100 time steps or until the temperature rose above 5° C (whichever came first), and eliminating all the kinetic energy again. This cycle was repeated until the temperature was close to absolute zero.

RESULTS AND DISCUSSION

Crystalline and Amorphous States

The average energy (kinetic plus potential) per atom after thermal equilibrium was plotted in Fig. 1 versus temperature which was the kinetic energy in units of ϵ/k multiplied by 120° (for Ar). A perfect fcc crystal at 0 K was assigned zero potential energy. The crystal line obtained by heating or cooling shows the stability of the crystalline state. Above about 200 K the crystal would become unstable and would melt. During melting, the total energy remained

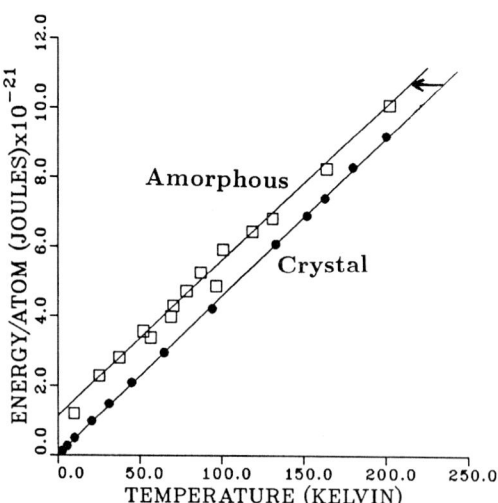

Fig. 1. Energy–temperature relations for crystalline and amorphous states.

constant (adiabatic) so the temperature dropped by about 25° C before the system became a liquid at thermal equilibrium. The cooling of the liquid state followed the amorphous line. Some of the points along the amorphous line were obtained by placing 108 atoms randomly into the same volume but avoiding too much overlap between atoms, and then allowing the system to relax until thermal equilibrium. It is seen that the points along the amorphous line have some scatter showing that the amorphous state is not unique.

Cavity Distribution

In addition to total energy, some structural details were revealed by the cavity distribution. A cavity is defined as the largest sphere which can fit within the interstices among four neighboring atoms. The atomic diameter was assumed to be the nearest neighbour distance, 1.122σ. The spherical cavity must be free space not overlapping with any other atom. An example is shown in Fig. 2 for the crystalline state at 2.7 K. The larger peak is for the tetrahedral holes and the smaller one for the octahedral holes. The number ratio is about 2 to 1. The hole diameter to atom diameter ratios are 0.225 and 0.414 at absolute zero (assuming no quantum effects). It is seen that there is some broadening even at 2.7 K. At 27 K, Fig. 3 shows considerably more broadening and the shift of the octahedral peak toward smaller cavity diameters. At 75 K the two peaks merged into one peak as shown in Fig. 4. The maximum is however skewed toward smaller cavity diameters. Finally at 109 K the distribution looks random (Fig. 5) indicating the possible disintegration of the crystal structure.

Rapid Quenching

To make sure that the crystal was melted, it was heated further to 152 K for which the cavity distribution resembled that of Fig. 5 at 109 K. However to our surprise the cavity distribution after rapid quenching from 152 K to 11 K revealed the two peaks as shown in Fig. 6. It shows that the crystal has not been melted even at 152 K. Later we found out that the abnormally high melting point was caused by the constant volume simulation of a high density system so that the pressure was very high ($\approx$ 3900 atmospheres). However, the characteristics of the melting process were found to be similar in a zero pressure simulation to be reported later.

Finally, at about 250 K the crystal was melted because the system changed from the crystal line to the amorphous line (see, Fig. 1). Rapid quenching after thermal equilibrium on the amorphous line revealed a broad peak in the cavity distribution indicating an amorphous structure. The time needed for the system to shift from the crystal line to the amorphous line depends on the temperature at which the crystal is superheated. As shown in Fig. 7, the time was about 6 ps after the crystal was superheated to 245 K. The time is longer for lower superheating and shorter for higher superheating. The structural evolution along the curve of Fig. 7 will now be discussed.

Incubation Period

While the temperature was dropping indicating that the kinetic energy was converted into potential energy, no permanent structural change could be detected for the first 4.65 ps along the curve in Fig. 7, because all the hot structure could be quenched into the original crystal structure with each atom returning to its original position. While the hot structure looked random like that of a liquid, the quenched structure was always the original perfect crystal.

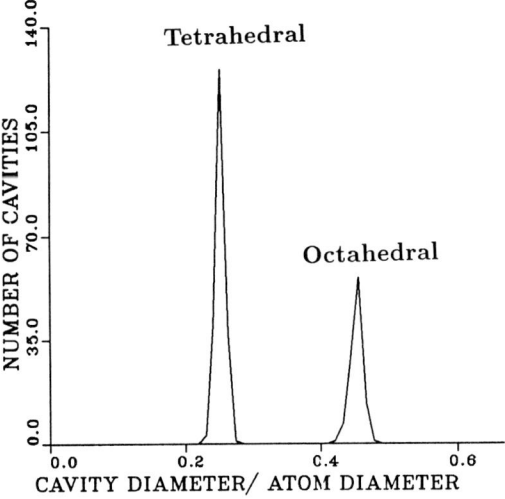

Fig. 2. Cavity distribution for the crystalline state at 2.7 K.

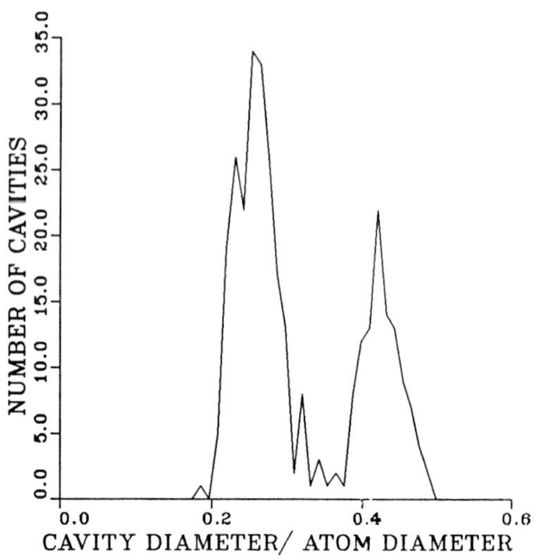

Fig. 3. Cavity distribution for the crystalline state at 27 K.

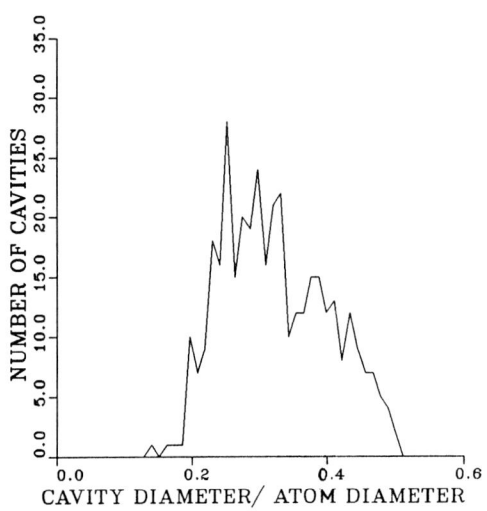
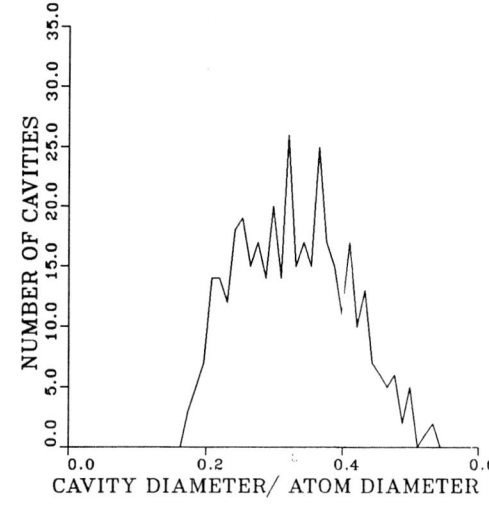

Fig. 4. Cavity distribution for the crystalline state at 75 K.

Fig. 5. Cavity distribution for the crystalline state at 109 K.

The large scatter in Fig. 7 is due to the fact that each point was an instantaneous state. If the temperature was averaged over one vibrational period (0.1 ps), its variation would be smoother. Since the structure can be quenched to recover the original crystal, it shows that every atom, while vibrating vigorously, has not traveled too far from its original position so as to forget the path to return.

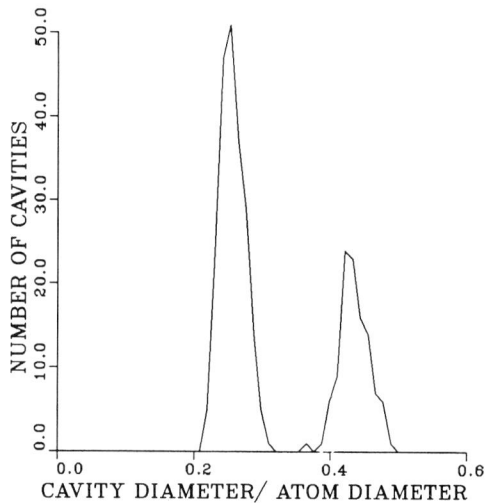

Fig. 6. Cavity distribution after quenching a crystal from 152 to 11 K.

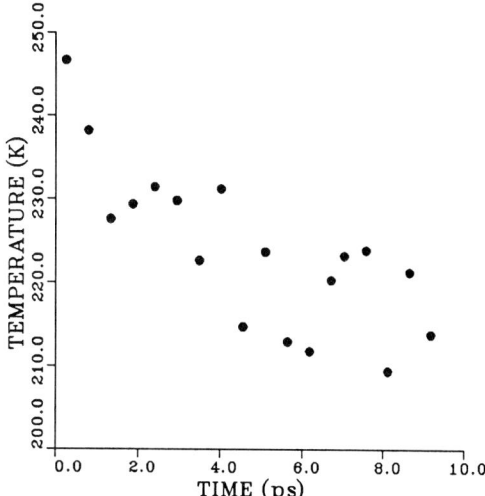

Fig. 7. Temperature decrease during adiabatic melting.

Point Defects

The first indication of disordering activity was the creation of a vacancy and a pair of split interstitials, in agreement with Stillinger and Weber (1984). This occurred after rapid quenching at 4.651 ps. Here the quenching times were further divided into 1 fs intervals. The interstitial pair was nearest neighbor to the vacancy indicating that the atom had moved too far in the <110> direction so that it was caught by its nearest neighbor to become a split interstitial pair leaving the original site of the displacing atom vacant. However, the creation of such point defects was only temporary because after another femtosecond, the system was quenched into the original crystal again, namely with each atom returning to the original position. Then nothing happened again until 4.88 ps.

Displacement Loops

At 4.88 ps the quenched structure showed four displaced atoms forming a closed loop in the (100) plane as shown in Fig. 8. The displacement vector was between nearest neighbors namely $(a/2)$<110> (a being the unit cell dimension of the fcc structure). Except for these displacements, the quenched structure was still a perfect crystal without any defects. It turned out that the displacement loop was also not permanent because after a few more femtoseconds, the structure was again the original crystal with no displaced atoms. However, from now on

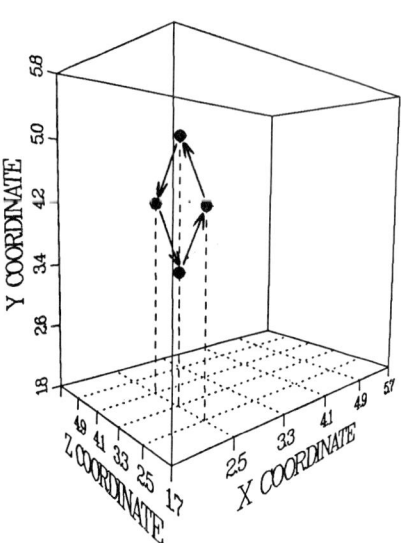

Fig. 8. Displaced atoms after quenching a crystal superheated for 4.88 ps.

the displacement loops appeared more frequently and the size increased also. For example, at 5.096 ps a loop involved 10 displaced atoms was found after quenching. This is shown in Fig. 9. All displacements were between nearest neighbours and they were all $(a/2)\langle 110\rangle$. The displacement marked 9 was also between nearest neighbors on account of the periodic structure. So instead of a loop, the displacement vectors connected together into many long chains. The displacement vectors are more clearly shown in Fig. 10. It is seen that the nine vectors are all $(a/2)\langle 110\rangle$ type and are evenly distributed among the three (100) planes.

The displacement loops were now more permanent. They would remain for some time or change partially. For example, at 5.264 ps the quenched structure revealed twelve displaced atoms seven of which existed before (filled circles) as shown in Fig. 11. While the displacement vectors among these twelve atoms still formed a single loop, sometimes there were two separate loops. In other words, the loop evolved by adding on smaller loops. If the small loop overlaps with the existing loop, they will form a larger loop or shrink into a smaller loop. If they do not overlap, two loops will appear and each evolves further with time.

The Point of No Return

The time 5.264 ps marked an important event because it was the last time the system could be quenched into a perfect crystal. A femtosecond later, namely at 5.265 ps, the system was quenched into a defect structure. The middle two layers (36 atoms) parallel to (001) plane of this structure before (still hot) and after quenching are shown in Figs. 12 and 13 respectively. At 5.264 ps the system was quenched into a perfect fcc crystal even though it contained loops of displacement vectors. But at 5.265 ps, the quenched structure was no longer crystalline. Examination of the consecutive close-packed planes revealed that there was a close-packed

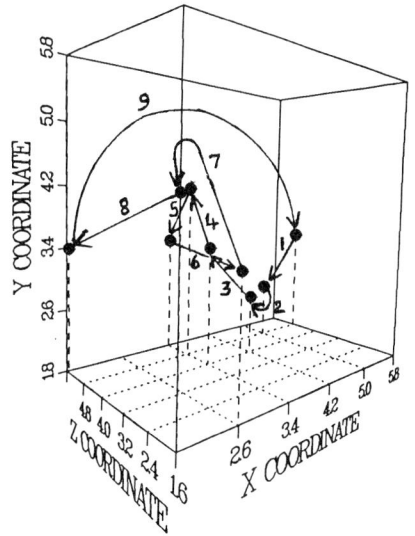

Fig. 9. Displaced atoms after quenching a crystal superheated for 5.096 ps.

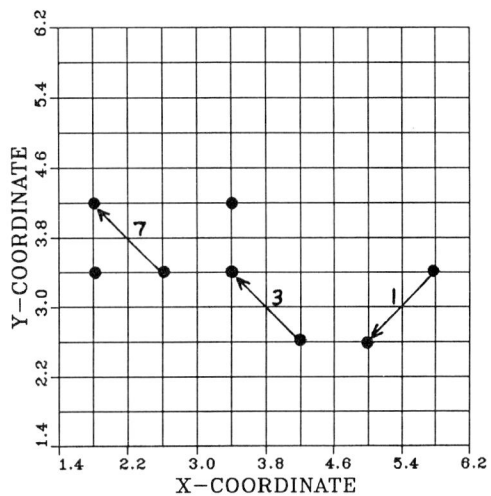

Fig. 10a. Displacement vectors for Fig. 9.

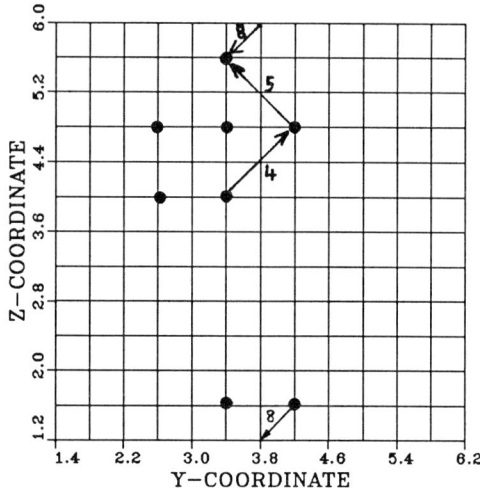

Fig. 10b. Displacement vectors for Fig. 9.

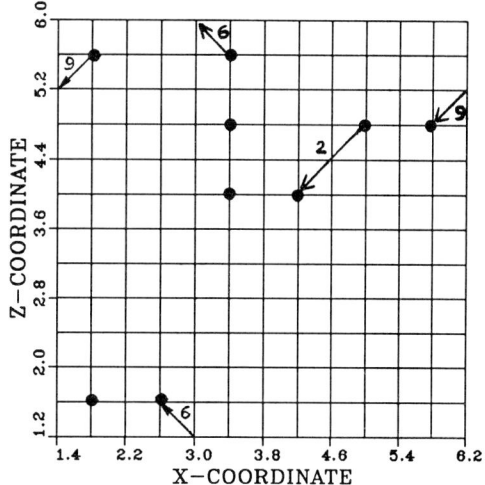

Fig. 10c. Displacement vectors for Fig. 9.

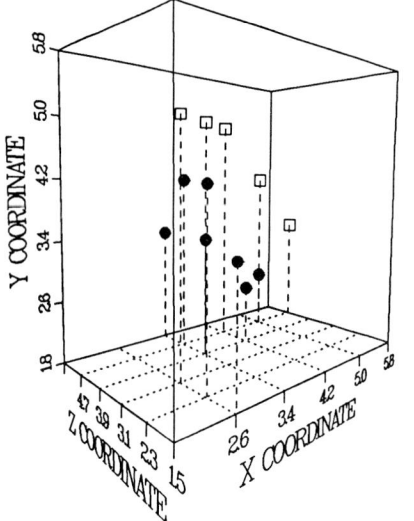

Fig. 11. Displaced atoms after quenching a crystal superheated for 5.264 ps.

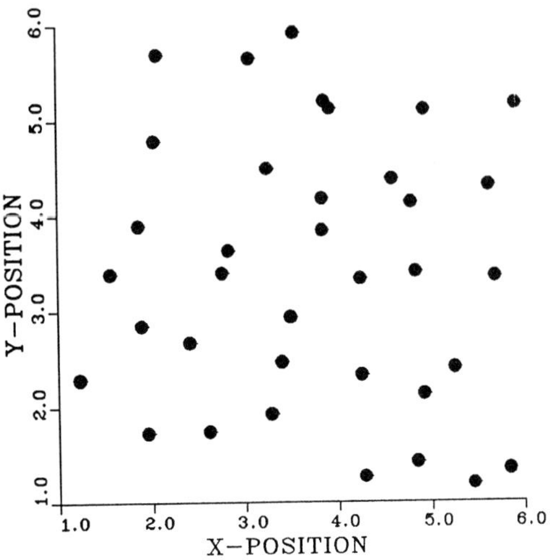

Fig. 12. Positions of atoms in the middle two layers of a crystal superheated for 5.265 ps.

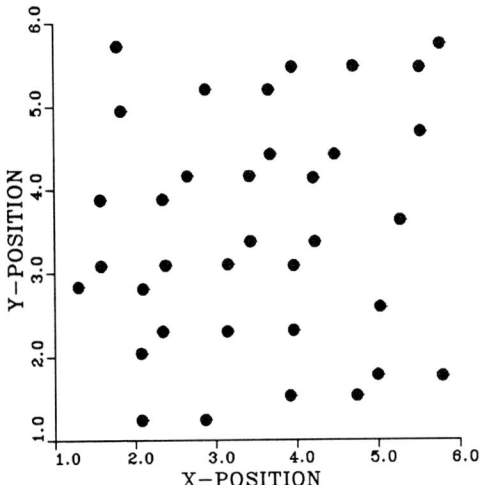

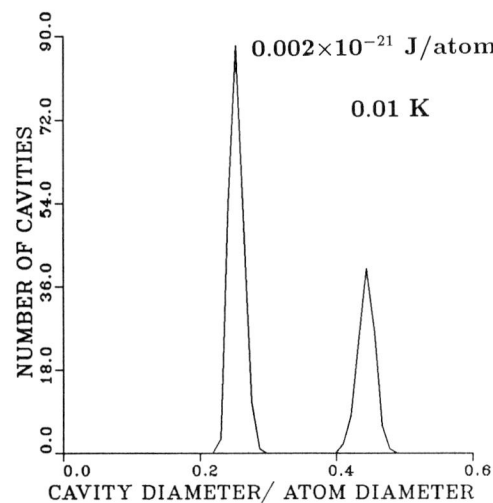

Fig. 13. Positions of atoms in the middle two layers of a crystal superheated for 5.265 ps and quenched.

Fig. 14. Cavity distribution for a crystal superheated for 5.265 ps and quenched immediately after the point of no return.

order in the (111) planes; however, there was no easily describable order in the other close-packed planes. Dislocation dipoles observed by Cotterill, Kristensen and Jensen (1974) were not found. From 5.265 ps onwards the quenched structure remained a defect structure so it was a permanent change. At this point the atoms apparently moved too far away from their original positions so as to forget the path to return. Yet only a femto second earlier they still remembered. This *point of no return* must represent some critical configuartion whose characteristics are not apparent at this stage of the investigation.

Completion of Melting

It turned out that right after the point of no return there was still some order in the quenched structure even though it looked amorphous. This order was revealed through the cavity distribution as shown in Fig. 14 for the structure quenched from the hot configuration at 5.265 ps. The low value of the potential energy confirmed this possibility. However, two more femto seconds later the potential energy of the quenched structure suddenly increases as shown in Fig. 15 with corresponding more disordering as shown by the atomic arrangement of the middle two layers (Fig. 16) and the cavity distribution (Fig. 17). The two distinct cavity peaks finally merged into one at 5.956 ps. The distribution was still skewed toward the smaller cavity and the potential energy at 0.3 K was 8.1×10^{-22} J/atom which was not quite that of the undercooled liquid yet. It took two or more picoseconds to melt completely.

CONCLUDING REMARKS

So far the melting phenomenon appears as follows: A superheated crystal does take time to melt, the higher the superheating the shorter will be the time. The disordering process begins by atoms displacing their nearest neighbors along <110> directions. The displacement vectors form a closed loop so that no point defects are permanently generated. Then more and more atoms are involved in the displacement process without creating any defects. In other words, the system is quenched into the original perfect crystal with only some atoms exchanging positions. Finally when the displacement activities reach some critical magnitude or frequency, the system passes over a *point of no return* after which the system can be quenched only into an amorphous structure. The *point of no return* shows also a sudden increase of potential energy of the quenched state. However, at this time the quenched amorphous structure still has some order in it as revealed by the cavity distribution. How to describe this order or disorder is still being investigated. Gradually, such order disappears with a simultaneous increase of potential energy and finally the crystal melts completely as indicated by the equilibrium temperature and the potential energy of a truly amorphous structure.

ACKNOWLEDGEMENT

This work is supported by the U.S. Office of Naval Research through N00014-85-K-0758. The use of supercomputers at the John von Neumann National Supercomputer Center, Princeton, NJ 08543 is gratefully acknowledged.

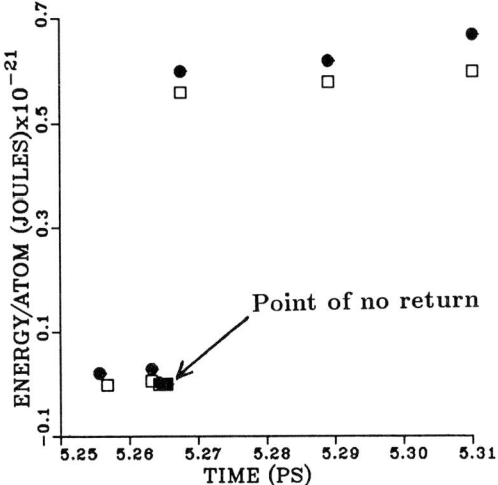

Fig. 15. Potential energy after quenching a crystal superheated for various times. Quenching was stopped when there was a change of less than $0.5°C$ during a relaxation period of 3 ps. Filled circles are the actual values obtained with the final (equlilibrium) temperatures usually less than $1\ K$. Squares show the same points extrapolated to $0\ K$.

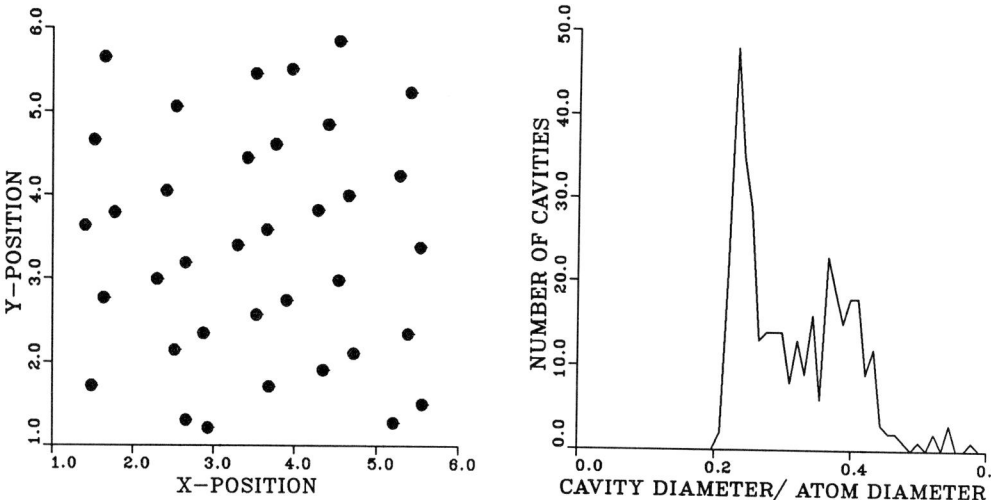

Fig. 16. Positions of atoms in the middle two layers of a crystal superheated for 5.267 ps and quenched.

Fig. 17. Cavity distribution for a crystal superheated for 5.267 ps and quenched.

REFERENCES

Born, M. (1939a). Thermodynamics of crystals and melting. J. Chem. Phys., 7, 591–603.

Born, M. (1939b). Melting as an order-disorder transition. J. Chem. Phys., 7, 810–817.

Cotterill, R. M. J., W. Damgaard Kristensen and E. J. Jensen (1974). Molecular dynamics studies of melting. II. Spontaneous dislocation generation and the dynamics of melting. Phil. Mag., 30, 245–263.

Cotterill, R. M. J. and J. U. Madsen (1980). Evidence against a crystal instability in melting. Nature, 288, 467–469.

Herzfeld K. F. and M. Geoppert-Mayer (1934). The theory of fusion. Phys. Rev., 46, 995–1001.

Jensen, E. J., W. Damgaard Kristensen and R. M. J. Cotterill (1973). Molecular dynamics studies of melting. I. Dislocation density and the pair distribution function. Phil. Mag., 27, 623–632.

Lennard-Jones, J. E. and A. F. Devonshire (1939a). A theory of melting and structure of liquids. Proc. Roy. Soc., A169, 317–338.

Lennard-Jones, J. E. and A. F. Devonshire (1939b). Theory of disorder in solids and liquids and the process of melting. Proc. Roy. Soc., A170, 464–484.

Ross, M. and G. Wolf (1986). High pressure melting curve of KCl: evidence against lattice instability theories of melting. Phys. Rev. Lett., 57, 214–217.

Stillinger, F. H. and T. A. Weber (1982). Hidden structure in liquids. Phys. Rev. A, 25, 978–989.

Stillinger, F. H. and T. A. Weber (1983). Dynamics of structural transitions in liquids. Phys. Rev. A, 28, 2408–2416.

Stillinger, F. H. and T. A. Weber (1984). Point defects in bcc crystals: structures, transition kinetics, and melting implications. J. Chem. Phys., 81, 5095–5103.

Verlet L. (1967). Computer "experiments" on classical fluids. I. Thermodynamic properties of Lennard-Jones molecules. Phys. Rev., 159, 98–103.

Cellular Solidification

D. VENUGOPALAN

Materials Department, University of Wisconsin – Milwaukee
Milwaukee, Wisconsin, USA

ABSTRACT

In this contribution, theoretical and experimental progress on stable cell configurations obtained in the single phase solidification of a binary alloy are reviewed. Important experimental observations such as the absence of near-marginal stable configurations and crystallographic preference of cell walls are highlighted. The paper concludes with suggestions for further work to resolve the difficulties in the theoretical description of the stable cellular solidification.

KEY WORDS

Solidification; non-equilibrium interface; solidification cells; free boundary problem.

INTRODUCTION

Cellular solidification refers to the particular mode of liquid-to-solid transformation in which the morphology of the interface is cellular as shown in Fig. 1. Cellular morphology is the precursor of dendritic morphology of the solid-liquid interface in directional solidification of metallic alloys in the presence of a positive temperature gradient in the liquid.

In directional solidification, the control variables are temperature gradient (G), growth (solidification) velocity (v) and alloy content (C_0) of the material. The growth variables determine the morphology of the solidification interface to be one of many observed morphologies, viz., planar, cellular, dendritic or irregular nonplanar configurations. Early work by Chalmers and coworkers (Tiller and others, 1953) was directed at the conditions for obtaining a nonplanar solidification interface and led to the derivation of the constitutional supercooling criterion. The three variables can be combined into a single parameter, α, given by Eqn. 1.

$$\alpha = \frac{GDk}{vmC_0(1-k)} \tag{1}$$

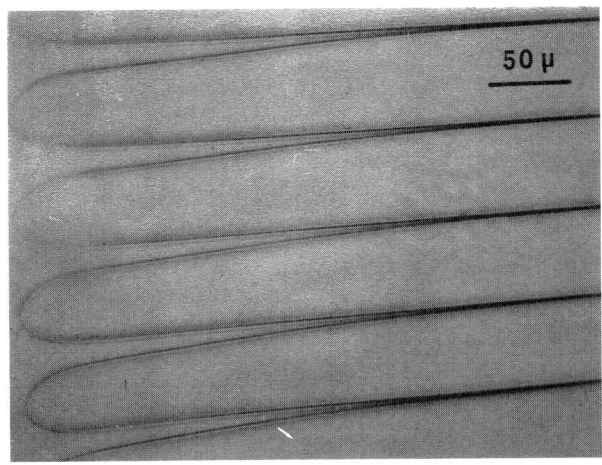

Fig. 1 Stable cellular interface in succinonitrile-salol

In the early analysis, a non planar interface is expected for α less than unity. It was later pointed out that surface tension effects lowered the critical value of α at which the transition from planar to nonplanar morphology is expected (Mullins and Sekerka, 1964). However, in most metallic systems the critical value of α is greater than 0.95 including surface tension effects. The dimensionless parameter, α, which was introduced in this form in literature by Kirkaldy (1980), is a measure of the deviation of the solidification conditions from the critical conditions at which the planar interface is marginally stable.

Although nonplanar interfaces are expected when α is less than unity, the regular patterns obtained with the cellular and dendritic morphologies are observed at values of α substantially lower than unity. Of the two, cellular morphology with smooth side walls and a well-defined and experimentally measurable length for the mixed solid-liquid portion of the solidifying sample, presents itself as the one as a true steady-state morphology which is more amenable to mathematical analysis over the entire interface. In this paper the past theoretical analyses and experimental observations in this field are summarized and an attempt is made to define some of the unsolved problems in cellular solidification.

EXPERIMENTAL OBSERVATIONS ON CELLULAR SOLIDIFICATION

Early experimental investigations on cellular interfaces in metals solidified in a temperature gradient furnace under steady-state conditions were made by interrupting steady-state growth by decanting or quenching techniques and observing the pattern formed by sectioning the solid (Rutter and Chalmers, 1953; Walton and others, 1955; Sharp and Hellawell, 1969; Burden and Hunt, 1974a; Morris and Winegard, 1969). Solidification cells were observed as a regular array of hexagons on the decanted surface (see Fig. 2(a)). The opacity of metals to light waves prohibits the direct observation of the solidifying interface. Jackson and coworkers (Hunt, Jackson and Brown, 1966) pioneered experimental research on solidification of thin films of transparent organic alloys which can be used as model

systems to study the solidification behavior of metallic materials.
The new technique permits the direct observation of the solid-liquid
interface during solidification in an optical microscope. A number
of recent experimental researchers have employed this technique for
observing interface morphology and pattern formation during solidification (Venugopalan, 1982; Venugopalan and Kirkaldy, 1982; Trivedi
and Sombonsook, 1985). The only reported comprehensive measurements
of configurational parameters (spacing and length) were made on the
model system of succinonitrile-salol (Venugopalan, 1982 and
Venugopalan and Kirkaldy, 1982). The experimental observations have
led to some general conclusions on morphological and constitutional
aspects of cellular solidification.

Transitions in Interface Morphology

From the reported experimental observations, it may be concluded that
cellular morphology is obtained only over a certain range of α. A
critical α ($\alpha_c \ll 1$) must be reached before regular arrays of cells
are observed. For values of α between α_c and unity, a "pox"
structure or irregular structures are observed (Walton and others,
1955; Bolling, Tiller and Rutter, 1956; Venugopalan and Kirkaldy,
1982). In Fig. 2 are shown a number of nonplanar morphologies
observed on decanted interfaces of tin-lead alloys. The irregular

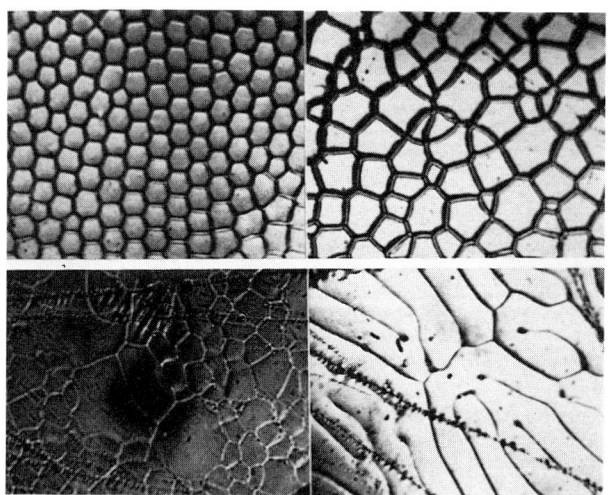

Fig. 2. Nonplanar interfaces in tin-lead alloys. (a) for $\alpha \ll 1$;
(b), (c) and (d) for α near unity. (Walton and others,
1955)

cellular structures of Figs. 2 (b-d) are obtained for α close to
unity while the regular array of cells shown in Fig. 2(a) is obtained
for α values much less than unity. Experiments on solidification of
thin film succinonitrile-salol specimens have shown that this
transition occurs at a typical α of 0.3 (Venugopalan, 1982). Such

data are not available for metallic systems. Fig. 3 illustrates a typical nonplanar morphology obtained in succinonitrile-salol alloys for marginal values of α. The observation of persistent intermediate irregular structures conclusively proves that for near marginal values of α, the solidification interface does not organize itself in a regular pattern. The implication of this is that theories (linear or nonlinear) dealing with marginal stability of a planar interface cannot be applied to stable cell configurations. It is unfortunate that the conclusive experimental evidence of this important phenomenon, which has existed for more than three decades, has largely been ignored in literature.

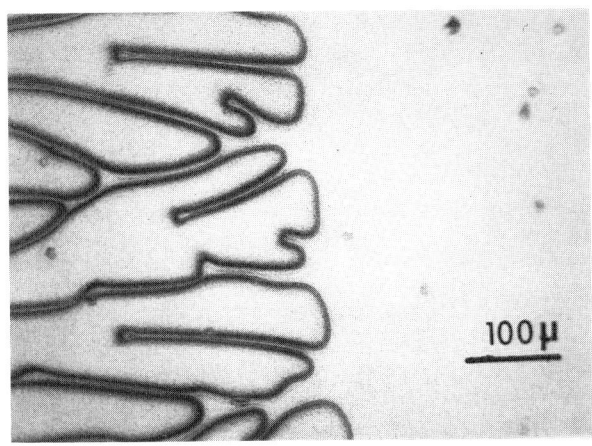

Fig. 3. Irregular interface in succinonitrile-salol typical of near-marginal structures (Venugopalan, 1982)

There is also a cell-dendrite transition marked by the appearance of side branches. This transition occurs typically at an α of 0.05. For lower values of α, a dendritic structure is obtained. In a recent analysis of experimental data on cell-dendrite transition in a number of metallic and model alloy systems, Tewari and Laxmanan (1987) show that the transition occurs over a wide range of α even in the same group of alloys. However, there is still no systematic experimental and theoretical studies of regular-irregular cell and cell-dendrite transitions on various alloy systems.

Configuration and Crystallographic Orientation of Cells

Cell walls are oriented along a preferred crystallographic direction. In cubic crystals this is the (100) direction (Morris and Winegard, 1969). The very strong tendency for preferred orientation of cell walls indicates that crystalline anisotropic effects are important in cellular solidification. This effect could be due to anisotropy in attachment kinetics, solid-liquid interfacial energy. The alignment of cell walls along preferred orientations is clearly seen from Fig. 4 which shows cellular morphology for the solid-liquid interface for two crystals of solid with different crystallographic orientations.

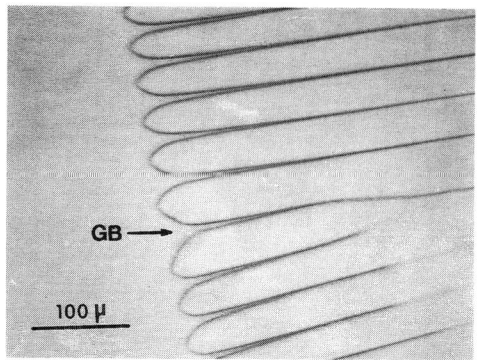

Fig. 4. Stable cellular growth of two crystals with different orientations. Note cell wall orientation and the differences in spacing and length of cells on either side of the grain boundary (Venugopalan, 1982).

Solidification cells are roughly elliptical near the tip near the irregular-regular cell transition and become increasingly paraboloidal as the cell-dendrite transition is approached. The spacing of cells observed in metallic and model systems depends on the growth parameters, viz. G, C_0 and v, and is of the same order of magnitude as the diffusion length, D/v (See Fig. 5). The radius of curvature of the cell tips decrease with increasing growth velocity for a given temperature gradient and alloy content. This quantity has been measured in a number of alloy systems. The length of cells also depends on the growth parameters and is typically an order of

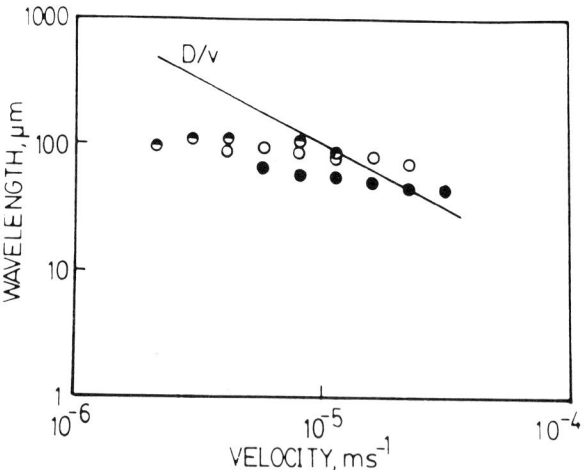

Fig. 5. Cell spacing (wavelength) observed in succinonitrile-2% Salol alloy for a number of temperature gradients.

magnitude larger than the spacing. The only measurements of cell length have been reported on succinonitrile-salol alloys by Venugopalan (1982). There is evidence that the magnitude of the configurational parameters also depend on the crystallographic orientation (See Fig. 4). The shape of the cells is such that a long, narrow region of intercellular liquid exists in contact with solid which is oriented along a preferred crystallographic direction. The region of liquid ends at the cell root whose radius of curvature is normally unresolved in the optical microscope.

Cells have been observed to persist for long times under steady-state growth conditions maintaining the same configurational parameters (spacing and length) and morphology. It has also been reported that the spacing and length assume unique values for a given set of growth parameters. Discrete mechanisms exist for obtaining the unique configuration as well as for equalizing the spacing across a given solidification front (Venugopalan and Kirkaldy, 1984). These observations on the configuration of the cells show that the cellular morphology is associated with true steady-state conditions.

Constitutional Aspects of Solidification Cells

There has been no report of experimental measurements of temperature and composition of the liquid/solid along the entire cellular interface. Thus there is no direct experimental evidence to confirm whether or not the interface is at local equilibrium. Experimental data on temperature and liquid/solid compositions at various positions along the interface could well lead to the resolution of some outstanding theoretical problems in cellular solidification. Most measurements on the cellular interface have been restricted to measurement of the cell tip temperature (Sharp and Hellawell, 1969; Burden and Hunt, 1974a; Jin and Purdy, 1974b). The tip temperature measured in Al-Cu alloys increased with increasing growth velocity initially and then at higher velocities decreased with increasing velocity. The latter relationship is observed generally beyond the cell-dendrite transition. The tip composition measured in an Fe-Ni alloy by Jin and Purdy (1974b) decreased with increasing velocity and remained constant at high velocities.

THEORETICAL TREATMENTS OF CELLULAR SOLIDIFICATION

Theoretical descriptions of cellular solidification have been aimed at predicting the morphological parameters such as the lateral spacing of the cells and constitutional conditions of the interface such as the interface temperature and composition. A complete theoretical description of the problem involves solving for the thermal and solute diffusion fields which have steady-state values in a moving frame of reference attached to the interface. It is expected of the correct theoretical description to predict all of the observations on stable solidification cells including the fact that such cells do not exist for near marginal conditions.

Initial attempts to apply the description of marginal instability of a planar interface to predict cell spacing have not been very fruitful (Mullins and Sekerka, 1964; Venugopalan and Kirkaldy, 1988). Observed cell spacings were too large by an order of magnitude compared with predictions of the linear theory. This is not surprising in the light of the observations made earlier. Later attempts to describe the cell problem focussed on the conditions in the vicinity of the cell tips (Bolling and Tiller, 1960; Donaghey and Tiller, 1968; Burden and Hunt, 1974b; Kirkaldy, 1959; Jin and Purdy, 1974a; Trivedi, 1970). Since the solutions are only obtained near the tip, the analyses are applied to dendrites as well as cells. In these analyses, the solute diffusion problem is solved in the tip region of an isolated cell or an array of cells whose shape is defined a priori. Mass conservation condition is satisfied at the tip. The thermal field is assumed to be represented by a constant temperature gradient. None of the analyses leads to a unique mathematical solution; but each analysis leads to a relationship between the liquid composition at the tip (hence the tip temperature from the phase diagram assuming local equilibrium) and either the tip radius of curvature or the cell spacing. An extra condition is thus required to select one solution from the infinite possible mathematical solutions.

A variety of suggestions have been made in literature to select a single solution out of all the possible ones. These criteria are based on one of the following: minimum tip undercooling, maximum intrusion criterion, marginal tip stability and minimum entropy production rate. All of these criteria provide reasonable agreement between experimentally measured tip temperature/composition and the calculated value for the selected solution. As mentioned elsewhere, all of these criteria except the marginal stability criterion are qualitatively equivalent in that the selected solution corresponds with maximizing segregation of solute. There is no theoretical basis for all but the minimum entropy production rate criterion. For example, there is no theoretical basis to require cell tips (or dendrite tips) to grow at the limit of marginal stability of the chosen shape.

As Kirkaldy (1959, 1980, 1985) has pointed out in literature, the lack of a unique solution is a characteristic of free boundary problems and frequently nature exhibits several of these possible solutions under the same steady-state conditions but with differing paths to the steady-state operating conditions. Such hysteresis effects are observed in hydrodynamics and in eutectic solidification in thin film specimens. While hysteresis effects are not observed in cellular solidification, an understanding of how nature chooses a particular solution from an infinite milieu is of interest and relevance to the cell problem.

Kirkaldy (1985) has made significant contributions to our understanding of the nature of the free boundary problems and the development of patterns characterized by morphological and constitutional parameters in dissipative structures in thermal, diffusive, hydrodynamic and life systems. His pioneering contributions to developing a theoretical structure for the resolution of the free boundary problems are rooted in irreversible thermodynamics applied to steady state phenomena and based on unique applications of the Second Law of thermodynamics. According to Kirkaldy, the observation of a unique solution characterized by a unique set of morphological and constitutional parameters (or a unique pattern) depend on the ability of the system to vary these parameters strictly according to the steady state conditions (irrespective of initial conditions). Kirkaldy points out that the presence or absence of a threshold in a thermodynamic function determines the ability of a dissipative system to access a continuum of mathematically valid steady state configurations in its search for a unique solution defined by the steady state operating conditions. It is noteworthy to mention here that Kirkaldy's theoretical structure predicts that near-marginal states are unable to overcome the threshold to organize themselves in any regular pattern forming state. This is consistent with the observations that regular solidification cells are formed only for values of   far removed from its critical value for the onset of nonplanar morphologies. From Kirkaldy's analysis, the steady state pattern is determined by an optimum in the entropy production rate. The optimum could be a minimum, a maximum or a saddle point configuration in the space defined by the configurational (or constitutional) parameters. The cellular solidification problem corresponds to a minimum in the entropy production rate.

More recent approaches to the problem attempt to solve the problem over the entire length of the interface rather than just at the tip. Kirkaldy (1980) argued that cells represent a morphology for which diffusional and thermal conditions require the interface to be out of local equilibrium. This is in sharp contrast to assumptions made in earlier work referred to above that the solid-liquid interface is always at local equilibrium. With the thermal field with a constant temperature gradient is imposed on the system, linear liquidus and solidus lines in the phase diagram and a nonlinear solute concentration profile required by the diffusion equation, the nearly parallel walls of cells can not maintain local equilibrium everywhere on the interface. Stable cellular interfaces cannot exist in local equilibrium (Kirkaldy and Venugopalan, 1988). As mentioned before, the constitution of the entire cellular interface is yet to be verified by experiment. Qualitatively, the absence of stable morphologies and the strong crystallographic preference for cell wall orientation suggest that nonequilibrium effects play an important role in cellular solidification. There is more than a hint that the crystallography of cell walls is related to kinetic stabilization of the nonequilibrium interface. The breakdown of the cell walls in the dendritic regime is analogous to the breakdown of the planar interface itself.

Kirkaldy's (1980) approximate analysis of the diffusion problem in the cell grooves yielded a relationship for the length of cells. Since the lateral solute segregation near the tips was not treated, no expression was obtained for the spacing. A substitute for minimum entropy production rate criterion (in the form of a minimum root radius) was used to obtain a unique solution. Significantly, no mathematical solutions existed for near marginal values of α.

A more rigorous treatment of the problem on the same basis as Kirkaldy's was conducted using numerical techniques on an assumed cell shape (Venugopalan, 1982 and Venugopalan and Kirkaldy, 1984). The two dimensional solute diffusion problem was solved in the liquid. The solution, which is described in detail in the above mentioned references, consisted of applying the Fourier solution to the diffusion problem following a standard procedure used in heat conduction (Arpaci, 1966). A bivariate analysis yielded an optimum length and spacing of cells. Figs. 6 and 7 show examples of the reasonable agreement obtained between calculated and experimental cell spacings and lengths. The complexity of the numerical methods, however, render the analysis a time consuming and tedious procedure. However, this is the only complete analysis in literature of the single phase binary cellular solidification. McCartney and Hunt (1981) applied a finite difference technique to the situation in which an eutectic is formed in the cell groove. While the problem is seemingly similar to the single phase cellular solidification problem, there are important differences in the boundary conditions preventing meaningful comparisons with this analysis.

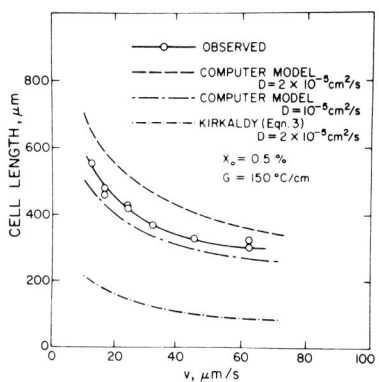

Fig. 6. Comparison of predicted and observed cell lengths in succinonitrile-salol (Venugopalan and Kirkaldy, 1982).

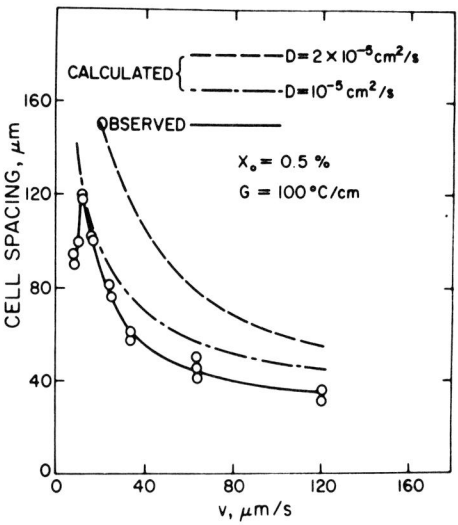

Fig. 7. Comparison of calculated and observed cell spacings in succinonitrile-salol (Venugopalan and Kirkaldy, 1982).

CONCLUSIONS

Much progress has been made in understanding the morphology of the solidification interfaces since the time of elucidation of the Constitutional Supercooling criterion by Chalmers and coworkers (Tiller and others, 1953). It is now understood that regular stable cellular structures are obtained only when the growth conditions are far removed from those for stable planar solidification. Thus marginal stability theories are clearly not applicable for stable cell formation. There is no elegant theoretical solution to the diffusional problem in the entire domain of liquid (present ahead of the cell tips and in the groove between cells). Numerical solutions to the problem are inelegant and tedious. It is understood that any theory of cellular solidification must be able to predict all of the experimental observations regarding stable cells.

Further work in the area of cellular solidification should be directed at (a) establishing experimentally the temperature and liquid composition along the cellular interface; (b) investigating the crystallographic preference of cell walls and mechanisms of stabilization of the cell walls; (c) a systematic investigation of the transitions between regular and irregular morphologies and (d) a more elegant and fruitful method to analyze the diffusion problem in the entire liquid region.

REFERENCES

Arpaci, V. S. (1966). Conduction Heat Transfer, Addison Wesley Pub. Co., Reading, Mass.
Bolling, G. F., and W. A. Tiller (1960). J. appl. Phys., 31, 2040-2045.
Bolling, G. F., W. A. Tiller, and J. W. Rutter (1956). Can. J. Phys., 34, 234-240.
Burden, M. H., and J. D. Hunt (1974a). J. Cryst. Growth, 22, 99-108.
Burden, M. H., and J. D. Hunt (1974b). J. Cryst. Growth, 22, 109-116.
Donaghey, L. F., and W. A. Tiller (1968). The Solidification of Metals, The Iron and Steel Institute, London.
Hunt, J. D., K. A. Jackson and H. Brown (1966). Rev. Sci. Instrum., 37, 805-810.
Jin, I., and G. R. Purdy (1974a). J. Cryst. Growth, 23, 29-36.
Jin, I., and G. R. Purdy (1974b). J. Cryst. Growth, 23, 37-44.
Kirkaldy, J. S. (1959). Can. J. Phys., 37, 739-754.
Kirkaldy, J. S. (1980). Scripta Metall., 14, 739-744.
Kirkaldy, J. S. (1985). Metall. Trans., 16A, 1781-1797.
Kirkaldy, J. S., and D. Venugopalan (1988). submitted to Scripta Metall.
McCartney, D. G., and J. D. Hunt (1981). Acta Metall., 29, 1851-1863.
Morris, L. R., and W. C. Winegard (1969). J. Cryst. Growth, 5, 361-375.
Mullins, W. W., and R. F. Sekerka (1964). J. appl. Phys., 35, 444-451.
Rutter, J. W., and B. Chalmers (1953). Can. J. Phys., 31, 15-20.
Sharp, R. M., and A. Hellawell (1969). J. Cryst. Growth, 6, 253-263.
Tewari, S. N., and V. Laxmanan (1987). Metall. Trans., 18A, 167-170.
Tiller, W. A., K. A. Jackson, J. W. Rutter and B. Chalmers (1953). Acta Metall., 1, 428-437.
Trivedi, R. (1970). Acta Metall., 18, 287-299.

Trivedi, R., and K. Sombonsook (1985). *Acta Metall.*, __33__, 1061-1074.
Venugopalan, D. (1982). *Cellular Instability in Binary Solidification*, Ph. D. Thesis, McMaster University, Canada.
Venugopalan, D., and J. S. Kirkaldy (1982). *Scripta Metall.*, __16__, 1183-1187.
Venugopalan, D., and J. S. Kirkaldy (1984). *Acta Metall.*, __32__, 893-906.
Venugopalan, D., and J. S. Kirkaldy (1988). submitted to *Scripta Metall.*
Walton, D., W. A. Tiller, J. W. Rutter and W. C. Winegard (1955). *Trans. Am. Inst. Min. Engrs.*, __203__, 1023-1026.

Channel Flow in Partly Solidified Alloy Systems

BY J. R. SARAZIN AND A. HELLAWELL

Department of Metallurgical Engineering
Michigan Technological University, Houghton, MI 49931, USA

ABSTRACT

Channel segregation is a type of macrosegregation which occurs during alloy solidification in situations where a two phase, solid + liquid, 'mushy zone' exists over a large temperature range for a long period of time.

Differences in thermal or solutal expansion or contraction may cause the interdendritic liquid within the mushy zone to differ in density from the bulk liquid and thus provide a driving force for convective flow. The convective flow patterns within the ingot give rise to channels, or streamers, of solute rich material, referred to as 'A' segregates or freckles in the foundry industry, which are left in the ingot after solidification. The defects are observed to run antiparallel to the direction of gravity in the columnar zone of a casting. Although different geometrical configurations are discussed, channels are particularly evident in base chilled configurations where the solute(s) rejected during solidification are less dense than the solvent.

Channel segregation has been examined in Pb alloys containing Sn and/or Sb as well as the transparent NH_4Cl-H_2O system. Channels are shown to originate from liquid perturbations arising from thermosolutal interactions at or near the dendritic interface, and assuming such a model, an analysis of the situation is presented.

KEYWORDS

Solidification, macrosegregation, 'A' segregate, channel, freckle, thermosolutal convection, perturbation, Rayleigh number.

INTRODUCTION

Channel segregation is a special kind of macrosegregation arising from density driven convective flow during alloy solidification. It occurs in situations where a two phase (solid + liquid) mushy zone exists over a large temperature range for an extended period of time. Thermal or solutal, expansion or contraction may cause the interdendritic liquid within the mushy zone to differ in density from the bulk and thus provide a driving force for convective flow.

The remnants of these convective flows are left in the ingot after solidification in the form of channels or streamers of solute rich material which run antiparallel to the direction of gravity in the columnar zone of a casting.

In large steel billet castings and other situations where heat flow is primarily horizontal through the mold wall they are referred to as 'A' segregates. Figure 1 shows the top section of a 5 ton killed steel ingot (1). Within the columnar zone, near the top, there are channels or 'A' segregates (dark) presumably rich in light elements such as sulphur, phosphorous and carbon. Channel segregation also occurs in situations where heat flow is vertically downward such as in electroslag refined (ESR) and vacuum arc remelted (VAR) ingots as well as directionally solidified Ni-base superalloy turbine blades where they are commonly referred to as freckles. Channel segregation can cause serious problems during subsequent processing, resulting in costly homogenization treatments or rejection of all or part of a cast ingot. Previous work has examined channel formation during directional solidification of base chilled ingots of Pb alloys containing Sn and/or Sb (2-9) as well as the transparent analogue system NH_4Cl-H_2O (4-7,10,11).

Figure 2a shows a vertical section through a Pb-10wt.% Sn ingot which was allowed to partially solidify with the growth direction vertically upward and then quenched from the top to reveal the solidification front. Near the center is a channel of Sn rich material beginning near the bottom and extending to the dendritic interface. Figure 2b is a cross section of a similar ingot at approximately the midpoint showing numerous freckles $\sim$ 1 mm in diameter against a background array of primary dendrite 'crosses'.

The NH_4Cl-H_2O eutectic system has been used for some time as an analogue to study the solidification of metals (12). Due to its low entropy of fusion it

Fig. 1. 'A' segregate channels, rich in sulphur, phosphorus and carbon, in top portion of a 5 ton killed steel ingot (1).

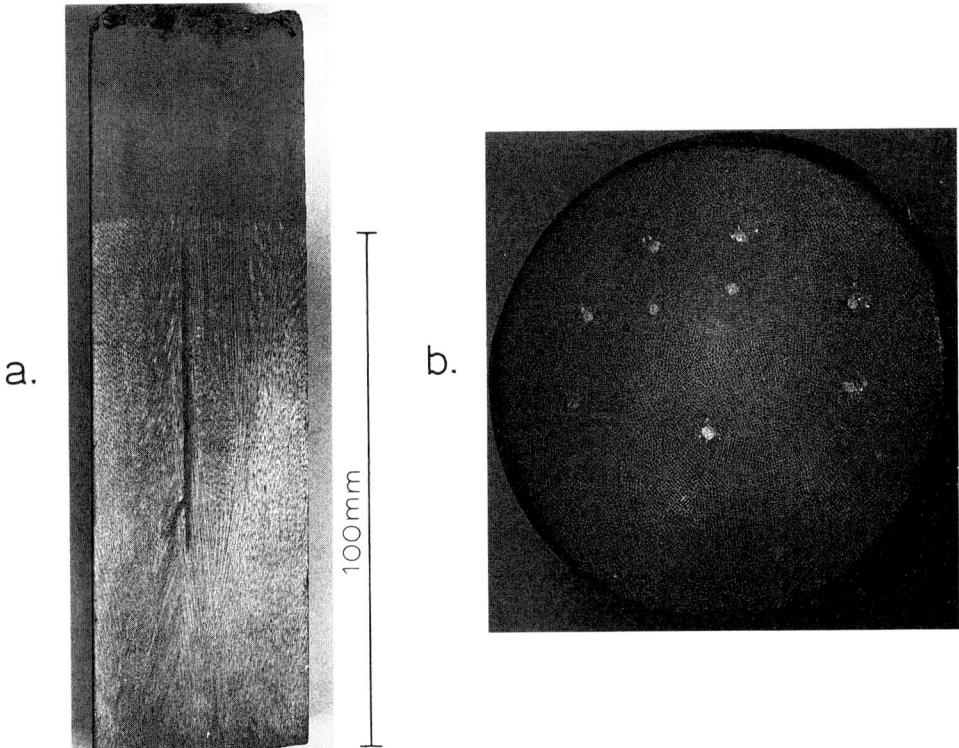

Fig. 2a) vertical and 2b) horizontal sections through a base chilled Pb-10wt.% Sn ingot to show the dendritic growth front and channels.

solidifies dendritically, in a way similar to metallic systems with the distinct advantage of a transparent liquid/opaque solid allowing direct observation of the solidification process. Figure 3a shows a top view of a solidifying NH_4Cl-35 wt.% H_2O casting. It can be seen that channel formation occurs in a similar manner to and on a similar scale as, the metallic systems. In Fig. 3b, a side view of the same casting, a streamlined plume of solute rich liquid can be seen rising from a channel on the dendritic interface. Observation is possible due to a change in refractive index with composition.

Although channels are more clearly found when the solute which is rejected during solidification is less dense than the solvent, similar convective flow patterns may exist during solidification or melting with a solute which is either less or more dense than the solvent. Figure 4.1 represents horizontal heat flow (mushy zone shaded) while Fig. 4.2 and 4.3 represent vertical heat flow parallel and antiparallel to gravity respectively. Figures 4.1-4.3a represent a less dense solute while 4.1-4.3b represent a more dense solute. The flow pattern shown in Fig. 4.1a would be representative of a killed steel ingot (Fig. 1) while Fig. 4.2a reflects the flow pattern observed in the directionally solidified Pb-base or NH_4Cl-H_2O castings (Fig. 2,3). Crystal growth using the Czockralski technique would exhibit the behavior shown in Fig. 4.3. It should be noted that no convection would be present in Fig. 4.2b since the more dense interdendritic liquid resides stably against the advancing interface.

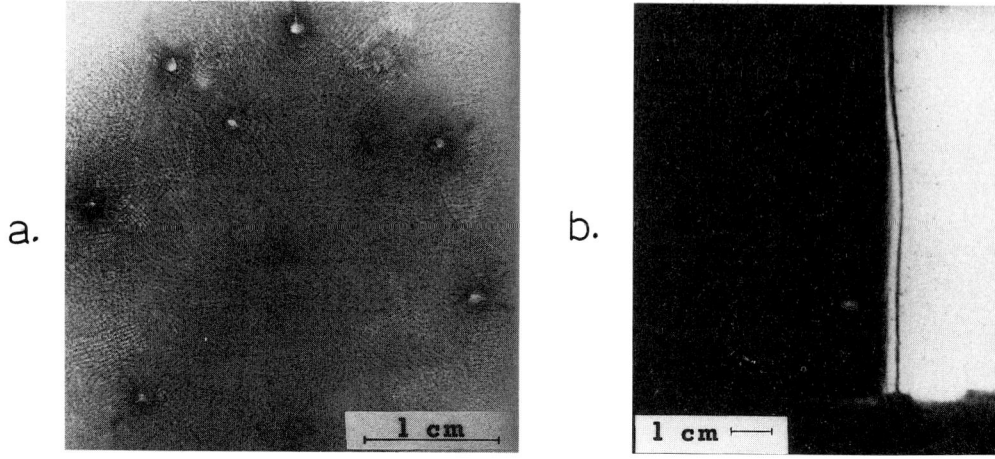

Fig. 3a) Top view and 3b) side view of a solidifying NH$_4$Cl-35 wt.% H$_2$O casting observed through the transparent bulk liquid showing channels and plume flow.

Model for Channel Formation

Consider the directional solidification of a base chilled ingot of composition C_o (Fig. 5, upper left) for the case where the equilibrium distribution is less than 1 and the solute is less dense than the solvent. The solute which is rejected into the liquid during solidification of the primary phase may build up a metastable layer of less-dense, solute-rich liquid in the interdendritic regions, extending a distance of approximately D/V into the bulk liquid, where D is the diffusion coefficient of solute in the liquid and V is the dendrite growth velocity (Fig. 5, upper right).

Examining the concentration profile in the vicinity of the dendritic interface (Fig. 5 lower left) shows a near constant bulk concentration of C_o, increasing as the interface is approached from the top, due to the presence of the rejected solute. The negative concentration profile (dC/dz) extends down into the mushy zone with a slope dependent on the temperature gradient and liquidus slope until the eutectic front is reached. Since the solute is less dense, the negative concentration profile within the mushy zone results in a positive density gradient (dρ/dz) (Fig. 5 lower right). However, the sample is sitting in a positive temperature gradient (dT/dz). As the density decreases slightly with temperature then the density gradient within the bulk liquid (dρ/dz) is negative or stabilizing. This results in a metastable condition which may persist for some time.

The formation of channels under these conditions has been well documented since the late 1960's (1,10). However, the origin of the channels or the exact conditions under which they arise, has not been clearly established. Two schools of thought have developed as to where the channels nucleate, one based on theoretical considerations (13,14,15) and the other based on direct experimental observation. The former group contends that channels originate at some imperfection or defect deep within the mushy zone and a channel develops upward by local melting, in a manner similar to a river system. Unfortunately, local acceleration of fluid flow is not possible without cooperative flows

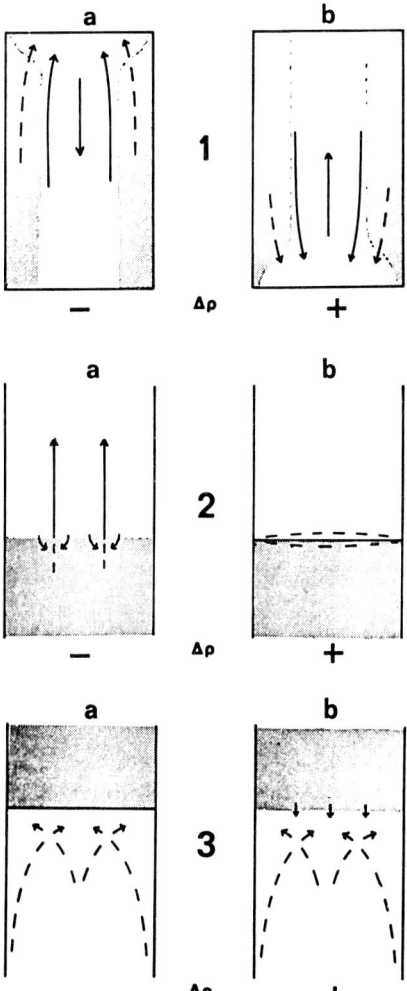

Fig. 4. Convective flow patterns within a casting for alternative heat flow and solute density configurations.

elsewhere within the system. Since the situation is one of a less dense volume of entrapped liquid within the mushy zone surmounted by a body of more dense, bulk liquid, which is thermally stabilized against convection, any motion must originate at or close to the dendritic front where the density gradient changes sign. An analogy has been drawn (11) to the draining of a swamp which is accomplished by creating a breach at the edge rather than by an internal trench or levy.

By direct observation of channel formation in the transparent NH_4Cl-H_2O system, Sample and Hellawell (6) have shown that channels originate from a liquid perturbation at or close to the dendritic interface. A schematic representation of the model is shown in Fig. 6. Figure 6a shows the situation, described previously, that is a less dense boundary layer with a thickness on the order of

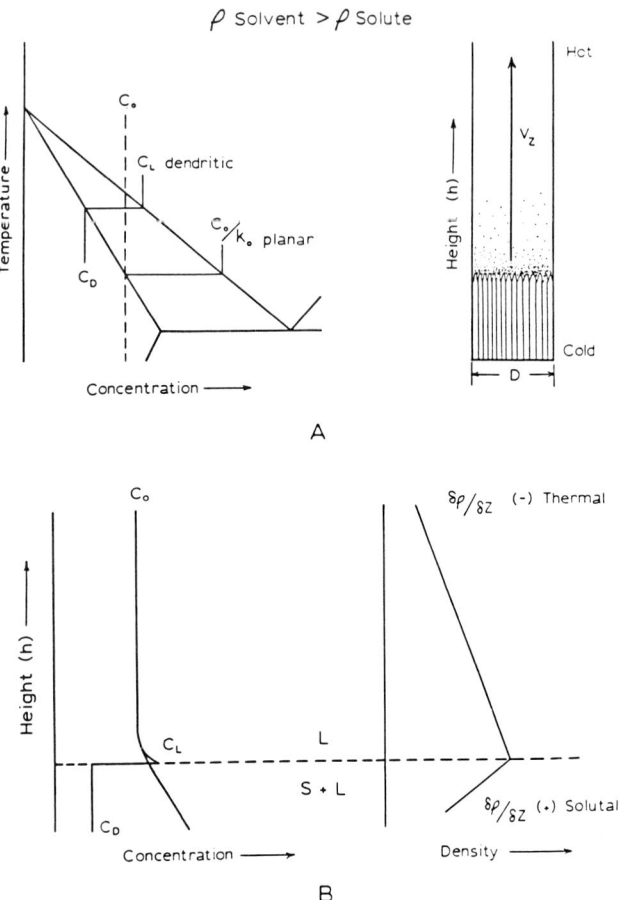

Fig. 5. Schematic concentration/density profiles above and below the dendritic growth front for vertical growth with a less dense solute in a positive temperature gradient.

D/V surmounted by the bulk liquid with the mushy zone below. Since a density inversion exists, the system is metastable and if conditions will allow, it can perturb as shown in Fig. 6b probably by double diffusive or thermal solutal interactions within the boundary layer. As the perturbation grows the concentration gradient and thus the temperature gradient below it is relaxed and the dendrite tips can be observed to accelerate forward to produce a small hillock on the interface as in Fig. 6c. As the perturbation continues to grow it begins to be fed not only by the boundary layer but the interdendritic liquid as well. Upward flow of the interdendritic liquid melts or erodes the dendrites, creating a channel backwards from the interface (Fig. 6d). For channel flow to persist, the upward flow from the channel mouth must be balanced by entrainment of the bulk liquid in the regions immediately adjacent to the channels, Fig. 6e. This results in a quasi-steady state flow pattern which persists until the dendritic interface reaches the top of the casting.

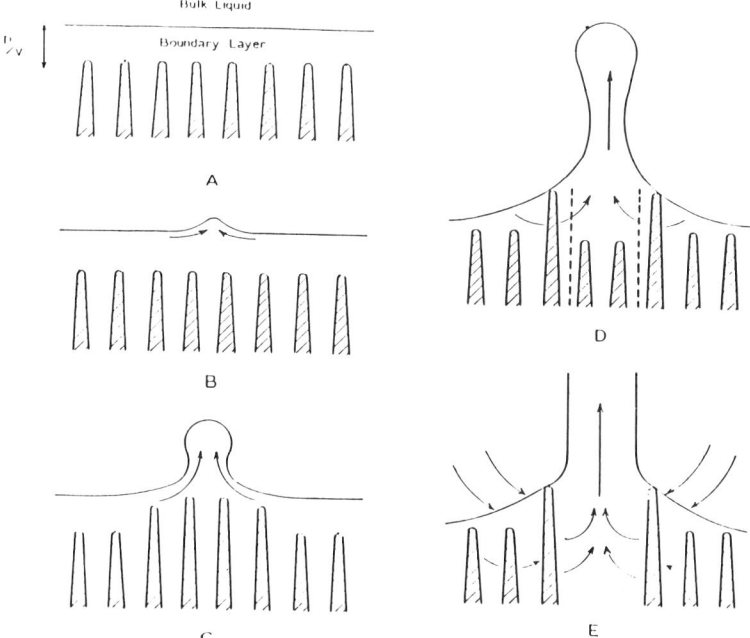

Fig. 6. Schematic model for channel formation by liquid perturbation at the dendritic growth front.

Experimental Work

Although the entire nucleation event occurs within seconds, careful observations of channel formation against the mold wall and simple experiments using the NH_4Cl-H_2O systems (4) have shown that channels nucleate by a liquid perturbation near the dendritic interface and that plume flow preceeds channel formation.

Experiments using potassium permanganate ($KMnO_4$) crystals as dies confirmed the channel flow patterns and local retrainment regions. Attempts were also made to artificially create channels (Fig. 7). An NH_4Cl-35 wt.% H_2O casting was allowed to solidify to point where channels were known to have formed. A fine 1 mm bore silica tube, sharpened on the end was then carefully twisted while inserting it into the mushy zone so as to drill out an artificial channel (Fig. 7a). Channels created in this manner failed to propagate and were soon grown over. The experiment was then repeated, this time the tube inserted to a point approximately 1 mm from the dendritic interface and a small amount of liquid drawn up so as to create an artificial plume (Fig. 7b). In each case a channel formed at the exact position where the plume had been created.

Although the NH_4Cl-H_2O system was very convenient to work with for modeling the mechanisms of formation, practical working conditions limited compositions to a narrow range around 35 wt.% NH_4Cl. In order to determine the effect of changing the fraction of solid within the mushy zone, the bottom chilled NH_4Cl-H_2O experiments were repeated using Pb alloys contain Sn and/or Sb (8). The relevant portions of the phase diagrams and regimes of channel formation are shown in Fig. 8.

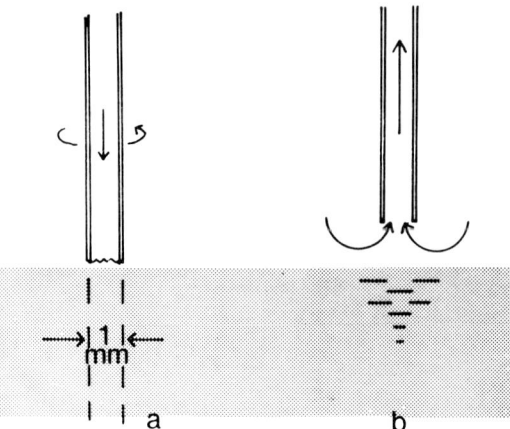

Fig. 7. Artificial formation of channels caused by (a) drilling into the mushy zone and (b) sucking up liquid to create an artificial plume.

In the metallic alloys two regimes of channel formation were observed. Firstly, at low solute contents numerous small channels were observed and secondly, at higher solute contents, a single, centrally located large channel was found (Fig. 9). The transition from the multiple small channels to the single large channel regime has been attributed to the increase in dendritic mesh permeability as the solute content is increased, allowing entrainment patterns on the ingot scale. It should be noted that the regimes of channel formation in the metallic systems are shifted to much higher fractions of solid or lower solute contents than in the analogue system. This is attributed to higher thermal diffusivity and lower kinematic viscosity (Table I) in the metallic system leading to easier establishment of streamlined flow. Within any given ingot in the multiple small channel regime, the number of channels is observed to decrease with distance up the ingot. It can be seen from Fig. 10, a Pb-2 wt.% Sb ingot sectioned at 2 cm intervals from the bottom, that the number of channels decreases with height and every channel present in the top section can be traced back to the bottom. This is consistent with the observations of the analogue castings where channels were observed to coalesce or die out during solidification.

Thermosolutal convection/Rayleigh number

The liquid perturbations leading to the formation of channels have been thought to arise from a phenomenon known as double diffusive or thermosolutal convection (17). This phenomenon is not unique to metallurgy and is believed to be responsible for convective flows in oceanographic, geological or even stellar contexts. Figure 11, after Schmitt (18), is a plot of logarithm of the Lewis number, (τ), defined as the ratio of the thermal to solutal diffusivity vs. Prandtl number (σ), defined as the kinematic viscosity over the thermal diffusivity, for a variety of double diffusive systems. Systems with high viscosity and low thermal diffusivity fall to the lower right in the figure while those with a low viscosity and high thermal diffusivity tend towards the upper left. The contours represent a normalized flow velocity.

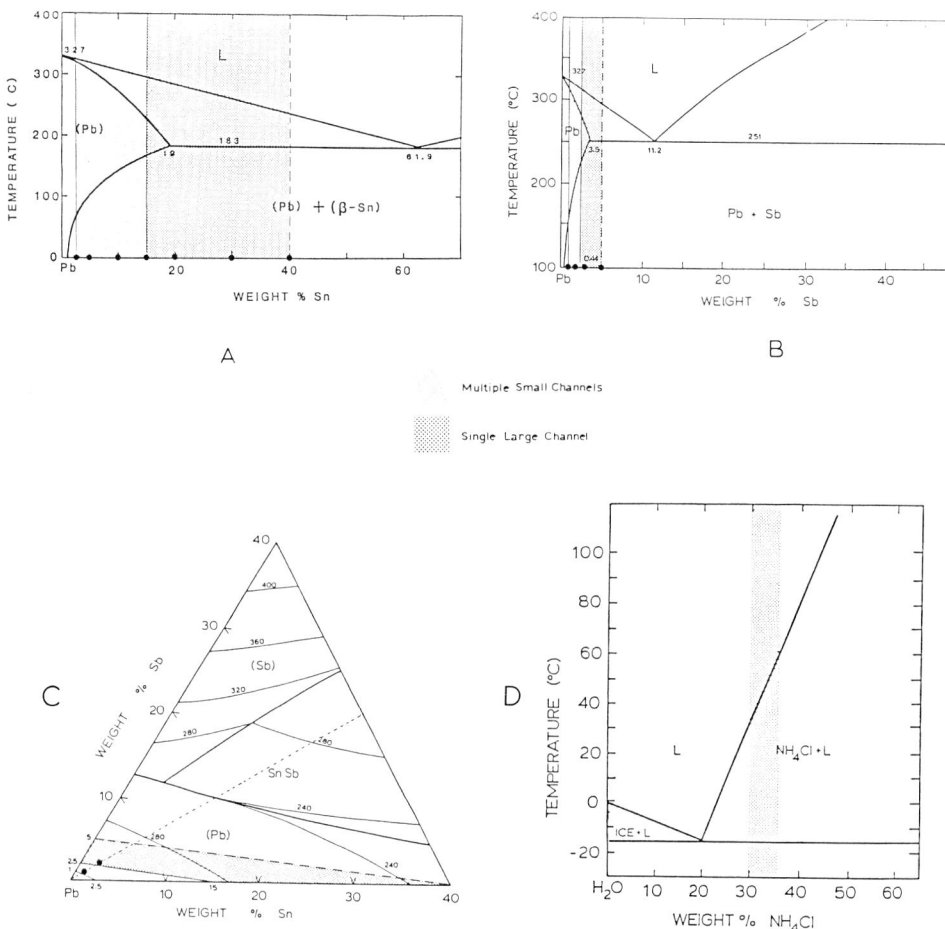

Fig. 8. Partial phase diagrams for (a) Pb-Sn, (b) Pb-Sb, (c) Pb-Sn-Sb and (d) NH$_4$Cl-H$_2$O showing compositional regimes (shaded) in which channels were observed.

The phenomenon of thermosolutal convection can arise in any situation where gradients of two or more properties have opposing effects on the vertical density gradient. In this case the properties of interest are heat and solute content. The mechanism of instability can be briefly described as follows: if a small parcel of fluid is displaced upwards its temperature equilibrates quickly but its solute content does not, since the diffusivity of heat is some 2 orders of magnitude greater than that of solute. The fluid parcel is thus solute rich and lighter than its surroundings, becoming increasingly buoyant as it continues to accelerate upwards.

If the solute were to be neglected and the density inversion in the liquid assumed to be due only to a temperature gradient, the classic Benard problem (19) analyzed by Lord Rayleigh in 1916 (20), then one can write a dimensionless thermal Rayleigh number (Ra_T) to describe the initial disturbance or perturbation.

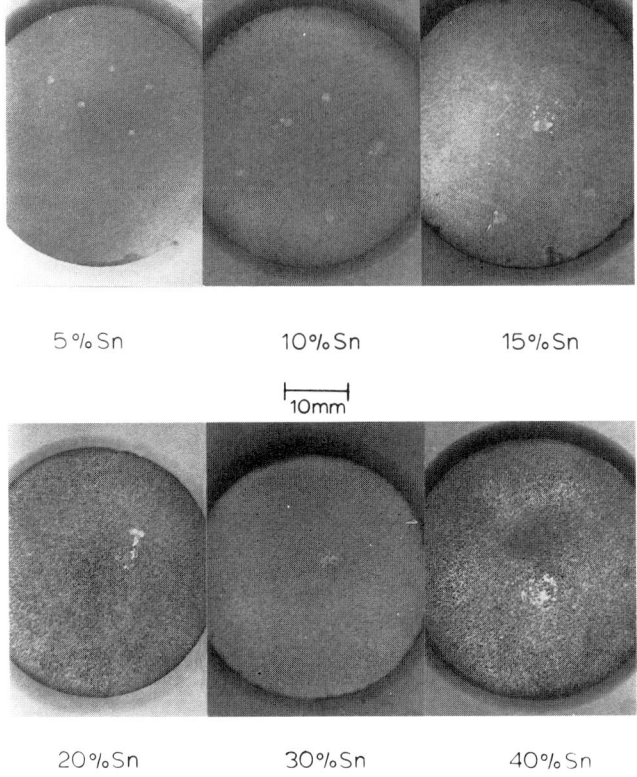

Fig. 9. Horizontal sections of Pb-Sn ingots, 60 mm from the bottom showing change in channel character with increasing concentration.

$$Ra_T = \frac{g\rho\alpha \ (dT/dz)}{\eta \ D_T/h^4} \qquad \text{Eq. 1}$$

g = gravitational constant
ρ = bulk liquid density
α = volume coefficient of thermal expansion
dT/dz = vertical temperature gradient
η = absolute viscosity
D_T = thermal diffusion coefficient
h = a characteristic linear dimension

Similarly, if the system were isothermal and the density inversion due only to a solute gradient then an equivalent solutal Rayleigh number (Ra_S) would be:

$$Ra_S = \frac{g\rho\beta \ (dC/dz)}{\eta \ D_S/h^4} \qquad \text{Eq. 2}$$

where α is replaced by the volume coefficient of solutal expansion β, dT/dz is replaced by the vertical solute gradient dC/dz and D_T replaced by the mass diffusion coefficient of solute in the liquid D_S.

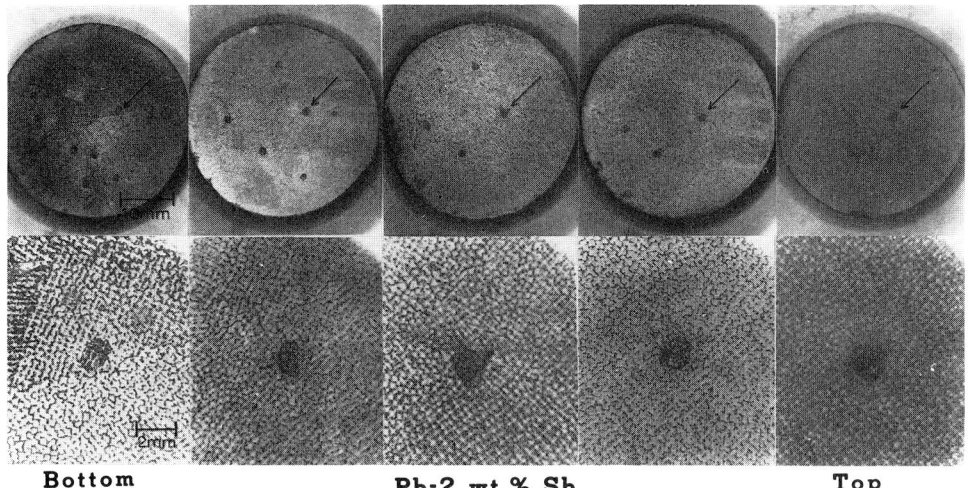

Bottom   **Pb-2 wt.% Sb**   **Top**

Fig. 10. A Pb-2 wt.% Sb ingot sectioned perpendicular to the growth direction at 20 mm intervals from the bottom.

Closer examination of equations 1 and 2 reveals that the numerators $g\rho\alpha(dT/dz)$ and $g\rho\beta(dC/dz)$ have equivalent units of pressure per unit area (Nm^{-4}). This can be thought of as the buoyant pressure exerted upward due to the density inversion. The units of denominators, $\eta\, D_T/h^4$ and $\eta\, D_S/h^4$ are the same and this can be thought of as the opposing or restraining pressure due to fluid viscosity and diffusion of heat or solute in the system.

When both thermal and solute gradients are present the Rayleigh numbers can be combined to yield an equivalent thermosolutal Rayleigh number ($Ra_{T/S}$) as follows:

$$Ra_{T/S} = \frac{Ra_S}{\tau} - Ra_T \qquad \text{Eq. 3}$$

where τ is the diffusivity ratio or Lewis number. The difference arises since the thermal effect is one of stabilization. $Ra_{T/S}$ may then be written as follows:

$$Ra_{T/S} = \frac{g\rho\,[\beta(dC/dz) - \alpha(dT/dz)]}{\eta\, D_T/h^4} \qquad \text{Eq. 4}$$

Analysis

Normally, the system is considered to perturb when $Ra_{T/S}$ exceeds some critical value. However, considerable disagreement exists in the literature as to the value of h, which is alternately taken as D_S/V (21,22), the height (16,24) or the radius (23) of the system. Since h is raised to the 4th power in the equation, these encompass an extremely large range of values making comparison of critical Rayleigh numbers meaningless from case to case.

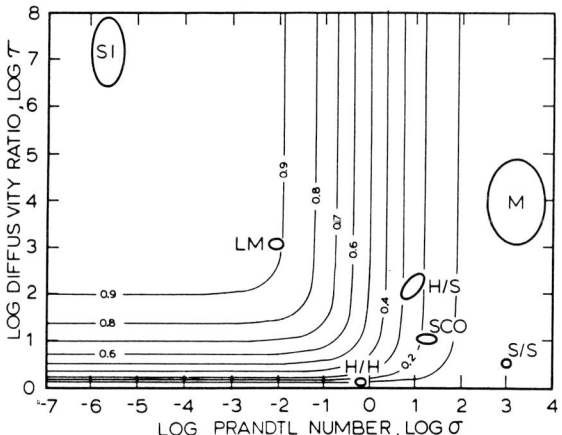

Fig. 11. A plot of Log τ (Lewis number) vs. Log σ (Prandtl number) for a variety of double diffusive systems including liquid metals (LM), aqueous heat/salt solutions (H/S), molten magmas (M), semiconductor oxides (SCO), aqueous salt/sugar solutions (S/S), heat humidity (H/H) and stellar interiors (SI). Contours correspond to a normalized flow rate after Schmitt (18).

In order to determine what might be a reasonable value for h, Sarazin and Hellawell (8,9,25) considered instability to occur when the buoyant forces exceeded the restraining effect of thermal diffusion and fluid viscosity or for a critical effective thermosolutal Rayleigh number ≅ 1. Back calculation of h from known data for Pb-Sn and NH_4Cl-H_2O revealed h to be on the order of the dendritic spacing in both cases. The term h appears, therefore, to represent a critical wavelength of a perturbation of the boundary layer. In retrospect it seems entirely reasonable that the perturbation wavelength should be related to the dendrite spacings since a periodic compositional variation would already exist ahead of the advancing dendritic interface, but to arrive at that result independently as it were is quite satisfactory.

The consequences of varying the primary dendrite spacing can be examined for the alloys previously studied. Kurtz and Fisher (26) have given an empirical expression for the dendrite spacing (λ), where:

$$\lambda = \frac{4.3 \, [\Delta T_o D \Gamma^{.25}]}{k^{.25} \, V^{.25} \, G^{.5}} \qquad \text{Eq. 5}$$

ΔT_o = temperature difference between liquidus and solidus
D = mass diffusion coefficient of solute in the liquid
Γ = Gibbs-Thompson coefficient
k = equilibrium distribution coefficient
V = growth velocity
G = temperature gradient

For a given composition, the primary dendrite spacing is proportional to $V^{-.25}$ $G^{-.5}$. Thus, if it were possible to control the thermal gradient and growth velocity, the dendrite spacing and perturbation wavelength for a given alloy could be varied.

Table I
Data for Metallic and Aqueous Systems (4,9,10,16)

	Pb-10 wt.% Sn	Pb-2 wt.% Sb	NH_4Cl-35 wt.% H_2O
Solvent: solute density ratio (solids)	1.55	1.69	1.53
Liquid density on liquidus, ρ, Kg m^{-3}	9.98×10^3	$\sim 10^4$	1.08×10^3
Dynamic viscosity, η, Kg m^{-1} s^{-1}	2.47×10^{-3}	$\sim 3 \times 10^{-3}$	1.03×10^{-3}
Kinematic viscosity, $\nu = \eta/\rho$, m^2 s^{-1}	2.47×10^{-7}	$\sim 3 \times 10^{-7}$	9.54×10^{-7}
Thermal diffusivity, κ, m^2 s^{-1}	1.08×10^{-5}	$\sim 1 \times 10^{-5}$	1.47×10^{-7}
Solutal diffusivity D, m^2 s^{-1}	3.0×10^{-9}	$\sim 3.0 \times 10^{-9}$	1.3×10^{-9}
Thermal expansion coeff., α, K^{-1}	1.15×10^{-4}	$\sim 10^{-4}$	6.0×10^{-4}
Solutal expansion coeff., β, wt.%$^{-1}$	5.2×10^{-3}	$\sim 7 \times 10^{-3}$	2.0×10^{-3}
Prandtl number, ν/κ	2.3×10^{-2}	$\sim 3.0 \times 10^{-2}$	6.81
Lewis number, κ/D	3.6×10^3	$\sim 3.0 \times 10^3$	107.7
Fraction liquid at eutectic, f_L	0.079	0.082	0.86
Temperature gradient, $\partial T/\partial z$, K m^{-1}	10^3	10^3	1.5×10^3
Interdendritic solute gradient, dc/dz, wt.%, m^{-1}	4.3×10^2	1.5×10^2	7.0×10^2
Flow rate to prevent channels, U, m s^{-1}	10^{-3}		10^{-2}
Interdendritic spacing, L, m	3×10^{-4}	3×10^{-4}	5.7×10^{-4}
Derived dimension h, m	6×10^{-4}	4×10^{-4}	4×10^{-4}

Future work will involve assembly of a separately controlled, 2 zone resistance furnace to permit relatively independent control of temperature gradient and growth velocity. This will allow determination of regimes of channel formation on a G-V plot such as the one shown schematically in Fig. 12 (10).

Increasing either the thermal gradient or the growth velocity for a given alloy would decrease the dendrite spacings and thus the perturbations' wavelength, making channel formation more difficult since a larger buoyancy pressure would be required.

Summary

1. Under conditions of positive temperature gradient upwards and vertical heat flow downwards, channel segregation has been observed in the transparent analogue NH_4Cl-H_2O system as well as Pb alloys containing Sn and/or Sb.

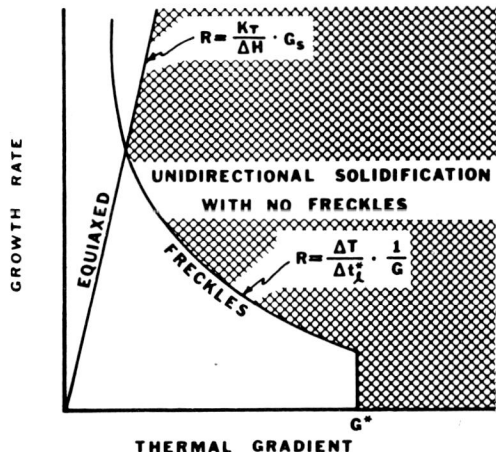

Fig. 12. Schematic plot of growth rate vs. temperature gradient for freckle formation after Copley et al (10).

2. The liquid perturbations leading to the formation of channels arise from thermosolutal interactions at or near the dendritic interface and have been assumed to occur when the thermosolutal Rayleigh number ($Ra_{T/S}$) exceeds 1.
3. The characteristic linear dimension or wavelength (h) in the calculation of Rayleigh numbers is on the order of the dendrite spacings.
4. Dendrite spacings are inversely proportional to the temperature gradient ($G^{.5}$) and the growth velocity ($V^{.25}$).
5. Increasing the thermal gradient or growth velocity for alloys in which channel formation has already been observed, should decrease the dendrite spacings and thus the perturbation wavelength and so make channel formation more difficult since a larger buoyancy pressure would be required.

Acknowledgments

This paper is based upon an ongoing research program supported by the National Aeronautics and Space Administration through the NASA-Lewis Research Center, grant #NAG-3-560.

References

1. R. J. McDonald and J. D. Hunt (1969). "Fluid motion through the partially solid regions of a casting and its importance in understanding A type segregation", Trans. TMS-AIME, Vol. 245, pp. 1993-1997.
2. N. Streat and F. Weinburg (1974). "Macrosegregation during solidification resulting from density differences in the liquid", Metall. Trans. Vol. 5, pp. 2535-2548.
3. N. Streat and F. Weinburg (1976). "Interdendritic fluid flow in a Pb-Sn alloy", Metall. Trans. B, Vol. 7B, pp. 417-423.
4. A. K. Sample (1984). M.S. Thesis: "Mechanisms of formation and elimination of channel segregation during alloy solidification", Michigan Technological University.
5. A. K. Sample and A. Hellawell (1984). "The use of analogue castings to study channel segregation", Modeling of Casting and Welding Processes II, TMS-AIME, pp. 119-125.

6. A. K. Sample and A. Hellawell (1984). "Mechanisms of formation and prevention of channel segregation during alloy solidification", Metall. Trans. A, Vol. 15A, pp. 2163-2173.
7. A. Hellawell (1985). "Channel segregation in alloy ingots", Proceedings 1st International Steel Foundry Congress, Chicago, P. 295-305.
8. J. R. Sarazin (1986). M. S. Thesis: "The influence of dendrite mesh permeability on the formation of segregation channels, Michigan Technological University.
9. J. R. Sarazin and A. Hellawell (1988). "Channel formation in Pb-Sn, Pb-Sb and Pb-Sn-Sb ingots and comparison with the system NH_4Cl-H_2O, Met. Trans. A, in press.
10. S. M. Copley, A. F. Giamei, S. M. Johnson, M. F. Hornbecker (1970). "The origin of freckles in unidirectionally solidified castings", Metall. Trans. Vol. 1, pp. 2193-2204.
11. A. K. Sample and A. Hellawell (1982). "The effect of mold precision on channel and macrosegregation in ammonium chloride-water analog castings", Metall. Trans. B, Vol. 31B, pp. 495-501.
12. K. A. Jackson, J. D. Hunt, D. R. Uhlmann, and T. P. Steward, III (1966). "On the origin of the equiaxed zone in castings", Trans. TMS-AIME, Vol. 236, pp. 149-158 (Feb. 1966).
13. M. R. Bridge, M. P. Stephanson, and J. Beech (1982). "Direct observations on channel segregate formation in aluminum alloys", Metals Technology, Vol. 9, pp. 429-433.
14. V. R. Voller, J. J. Moore, and N. A. Shah (1983). "Modification of mathematical analysis and related physical descriptions used to describe channel segregation", Metals Technology, Vol. 10, pp. 81-84.
15. M. Simpson, M. Yerebakan, and M. C. Flemings (1985). "Influence of dendritic network defects on channel segregate growth", Metall. Trans. A, Vol. 16A, pp. 1687-1689.
16. S. R. Coriell, M. R. Cordes, W. J. Boettinger, and R. F. Sekerka (1980). "Convective and interfacial instabilities during directional solidification of a binary alloy", J. Crystal Growth, Vol. 49, pp. 13-28.
17. J. S. Turner (1973). "Buoyancy effects in fluids", Cambridge Univ. Press.
18. R. W. Schmitt (1983). "The characteristics of salt fingers in a variety of fluid systems including stellar interiors, liquid metals, oceans and magmas", Phys. Fluids 26 (9), pp. 2373-2377.
19. H. Benard (1901). "Les Tourbillons cellulaires dans une nappe liquide", Revue generale des Sciences pures et appliquees, 11, 1261-1271 and 1309-1329.
20. Lord Rayleigh (1916). "On convective currents in a horizontal layer of fluid when the higher temperature is on the under side", Phil. Mag. 32, 529-546.
21. D.T.J. Hurle, E. Jakeman, and A. A. Wheeler (1982). "Effect of solutal convection on the morphological stability of a binary alloy", J. Crystal Growth, 58, pp. 163-179.
22. G. B. McFaddan, R. G. Rehm, S. R. Coriell, W. Chuck and K. A. Morrish (1984). "Thermosolutal convection during directional solidification", Metall. Trans. A., Vol. 15A, pp. 2121-2137.
23. J. D. Verhoeven, J. T. Mason and R. Trivedi (1986). "The effect of convection on the dendritic to eutectic transition", Metall. Trans. A, Vol. 17A, pp. 991-1000.
24. H. E. Huppert and J. S. Turner (1980). "Ice blocks melting into a salinity gradient", J. Fluid Mechanics, Vol. 100, Part 2, pp. 367-384.
25. J. R. Sarazin and A. Hellawell (1987). "Channel flow in partly solidified systems", Proceedings of Third International Conference on Solidification Processing, Sheffield, England, pp. 70-73.
26. W. Kurtz and D. J. Fisher (1984). "Fundamentals of solidification", Trans. Tech. Publications, Switzerland, p. 86.

Dealloying Reactions as Cellular Phase Tranformations

D. J. YOUNG

School of Chemical Engineering and Industrial Chemistry
The University of New South Wales
P.O. Box 1, Kensington, NSW, 2033

ABSTRACT

Alkaline dissolution of aluminium from $CuAl_2$ in the temperature range 274-380 K is shown to be a selective dissolution process leading to the formation of a porous copper residue. The morphology of the residue is one of cylindrical copper rods lying parallel to the direction of the reaction, and surrounded by continuous pore space. The rate of the process is found to be controlled by diffusion through the liquid phase occupying the pores. However the size and spacing of the rods is controlled by diffusion along the boundary between $CuAl_2$ and its reaction products. Cellular phase transformation kinetics accurately describe the relationship between spacing and reaction rate, and correctly predict the pore-size distribution of the copper residue. When the residue is aged for long times in alkali, its structure coarsens. This occurs as a result of the slow dissolution and reprecipitation of copper. Some qualitative observations on the dealloying of $NiAl_3$ and Ni_2Al_3 are also reported.

KEYWORDS

Dealloying, selective dissolution, cellular phase transformation, pore measurements, ripening, catalysts.

INTRODUCTION

Dealloying reactions have been known for many years (Bengough and May, 1924; Raney, 1925) but the mechanisms of these processes are only now being understood. A good example is provided by the room temperature leaching in alkali of aluminium from $CuAl_2$ to leave an almost pure copper residue:

$$CuAl_2(s) + 2OH^-(aq) + 6H_2O = Cu(s) + 2Al(OH)_4^-(aq) + 3H_2(g)$$

Although the external dimensions of the material are unchanged, it suffers the loss of two-thirds of its atoms, about 50% of its mass and 74% of its volume. The residual copper is obviously highly porous, a feature of practical interest as the associated high surface area provides catalytic activity (Friedrich, Wainwright and Young, 1983). From a more fundamental point of view, the question of interest is how the material achieves such a radical change in structure and morphology at room temperature.

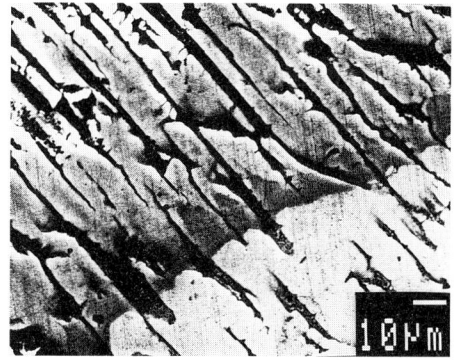
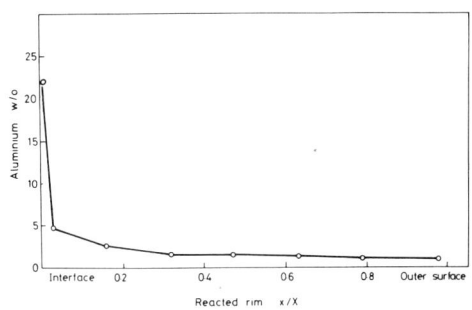

Fig. 1. Cross-sectional view of interface between the de-alloying product and as-yet unreacted Cu-Al alloy.

Fig. 2. Microprobe analysis for aluminium in a leached copper rim produced at 298 K in 20 w/o NaOH solution.

This question has traditionally been answered in two ways. One possibility is that both noble and reactive constituents dissolve simultaneously, and the more noble constituent subsequently re-precipitates to form the new structure. The difficulty with this mechanism lies in the proposal that a metal dissolves from a low activity site and diffuses to and precipitates at a high activity site. As will emerge subsequently, this mechanism cannot account for the primary dissolution process in the systems being considered here. The alternative possibility is that only the reactive alloy constituent dissolves, leaving an alloy surface region enriched in noble metal. The difficulty here is in understanding the mass transfer processes within this surface zone. This paper is concerned with an analysis of these processes, and its application to the prediction of reaction product morphologies in dealloyed $CuAl_2$.

The kinetics and macroscopic morphology of the dealloying reaction are first examined and interpreted as evidence for a selective dissolution mechanism. Pore size and copper crystallite size measurements are then related to reaction rates via the theory of cellular phase transformations. Measurements made over extended periods of time are then used to examine the aging, or ripening, of the porous copper structure. The dealloying kinetics, their relationship with the resultant pore structure and the subsequent ripening of this structure serve to identify the mass transport processes occurring within the two-phase (copper plus liquid-filled pores) reaction product layer. Certain key deductions as to the nature of the porous copper microstructure are then confirmed by direct electron microscopic examination. Finally, some qualitative comparisons are made with the dealloying behaviour of $NiAl_3$ and Ni_2Al_3.

DEALLOYING KINETICS

Portions of a cast alloy Cu-50 wt. pct. (w/o) Al, have been exposed to aqueous NaOH solutions at temperatures of from 274 to 323 K and subsequently sectioned and examined metallographically (Young, Wainwright and Anderson, 1980; Friedrich, Young and Wainwright, 1981a,b). A rim of reaction product is formed on the surface of a shrinking core of parent material. The general appearance of this rim is shown in Fig. 1 where the microstructure of the as-yet unreacted alloy is also visible. It is clear that the dendritic structure of the parent alloy is preserved in the reaction product: the $CuAl_2$ dendrites are converted to copper whilst the interdendritic $Al-CuAl_2$ eutectic leaches to yield void space. The interface between the reaction product and the parent alloy is very sharp. Microprobe analysis results shown in Fig. 2 identify the interface with a step

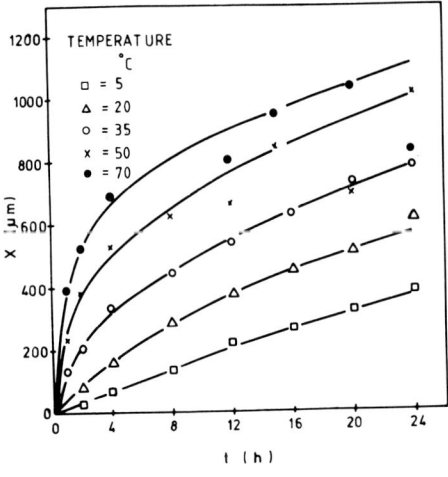

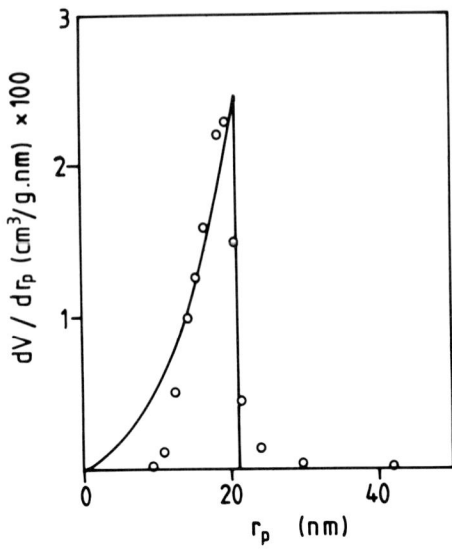

Fig. 3. Kinetics of $CuAl_2$ dealloying in 20 w/o NaOH solution.

Fig. 4. Measured pore size distribution in copper dealloyed at 274 K for 90 min. Continuous line calculated from Eqs. (1) and (3).

function change in composition. It is seen from the aluminium concentration profile that almost all of the aluminium dissolution process occurs at this interface. Since dissolution at the reaction front leads nonetheless to preservation of the dendritic structure, it is concluded that the process is one of selective dissolution.

Measurements of product rim thickness, X, as a function of time, t, show that the reaction kinetics are fast and parabolic

$$X^2 = k_p t \qquad (1)$$

as seen in Fig. 3 for short-term experiments. Much longer experiments (Tomsett, Young and Waingwright, 1984) have shown that this relationship is preserved for hundreds of hours, suggesting that the process is diffusion controlled. The deduction is supported by the observation that the sharply defined reaction front cuts across the alloy phases: despite their difference in aluminium activity, these phases leach at the same rate. Thus mass transfer via diffusion to or from the reaction front is rate-determining. Solid-state diffusion, either in the product layer or in the underlying alloy, can be ruled out as being too slow to support the observed rates (Friedrich, Young and Wainwright, 1981b). The remaining possibility of diffusion within the liquid-filled pore space is quantitatively reasonable. Thus the value of D calculated from Eq. (1) is of order 10^{-6} cm^2 s^{-1} and the activation energy for k_p is 29 ± 2 kJ.

The detailed microstructure of leached copper residues has been examined using gas adsorption, mercury intrusion porosimetry and X-ray line broadening (Friedrich, Young and Wainwright, 1981; Tomsett, Young and Wainwright, 1984; Tomsett, Wainwright and Young, 1984). . Dealloying at room temperature produces material with a surface area of around 20 m^2/g, average pore radius of order 15 nm and a copper crystallite size of order 10 nm. Evidently the microstructural rearrangement accompanying leaching takes place on a very fine scale. The mechanism whereby this rearrangement occurs is now considered.

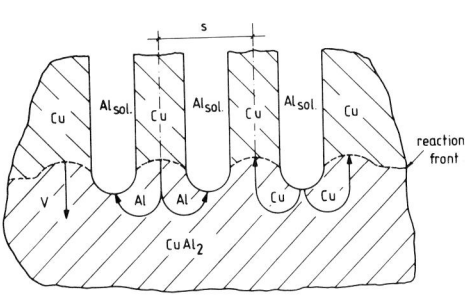
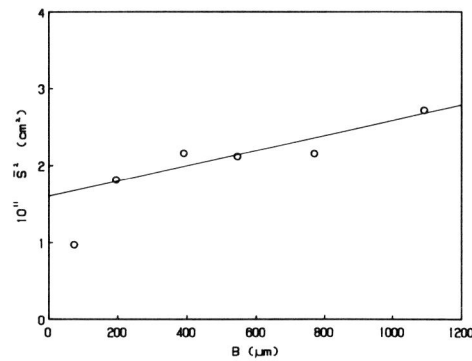

Fig. 5. Schematic representation of a CuAl$_2$-reaction product interface.

Fig. 6. Average spacing squared as a function of sample radius.

PORE DEVELOPMENT DURING DEALLOYING

Since the CuAl$_2$ dealloying reaction is supported by liquid phase diffusion, the pore space which the liquid occupies must be continuous from the sample exterior to the reaction front. Gas adsorption and desorption on the pore walls has been found (Friedrich, Wainwright and Young, 1983) to be characteristic of non-intersecting pores and reveals that the pore size distribution is rather narrow, as shown in Fig. 4, when the dealloying reaction is carried out at a low enough temperature. An array of non-intersecting, uniform pores providing a continuous pathway from one side of the product layer to the other can be represented as shown in Fig. 5. Such a structure can arise through microsegregation of the alloy constituents toward the tips of the advancing phases, solid copper and aqueous solution of aluminium.

Cellular phase transformations of this sort are familiar as eutectic and eutectoid reactions, as discontinuous precipitation, and have also been observed (Young, Smeltzer and Kirkaldy, 1976; Fridberg and Hillert, 1977) in cases of reaction of the parent phase with an externally supplied gaseous reactant. Selective dissolution reactions appear to fall within this latter class. The structure has a spacing, S, defined in Fig. 5, which is related to the velocity with which the interface advances, v, according to the mechanism of lateral segregation. If volume diffusion within the parent phase provides the means of segregation, then (Zener, 1946; Hillert, 1957; Cahn, 1959; Puls and Kirkaldy, 1972)

$$v = \frac{2D}{a\, f^{Cu} f^P RT} \frac{\sigma^{CuP}}{c^{Cu}-c^P} \frac{1}{S^2} \tag{2}$$

The product phases have been identified as copper and pore (P), D is the diffusion coefficient in CuAl$_2$, σ^{CuP} is the surface energy of the copper-pore boundary, a is a geometric constant of order unity, f^{Cu} and f^P are the ratios S^{Cu}/S and S^P/S, c^{Cu} and c^P are solute concentrations within the product phases. If diffusion along the interface between the parent and product phases is the mechanism of segregation, then (Cahn, 1959; Turnbull, 1955; Shapiro and Kirkaldy, 1968; Hillert, 1969; Sundquist, 1968)

$$v = \frac{24\, D_B \delta K}{a\, RT} \frac{f^{Cu} f^P \sigma^{CuP}}{c^{Cu}-c^P} \frac{1}{S^3} \tag{3}$$

Here D_B is the boundary diffusion coefficient, δ is the boundary width and K is the equilibrium constant for partition of solute between bulk and boundary. In order to test the applicability of these expressions, it is necessary to measure pore spacings as a function of the selective dissolution rate.

Because the rate varies with extent of reaction, as shown by Eq. (1), the value of S is also predicted to vary. The magnitude of S was evaluated by gas adsorption measurements, and therefore represents a volume average for the entire sample. Since the leaching kinetics are parabolic, the average leaching rate over the complete course of reaction is lower for a large alloy sample than for a small one. Thus it is possible to investigate reaction morphology as a function of average dealloying rate by using alloy samples of discrete size ranges. The average quantities are now related.

By definition the average quantity S is given by

$$\overline{S^2} = \frac{\int S^2 \, dV}{\int dV} \qquad (4)$$

where V represents volume. For spherical samples of radius B

$$\overline{S^2} = 3 \int_0^B S^2 r^2 dr / B^3 \qquad (5)$$

where r is the radial coordinate within the sphere. Differentation of Eq. (1) and substitution into (2) yields for a spherical sample,

$$S^2 = \frac{2 \kappa D}{k_p}(B-r) \qquad (6)$$

where

$$\kappa = 2\sigma^{CuP}/f^{Cu}f^P a \, RT \, (c^{Cu}-c^P).$$

Combination of Eqs. (5) and (6) followed by integration leads to

$$\overline{S^2} = \frac{\kappa D}{2k_p} B \qquad (7)$$

The corresponding expression for boundary diffusion is found to be

$$\overline{S^3} = \frac{\kappa' D_B \delta K}{2k_p} B \qquad (8)$$

where

$$\kappa' = 24 f^{Cu}f^P \sigma^{CuP}/aRT(c^{Cu}-c^P)$$

Rather simple dependencies of S on B are predicted from these two descriptions.

Alloy samples having discrete particle size ranges were prepared, leached to completion in 20 w/o NaOH solution, and the resultant average pore sizes measured by gas adsorption (Tomsett, Young and Wainwright, 1984). The results are shown in Table 1. In calculating S, both the pore diameter and intervening copper width are necessary. Knowing the pore volume and size from adsorption measurements, one can calculate the copper width on the basis of the structure's geometry. We proceed on the basis that the axes of the structure elements are arrayed on a square grid. In principle, either the metal or the pore space could constitute the continuous phase. In the case of $CuAl_2$ it is the pores which are continuous and the copper forms discrete rods as discussed later. It is this model which leads the copper rod diameters in Table 1.

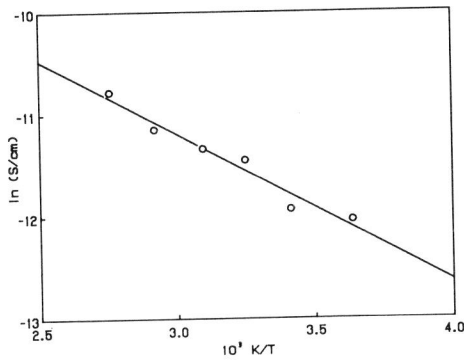
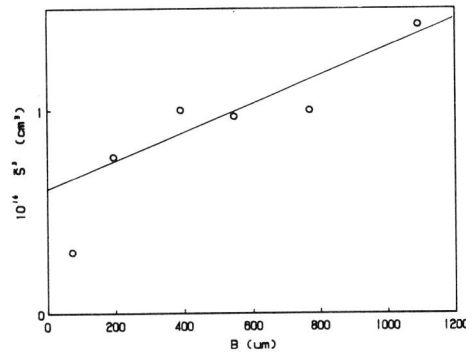

Fig. 7. Average spacing cubed as a function of sample radius.

Fig. 8. Arrhenius plot according to Eqns. (9) and (10).

Plots of S^2 and S^3 vs B are shown in Figs. 6 and 7. The first point shown in these diagrams is disregarded because the leaching kinetics at this early stage were found to be under chemical reaction control. The remaining data conform reasonably well with the predicted straight line relationships, but no statistically significant difference can be found between the two fits. The value of D is calculated from the slope of the S^2 vs B plot using Eq. (7) and $a=1$, $f^{Cu} = 0.36$, $f_2 = 0.64$, $\sigma^{Cu} = 300$ erg cm^{-2}, $C^{Cu} = 0.13$ mol cm^{-3}, $C_1 = 0_2$ and $k = 3.97 \times 10^{-8}$ cm^2 s^{-1}. The resulting estimate at 298 K is $D = 1 \times 10^{-11}$ cm^2 s^{-1}. This value seems very high for solid-state volume diffusion in CuAl$_2$ at room temperature, even given the probable error in the estimate of σ^{Cu}.

If alloy segregation occurs via boundary diffusion, then Eq. (8) applies. The slope of the S^3 vs B plot leads to the estimate at 298 K of $D_B \delta K = 1.1 \times 10^{-16}$ cm^3 s^{-1}. Using the estimates $\delta = 1$ nm and $K = 2$, it is found that $D_B = 6 \times 10^{-10}$ cm^2 s^{-1}. No independent information is available for the diffusion properties of the CuAl$_2$/Cu boundary, and the physical reasonableness of this value cannot be assessed.

The dependence of S on dealloying temperature for fixed sample sizes is found by logarithmic differentiation of Eqs. (7) and (8)

TABLE 1. Surface and pore structure data for alloy samples leached to completion at 293 K

Surface radius (μm)	Surface area (m$_2$/g)	Pore radius* (nm)	Pore volume (cm^3/g)	Interpore distance (nm)
50-90	31.4	11.8	0.214	21
180-210	25.3	15.8	0.203	29
350-420	23.5	17.1	0.197	32
500-590	24.0	16.9	0.197	31
700-840	23.9	17.1	0.197	32
1000-1180	21.4	19.1	0.195	36

* Value of r for which dV/dr is maximum in pore-size distribution.

$$\frac{\partial \ln \overline{s^2}}{\partial (1/T)} = - (Q_V - E_A)/R \tag{9}$$

$$\frac{\partial \ln \overline{s^3}}{\partial (1/T)} = - (Q_B - E_A)/R \tag{10}$$

where the linear term in T has been neglected. Here Q_V, Q_B and E_A are, respectively, the activation energies for volume diffusion, boundary diffusion and the dealloying reaction. The results in Table 2 for alloy samples of the same size, leached to completion at different temperatures, have been used to construct the Arrhenius plot shown in Fig. 8. From the slope of this plot and the measured value of E_A = 29 kJ, it is calculated that if volume diffusion is operative, then Q_V = 52 ±2 kJ; whereas, if boundary diffusion is in effect, then Q_B = 65 ± 4 kJ.

Activation energies for volume diffusion in $CuAl_2$ and Cu have been reported (Smithells, 1967) as 127 and 211 kJ, respectively. Clearly, the value estimated from the application of the volume diffusion model to the present results is quite inconsistent with these values. However, the value of Q_B estimated from the boundary diffusion model seems quite reasonable.

The volume diffusion model has been found to lead to an extraordinarily low estimate of the activation energy, and a remarkably high value for the diffusion coefficient. For these reasons, the hypothesis that alloy segregation occurs via volume diffusion is rejected. Conversely, the boundary diffusion model yields an activation energy of the expected magnitude and a diffusion coefficient which there is no reason to reject. In addition, this model correctly predicts the relationship between the structure's spacing and its rate of formation. It is therefore concluded that alloy segregation at the dealloying reaction front occurs via boundary diffusion.

This reaction mechanism and the associated reaction product geometry can be used to describe more accurately the pore size distribution within the dealloying residue. Because the dealloying reaction rate decreases with extent of reaction according to Eq. (1), it follows from Eq. (3) that S, and hence pore size, increases with leaching depth. Thus the pores at the sample exterior should be

TABLE 2. Effect of dealloying temperature on residue structure in completely leached samples with B = 0.5-0.59 mm

Extraction temperature (K)	Surface area (m^2/g)	Pore radius (nm)	Interpore distance (nm)
275	25.4	15.1	28
293	24.0	16.9	31
308	22.6	27.5	50
323	18.5	30.5	56
343	16.9	37.5	67
363	12.7	53.8	98

* See Table 1.

the most finely spaced, and those at the reaction front the coarsest. The distribution of pore sizes is readily calculated from Eqs. (1) and (3), using parameter values listed earlier. The results of such a calculation are compared with a pore size distribution measured by mercury intrusion porosimetry in Fig. 4. Satisfactory agreement was obtained using $D_R \delta K = 2 \times 10^{-16}$ cm^3 s^{-1}, a value slightly higher than the 1×10^{-16} cm^3 s^{-1} deduced from the kinetic experiments described in the previous section.

The preceding analysis assumes that once a dealloyed structure is formed, its morphology is thereafter unchanged with time. Although this is a reasonable approximation for the short times considered above, it is most unlikely to hold for long exposure times at relatively high temperatures. The intrinsic instability of a high surface area structure exposed to a medium in which its solubility is not zero is expected to lead to a coarsening of the morphology with time. The kinetics of this coarsening process have been studied for the case of copper.

AGING OF DEALLOYED COPPER

Large rectangular prisms of Cu-50 w/o Al alloy were exposed to 20 w/o NaOH solution at temperatures of 274 to 343 K for varying periods of time and the exterior surfaces of the partially dealloyed prisms then examined using X-ray diffractometry (Tomsett and others, 1987). Broadening of the copper (111) and (200) lines was used to determine the average crystallite size at the surface. The results shown in Fig. 9 indicate that the copper structure at the sample exterior is coarsening as the time of exposure to the alkali solution continues. In order to determine whether the dealloying process itself is involved in the coarsening phenomenon, the Cu-50 w/o Al alloy was dealloyed for a period of time and the dealloyed rim then mechanically removed from the unreacted alloy core. The separated residue was then exposed to alkali solution for further periods of time. Copper crystallite size measurements in pre-leached and separated material are compared in Fig. 10 with measurements made at the external surface of large samples within which the leach reaction was continuing. It is seen that the coarsening process occurs at the same rate and reaches the same apparent steady-state in the two different cases. Evidently the dealloying reaction itself plays no part in the crystal enlargement process, and it follows that the

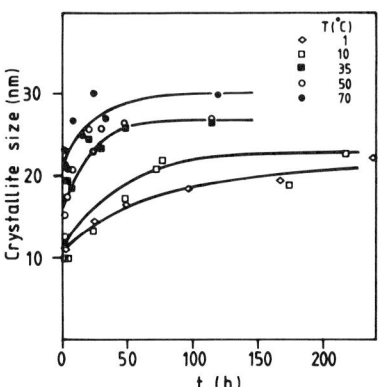
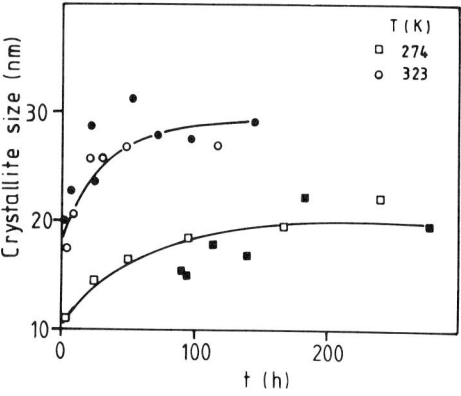

Fig. 9. Copper crystallite sizes at the surface of Cu-50 w/o Al pieces leached to different depths.

Fig. 10. Comparison of crystallite size of leached rims, aged after removal from alloy (closed symbols), with those of the external surface of partially leached samples (open symbols).

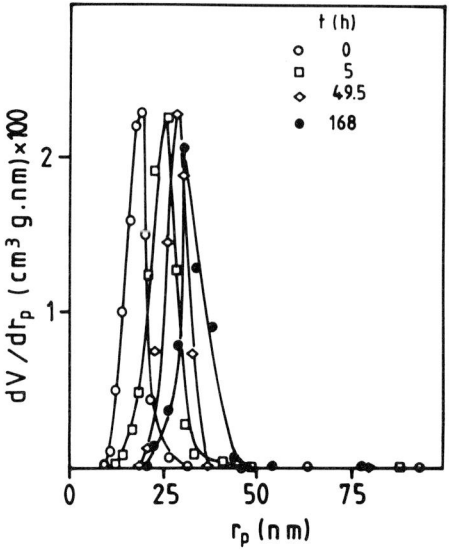 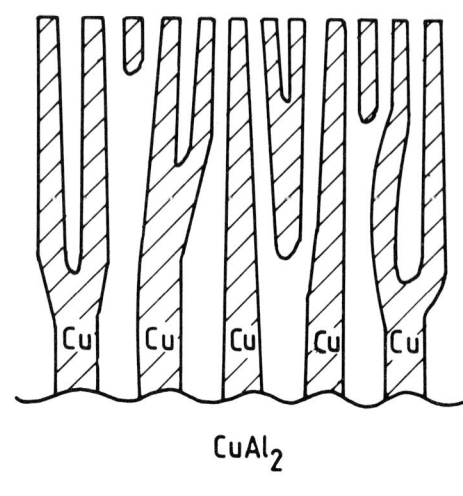

Fig. 11. Pore coarsening with aging time at 283 K after initial dealloying at 274 K.

Fig. 12. Schematic representation of faults in copper rod structure.

morphological changes result solely from the interaction between the porous copper and the alkaline environment.

Dealloyed copper residues were produced by leaching at 274 or 323 K. The residues were mechanically removed from their alloy cores and then subjected to lengthy periods of exposure to alkali solution. The morphological evolution of these structures was observed by measuring pore size distributions after different aging times. Typical results are shown in Fig. 11. A large number of observations revealed that the rate of pore size enlargement

(a) increased with aging temperature,
(b) increased with increasing alkalinity of the aging solution,
(c) was initially greater for a more finely spaced starting structure.

The magnitude of the changes during aging indicates that a fairly rapid mass transfer process is occurring. The only mechanism available for this mass transfer is liquid diffusion, and it is therefore concluded that a process of copper dissolution and reprecipitation is in operation. This conclusion is supported by the observations of the effects of temperature, alkalinity and initial structure spacing on the rate of the process. Chemical analysis was used to confirm that small concentrations of copper did in fact accumulate in the alkali solution during aging.

The driving force in the aging process is, of course, the surface free energy. How this works in detail can be seen from examination of the morphology of a freshly dealloyed material, prior to any aging. As discussed earlier, the pore spacing is initially controlled by the dealloying rate, and therefore increases with depth of reaction. Because the parent $CuAl_2$ phase is close to homogeneous, the pore volume of the leached residue is everywhere the same, and an increase in spacing necessarily implies a decrease in the number density of copper rods through the product rim. Such a decrease is presumably achieved by rod confluence or termination, as shown schematically in Fig. 12. If it is assumed that the terminations are hemispherical tips, it is seen that copper will preferentially dissolve from the high curvature tips and reprecipitate on the lower curvature sides of adjacent rods. Reprecipitation will be even more favoured at points of rod confluence where a negative radius of curvature is found. Similar faulted structures have been observed in eutectics with lamellar and rod geometry and found to result in coarsening (Graham and Kraft, 1966; Cline, 1971). The description developed by Cline (1971) for eutectic coarsening is now applied to the aging of porous copper residues.

Using the Gibbs-Thomson equation to describe the effect of curvature on solubility, and a steady-state solution of the diffusion equation to describe copper dissolution at the tips, Cline (1971) showed that under the approximations of low supersaturation and small numbers of copper tips, the rate of recession of a tip was given by

$$\frac{dl}{dt} = \frac{-2\sigma^{CuP} V_m^2 DC^*}{RT(1-\bar{r}/S)} \cdot \frac{1}{\bar{r}^2} \qquad (11)$$

where V_m is the molar volume of solid copper, C^* is the solubility of copper in the liquid at equilibrium with a flat surface, D is the diffusion coefficient of the dissolved copper, and $\bar{r}$ is the local average rod radius. Assuming then that the volume fraction of copper in the structure is invariant, it may be shown that the number of copper rods per unit area of a surface normal to the rod axes, N, varies with aging time according to

$$N = \frac{N_o}{\frac{1}{2}\epsilon f_o t + 1} \qquad (12)$$

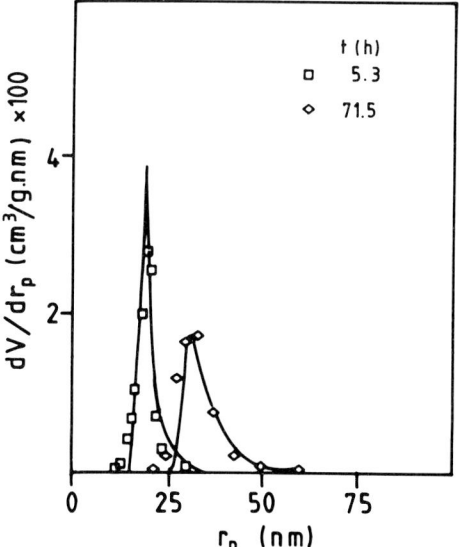

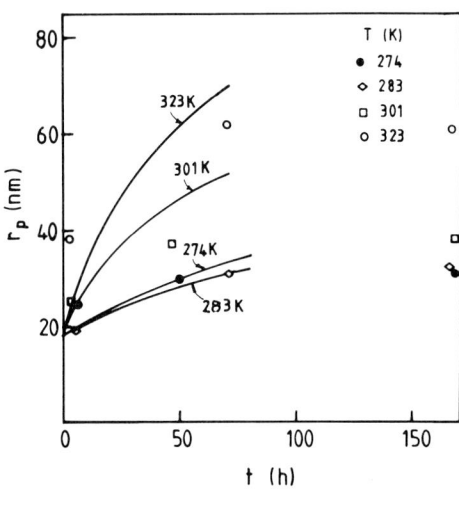

Fig. 13. Experimental pore size distributions of aged rims and those calculated from Eq. (12). Initial dealloying T = 274 K, aging T = 283 K,

Fig. 14. Comparison of experimental r_p(max) with values calculated from Eq. (12).

where

$$\epsilon = \frac{-8\sigma^{Cu} V_m^2 \, DC^*}{RT(1-\bar{r}/S)(f^{Cu})^2} \qquad (13)$$

Here N_o is the value of N resulting from dealloying (i.e. prior to any aging) and f_o the volume density of faults at this time.

The applicability of this model to the aging process was tested by applying Eqn. (12) to the initially formed pore size distribution, which defines N_o as functions of position. Aged pore size distributions calculated in this way are compared with measurement in Fig. 13. Agreement is excellent at 283 K and reasonable at 301 K. The quantity ϵ, defined in Eq. (13), was treated as an adjustable parameter in optimizing the agreement between calculated and measured distributions at each temperature. A useful measure of the extent of agreement achieved is provided by comparing the calculated and measured values of r for which the pore volume differential, dV/dr, is maximum. Fig. 14 demonstrates that good agreement is found only at low temperatures and relatively short times. However, in the advanced stages of aging the pore structure ceases to be a parallel sided one and neither the measurements nor the model can be valid.

Using the values of ϵ found in the fitting process, it is possible to calculate D for copper in the alkali solution. A value of 3×10^{-6} cm^2 s^{-1} is found for a temperature of 274 K and 2×10^{-5} cm^2 s^{-1} for 301 K. Uncertainties in the solubility of copper in alkaline solutions make asessment of these estimates difficult, but the order of magnitude seems reasonable for aqueous diffusion.

It is concluded that dealloyed copper residues age by a process of dissolution and precipitation. Under relatively mild conditions, this process can be described quantitatively using the approximation that dissolution is significant only at copper rod ends.

MICROSCOPIC VERIFICATION

Replica and thin foil microscopy have now been used by Szot and others (1987) to examine directly the morphologies of dealloyed copper residues. A Cu-50 w/o Al alloy was leached in 20 w/o NaOH solution either at 274 K for 12 h to produce a rim 0.06 mm deep, or at 366 K for 3 h to produce a rim 0.8 mm deep. The external surface of the material dealloyed at 274 K was examined by replica microscopy using a two-stage technique. Remnants of the copper frequently adhered to the replica and the surface topography was thereby revealed. Thin foil preparation was achieved by vacuum impregnating the structures with polyester resin and then thinning a section by mechanical polishing, dimpling and ultimately ion beam milling.

A replica from an external surface of a sample dealloyed at 274 K revealed an array of approximately uniformly sized pieces. These were identified by selected area diffraction and x-ray microanalysis as pure copper. The particles were spaced at intervals of about 40 nm. Although several were polycrystalline, many were single crystals. A stereo pair of transmission micrographs of a foil made from a sample dealloyed at 266 K is shown in Fig. 15. The copper objects are seen to be curved rods or fibres of circular cross-section. The rods were remarkably uniform in diameter, and oriented approximately parallel to the reaction direction. Micro-diffraction experiments showed that each rod was a single-crystal. Selected area diffraction experiments showed that the copper rods in a field of view all had the same crystallographic orientation.

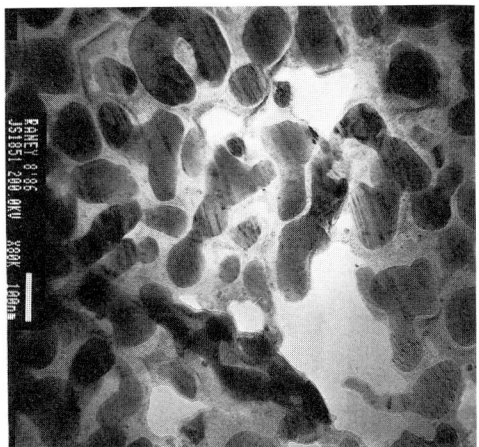 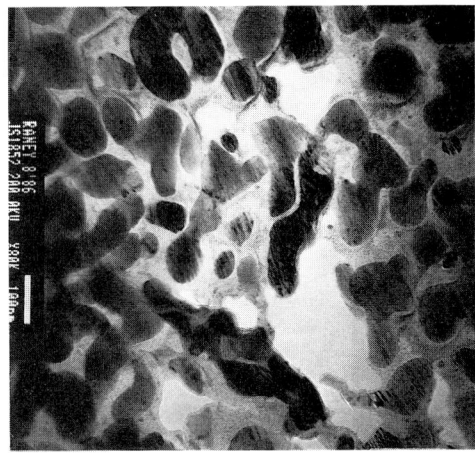

Fig. 15. Stereo pair (5° separation) of copper residue produced at 366 K.

It is concluded that the dealloyed copper residue consists of single-crystal fibres surrounded by continuous pore space. The fibres are curved but have a rather well defined average direction, normal to the dealloying reaction front. The rod spacing can be calculated from their observed diameter and the known pore volume using an assumed geometry in which the rod centres are aligned on a square grid.. The rod diameter was measured at a number of positions within the leached rim and was found to be 75±10 nm. The corresponding pore width is therefore 75±20 nm. The average pore diameter was measured by mercury porosimetry as 65±15 nm, in excellent agreement. The observations that the copper residue consists of uniform rods and that the assumption of uniform rod spacing leads to a calculated pore size in agreement with measurement, indicate that the residue morphology is indeed cellular.

The structure produced by dealloying at 266 K is substantially coarser than the surface structure produced at 274 K. This difference reflects both the ripening process known to occur at high temperatures and the temperature effect on the diffusional processes which support the initial transformation. Microscopic information on the variation in rod size with position in a non-aged residue has not yet been obtained.

DEALLOYING OF NICKEL ALUMINIDES

Single-phase samples of $NiAl_3$ and Ni_2Al_3 have been dealloyed in 20 w/o NaOH solutions, and the reaction kinetics and morphology determined (Bakker, Young and Wainwright, 1988). At temperature of 274 to 323 K, $NiAl_3$ dealloyed to produce a fragile nickel product which failed to form a coherent layer, allowing the reaction to continue at a uniform rate. At these temperatures Ni_2Al_3 was unreactive, but at 343-380 K it leached according to parabolic kinetics, producing a strong, tightly adherent rim of residual material.

This difference in behaviour can be ascribed, in part, to the differing pore volumes of the dealloying residues. Pore volumes calculated on the basis of 100% aluminium dissolution and no shrinkage in the residue and expressed as fractions of the external volume are 0.81 for $NiAl_3$ and 0.60 for Ni_2Al_3 compared to 0.74 for $CuAl_2$. It appears that the nickel residue formed from $NiAl_3$ is too disperse to provide a coherent product layer.

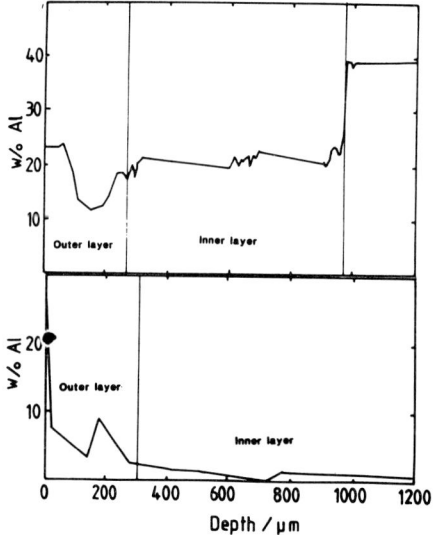
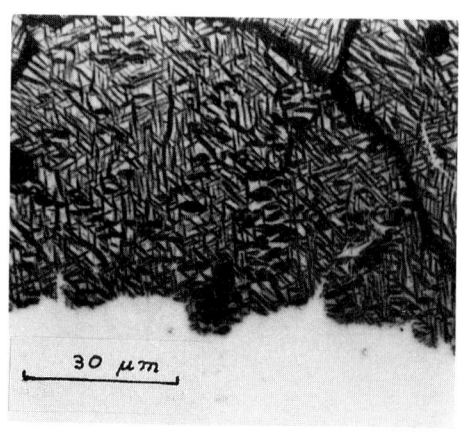

Fig. 16. Microanalysis of rims formed on Ni_2Al_3 after dealloying for 200 h at 380 K (upper plot) and after reaction to completion (300 h at 380 K).

Fig. 17. Interface between Ni_2Al_3 and inner reaction product layer after reaction for 200 h at 380 K.

The phase Ni_2Al_3 leached according to very slow parabolic kinetics. An approximate calculation of the corresponding liquid phase diffusion coefficient yields values of 10^{-9}, 10^{-8} and 10^{-6} cm^2 s^{-1} at temperatures of 343, 363 and 380 K respectively. These values, and the corresponding apparent activation energy of 110 kJ are quite unlike the values expected for aqueous diffusion. It is concluded that liquid phase diffusion is rapid enough to support the observed reaction rates but does not control them.

The reaction product layer on Ni_2Al_3 was more complex than any observed before. The microprobe analysis profile shown in Fig. 16 shows that two sub-layers are formed. The inner layer of the rim has an approximately uniform composition of about 22 w/o Al. In the outer layer, the aluminium content reaches a minimum of about 12 w/o and then rises to a maximum of around 24 w/o at the external surface. No single Ni-Al phase exists with an aluminium content of 22 w/o Al or of 12 w/o Al and the two sublayers must each contain more than one phase. This is confirmed in the case of the inner layer by its visible microstructure shown in Fig. 17.

X-ray diffraction of the leached layer revealed the presence of Ni and Ni_2Al_3 only, but the possible presence of amorphous $Al_2O_3 \cdot xH_2O$ must be recognized. Microprobe analysis of a sample approaching complete leaching showed that the aluminium content of the inner layer approaches zero, whilst that of the outer layer still has a high value at the rim surface. It is tentatively concluded that partial leaching of Ni_2Al_3 leads to the formation of a composite residue consisting of Ni_2Al_3 plus Ni. The outer part of the rim is more extensively reacted, producing a higher Ni/Ni_2Al_3 ratio. However, reprecipitation of an alumina hydrate near the rim exterior leads to an increase in apparent aluminium content. Prolonged leaching leads to reaction of the remnant Ni_2Al_3 material, leaving only nickel and reprecipitated Al_2O_3 present.

A reaction which converts Ni_2Al_3 into a mixture of Ni_2Al_3 plus Ni whilst preserving the detailed microstructure of the parent alloy is obviously one of selective dissolution. What is not yet clear is why some of the original Ni_2Al_3 is more readily leachable than the rest. Reliable information on pore morphologies is not yet available, and the intriguing question of what controls each of the leaching rate and the resultant pore spacing can not yet be addressed.

CONCLUSIONS

Dealloying of $CuAl_2$ in alkaline solution occurs by selective dissolution of aluminium. The rate of the process is controlled by diffusion through the liquid phase occupying the pores and parabolic kinetics result.

The residue morphology is cellular, that is it is one of uniformly spaced copper rods which have a circular cross-section and are aligned parallel to the reaction direction. Direct microscopic observation has confirmed this to be the structure, in agreement with earlier deductions based on porosimetry results.

Rearrangement of the parent material to accommodate the aluminium loss occurs almost entirely at the reaction front when temperatures are low. Under these conditions, cellular phase transformation kinetic theory succeeds in relating rod spacing to dealloying rates, and in accurately predicting the pore size distribution within the dealloyed residue. Diffusion along the parent-product phase boundary provides the mechanism for segregation of copper and aluminium toward their respective product phases.

When the copper residue is aged in alkaline solutions for long times, its structure coarsens. This ripening process is found to occur via dissolution and reprecipitation, a process driven by differences in surface curvature. Providing the aging conditions are not too severe, the process can be successfully modelled by eutectic coarsening theory.

Dealloying of $NiAl_3$ and Ni_2Al_3 has been examined briefly. The small volume fraction of solid in the residue of $NiAl_3$ does not permit formation of a coherent product, whereas the large value for the corresponding volume fraction in Ni_2Al_3 leads to the production of a strong, coherent residue. This latter phase dealloys according to a mechanism of selective dissolution, but the mechanisms controlling the reaction rate and the product morphology have not yet been determined.

ACKNOWLEDGEMENTS

Various aspects of this work have been supported by the National Energy Research, Development and Demonstration Program or the Australian Research Grants Committee.

REFERENCES

Bakker, M.L., Young, D.J., and Wainwright, M.S. (1988). J. Mat. Sci., in press.
Bengough, G.D., and May, R. (1924). J. Inst. Met., 32, 169.
Cahn, J.W. (1959). Acta Metall., 7, 18.
Cline, H.E. (1971). Acta Metall., 19, 481.
Fridberg, J., and Hillert, M. (1977). Acta Metall., 25, 19.
Friedrich, J.B., Young, D.J., and Wainwright, M.S. (1981a). J. Electrochem. Soc., 128, 1840.
Friedrich, J.B., Young, D.J., and Wainwright, M.S. (1981b). J. Electrochem. Soc., 128, 1845.
Friedrich, J.B., Wainwright, M.S., and Young, D.J. (1983). J. Catal., 80, 1.
Graham, L.D., and Kraft, R.W. (1966). Trans. Met. Soc. AIME, 236, 94.
Hillert, M. (1957). Jernkont. Ann., 141, 757.
Hillert, M. (1969). Monograph and Report Sereies No. 33, Institute of Metals, London.
Puls, M.P., and Kirkaldy, J.S. (1972). Metall. Trans., 3, 2777.
Raney, M. (1925). U.S. Patent, 1 563 587.
Shapiro, J.M., and Kirkaldy, J.S. (1968). Acta Metall., 16, 579.
Smithells, C.J. (1967). Metals Reference Book, Vol. 2., Butterworths, London.
Szot, J., Young, D.J., Bourdillon, A. and Easterling, K.E. (1987). Phil. Mag. Lett., 55, 109.
Sundquist, B. (1968). Acta Metall., 16, 1413.
Tomsett, A.D., Curry-Hyde, H.E., Wainwright, M.S., Young, D.J., and Bridgewater, A.J., (1987). J. Appl. CAtal., 33, 119.
Tomsett, A.D., Wainwright, M.S. and Young, D.J. (1984). J. Appl. Catal., 12, 43.
Tomsett, A.D., Young, D.J., and Wainwright, M.S. (1984). J. Electrochem. Soc., 131, 2476.
Turnbull, D. (1955). Acta Metall., 3, 55.
Young, D.J., Smeltzer, W.W. and Kirkaldy, J.S. (1976). J. Electrochem. Soc., 123, 1758.
Young, D.J., Wainwright, M.S., and Anderson, R.B. (1980). J. Catal., 64, 116.
Zener, C. (1946). Trans. AIME, 167, 550.

Modelling of Oxidation and Sulfidation Reactions

W. W. SMELTZER

Department of Materials Science and Engineering,
McMaster University, Hamilton, Ontario, Canada, L8S 4M1.

ABSTRACT

Models common to those for phase transitions correlating diffusion with thermodynamics are advanced to describe the growth of oxides and sulfides on several transition metals and alloys in single oxidant gases at temperatures ranging from 400° to 1400° C. Ambipolar lattice and short-circuit diffusion describe reactant transport and growth of superficial scales; these diffusion properties are combined with theory for general and cellular precipitation to explain development of internal oxidation zones beneath scales on alloys.

KEYWORDS

Oxidation and sulfidation; transition metals and alloys; solid-state reactions.

INTRODUCTION

Many transition metals when exposed to a gas oxidant at elevated temperatures form compact uniformly thick films ($\leq 0.1~\mu$) or scales ($\geq 0.1~\mu$) of plasticity sufficient for oxide retention to the receding metal interface. If diffusion occurs by interstitials or vacancies in the superficial reaction product layer, the reaction kinetics are parabolic and amenable to description by theory for ambipolar diffusion of the reactants (Wagner, 1933). An extension of this theory to ternary diffusion has been formulated to account for growth of single-phase ternary oxide scales on binary alloys (Wagner, 1969).

Metals and alloys are frequently covered during early exposure by microcrystalline films resulting from the growth processes and the metal-film system minimizing strain. Ion transport across films at low temperatures proceeds under the influence of an internal electric field. Thermal energy, however, is sufficient to promote oxide growth by chemical diffusion at elevated temperatures because the ratio of boundary to lattice diffusion is as high as 10^4-10^6 with corresponding ratios of activation energies ranging from 0.4-0.7. An extensive body of research now demonstrates that short-circuit diffusion of reactants by oxide grain boundaries can determine the oxidation rates of metals and alloys at moderately elevated temperatures.

Alloy oxidation is more complicated since composition variations lead to internal oxidation (subscale growth) involving oxide precipitation in the alloy beneath the superficial scale. Criteria defining limiting alloy compositions to prevent internal oxidation (sulfidation) are significant to design of heat-resistant alloys. Present research on the transition from internal oxidation to exclusive growth of superficial scales emphasize mechanisms of general and cellular precipitation in controlling oxide growth and morphological development.

The purpose of this contribution is threefold: first, to describe theory for growth of oxide and sulfide scales on metals and alloys by ambipolar lattice diffusion of the reactants; second, to extend this theory to account for growth of scales by ambipolar short-circuit diffusion and, finally, to outline features of alloy internal oxidation. Within this context, an attempt is made to illustrate that advances in the modelling of oxidation and sulfidation reactions are allied to advances in the field of phase transitions.

SCALE GROWTH BY LATTICE DIFFUSION

Metal and oxygen are assumed to migrate across a reaction product layer as the rate determining step by independent migration of ions and electrons under the influence of the electrochemical potential gradient. Thus, the flux of electrically charged species i is

$$J_i = C_i v_i = \frac{-C_i v_i}{q_i} \left(\frac{du_i}{dx} + q_i \frac{dV}{dx} \right) \quad [1]$$

where C_i is concentration, v and u are particle velocity and mobility and the gradient of the electrochemical potential is expressed in terms of the species chemical potential u_i and the electrical potential $q_i V$ where q_i is the particle charge. For ambipolar diffusion (charge transport with no net electrical current) the flux can be expressed by migration of the reactants under the influence of an elemental chemical potential gradient. This condition leads to parabolic kinetics for scale growth

$$X^2 = k_p t + B \quad [2]$$

where X is the oxide layer thickness at time t, k_p is the parabolic oxidation rate constant and B is a constant. If the flux is regarded as divergence free and the scale is a metal conducting semiconductor (transport number of electrons ~ 1), k_p is proportional to the parabolic oxidation rational rate constant (Wagner, 1933, 1951)

$$k_r (eq/cm\ s) = \frac{RT}{|z_o|F^2} \int_{a_o(o)}^{a_o(x)} t_M 6 d\ln a_o = \bar{C}_{eq} \int_{a_o(o)}^{a_o(x)} \frac{Z_M}{|Z_o|} D_M d\ln a_o \quad [3]$$

Here, t_M, 6 and D_M represent the cation transport number, electrical conductivity and the metal diffusivity of the oxide, respectively, Z is a valency F is the Faraday, $\bar{C}_{eq}$ is the average oxide composition in equivalents and a_o is the oxygen activity at the inner (o) and outer (x) scale interfaces.

Interpretation of scaling by means of the Wagner theory is treated in depth in several texts and recent reviews, as example (Smeltzer, 1987). Accordingly, discussion is confined to two illustrative examples, oxidation of nickel and sulfidation of iron.

Nickel oxide exhibits an extremely narrow range of nonstoichiometry 10^{-3} at.% brought about by metal vacancies predominantly ionized to a doubly charged state. Assuming a constant vacancy mobility, eqn[3] becomes

$$k_r = 3\bar{c}_{eq} D_{Ni}^0 (P_{O_2}^{(x)})^{1/6} [1 - (P_{O_2}^{(o)}/P_{O_2}^{(x)})^{1/6}] \qquad [4]$$

where D_{Ni}^0 is the nickel diffusivity in NiO at $P_{O_2} = 1$ atm. Experimental and calculated values of the rate constant at temperatures higher than 500°C are illustrated by the Arrhenius plots in Fig. 1. Good agreement exists above 1000°C but the experimental values are larger at lower temperatures, the divergence from the calculated values increasing with decreasing temperature. This behavior and the influence of nickel crystal faces on the oxidation rate is be shown subsequently to be brought about by more rapid short-circuit diffusion of nickel in the polycrystalline scales.

Sulfidation of transition metals forming metal conducting sulfides usually proceed more rapidly than the corresponding oxidation reactions because the

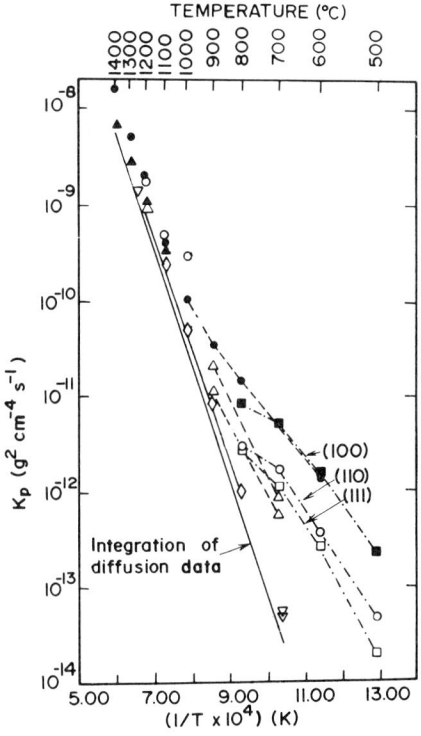

Fig. 1. Arrhenius plots of parabolic oxidation constants at $P_{O_2} = 1$ atm for NiO layer growth on polycrystalline, crystal faces and pre-oxidized polycrystalline nickel (Elrefaie, Manolescu and Smeltzer, 1985).

sulfides exhibit larger defect concentrations and mobilities. An extensive review of the defect and transport properties of sulfides and sulfidation of metals is available (Mrowec and Przybylski, 1984). Sulfidation of iron is utilized as an illustrative example of the application of the Wagner theory. FeS exhibits nonstoichiometry extending up to 25 at.% metal deficiency due to iron vacancies which exhibit mutual repulsive interactions. In this case, eqn.[3] leads to

$$k_r = \frac{\bar{D}\Delta\delta}{\bar{V}} \qquad [5]$$

where $\bar{D}$ is the chemical diffusivity, $\Delta\delta$ is the change in nonstoichiometry across the scale and $\bar{V}$ is the equivalent volume of FeS (Fryt and coworkers, 1979). The increasing influence of sulfur pressure on the parabolic sulfidation rate at temperatures 600-980°C is shown in Fig.2. The predicted and experimental values of the rate constants agree at sulfur pressures up to 10^{-2} atm; at larger sulfur pressure, the rate constants increase by a factor of 2 because the scale texture changes from an 'a' to 'c' crystallographic orientation (FeS is of NiAs structure) which supports more rapid iron diffusion.

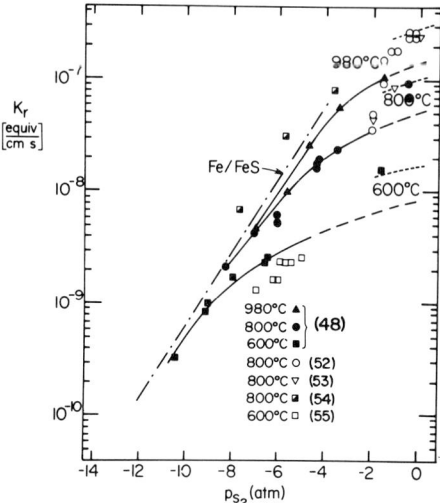

Fig.2. Dependence of the sulfidation parabolic rational rate constant on sulfur pressure at 600, 800 and 980°C. Curves correspond to predicted values calculated by eqn.[5] (Fryt and coworkers, 1979).

Growth and morphological development of scales on alloys are so complicated that theoretical analysis has been mainly confined to binary alloys. A diffusion model (Wagner, 1969), which is an extension of his model for a pure metal, describes growth of a ternary homogeneous oxide scale whereby one calculates from the diffusion equations the parabolic oxidation rate constant, k, the composition and oxygen activity profiles across the scale knowing the cation diffusivities in the oxide as functions of composition and oxygen activity. Since diffusion defines the flux of each metal component across a scale, the total metal flux is given by

$$D_A(1-\xi)(\frac{-\partial \ln a_{AO}}{\partial \xi}\frac{d\xi}{dy} + \frac{Z_A}{Z_O}\frac{d\ln n_O}{dy}) + D_B\xi(\frac{-\partial \ln a_{BO}}{\partial \xi}\frac{d\xi}{dy} + \frac{Z_B}{Z_O}\frac{d\ln a_O}{dy}) = k \quad [6]$$

and the composition profile by

$$yk = \frac{d\xi}{dy}(D_B\xi(\frac{-\partial \ln a_{BO}}{\partial \xi}\frac{d\xi}{dy} + \frac{Z_B}{Z_O}\frac{d\ln a_O}{dy})] = k \quad [7]$$

This theory has been used to interpret oxidation and sulfidation kinetics of binary transition metal alloys forming NaCl structural type ternary solid solution scales. The scaling model is depicted in Fig.3. An example of the calculated and experimentally determined values of the composition profile across a scale is shown in Fig.4 for the case of Co-Ni alloy oxidation (Narita and coworkers, 1982).

The evolution and application of this theory is the topic of reviews; one recently (Bastow and coworkers, 1985) includes a critical appraisal of theory, its application and attempts by investigators to obtain a more rigorous analysis based upon properties of point defects and their diffusion. Oxidation models have been developed using these principles for growth of multi-layer scales on

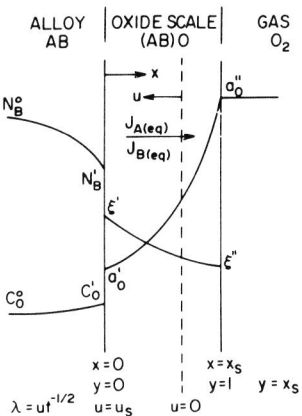

Fig. 3 Ternary diffusion model for growth of an oxide scale on a binary alloy with notations used in text.

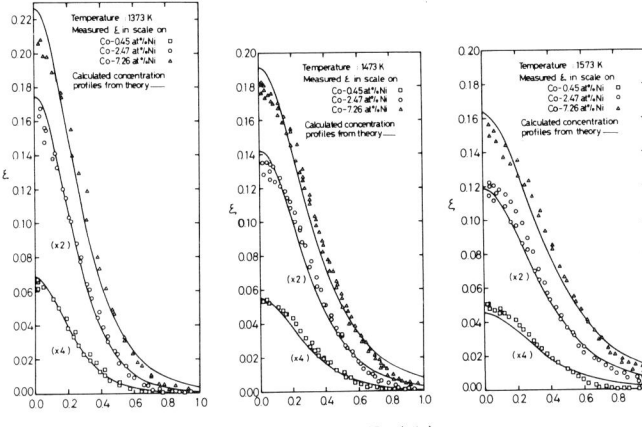

Fig. 4. Calculated and experimentally determined values for nickel composition as a function of normalized distance, y, in (CoNi)O scale formed on Co-Ni alloys containing 0.45, 2.44 and 7.26 at. % Ni under 1 atm oxygen pressure at 1373, 1473 and 1573 K (nickel compositions for the Co-0.45 Ni and the Co-21.47 Ni alloys were plotted by multiplying two and four times their original values, respectively.

pure metals and alloys but not to any degree of sophistication as a result of the diversity of oxidation phenomena (Smeltzer, 1987).

SCALE GROWTH BY SHORT-CIRCUIT DIFFUSION

The evidence for short-circuit diffusion during oxidation of metals and alloys is a topic of past reviews, as an example (Atkinson, 1985). These oxide films and thin scales formed at temperatures $\leq 1/2$ the oxide melting temperature may contain grains of sufficiently small size for boundaries and dislocations to act to short-circuit lattice diffusion. Oxidation kinetics, accordingly, are characterized by short-circuit diffusion which undergoes a gradual transition with increasing oxide thickness and at higher temperatures to diffusion ultimately controlled by migration of lattice point defects.

If oxidation proceeds by short-circuit diffusion,

$$\frac{dX}{dt} = \Omega D_{eff} \frac{\Delta C}{X} \qquad [8]$$

and

$$D_{eff} = D_L(1-f) + D_B f \qquad [9]$$

the following modified parabolic rate equation describes growth of the oxide layer

$$X^2 = 2\Omega D_L \Delta c \int_0^t (1 + \frac{D_B}{D_L} f) dt = k_p \int_0^t (1 + \frac{D_B}{D_L} f) dt = k_p(eff) t \qquad [10]$$

In these expressions, X is the layer thickness, Ω is the volume of oxide formed by a diffusing particle, D_B and D_L are the boundary and lattice diffusivities, respectively, and $\Delta C/X$ is the concentration gradient across the oxide layer.

Since k_p = $2\Omega D\Delta C$ (the parabolic rate constant) represents lattice diffusion of metal or oxygen by point defects, eqn. [10] describes an oxidation curve conforming to limiting parabolic relationships. The effective rate constant contains terms for boundary and lattice diffusion which is constant if the oxide grain size remains unchanged during early exposure and, at long times, the classical parabolic oxidation relationship is obtained for lattice diffusion if the large density of oxide boundaries is decreased by grain growth. The time dependent oxidation rate parameter, kp (eff), may also be defined by an expression similar to eqn. [3] in order to determine the influence of oxygen pressure on the oxidation rate because short-circuit diffusion is also ambipolar. Several empirical oxidation curves have been generated to describe the reaction kinetics by assuming constant values for the diffusivities and that decay of diffusion sites in the low resistance paths follows first order or grain growth mechanisms (Smeltzer and Young, 1975).

Two topics are discussed to illustrate reactant short-circuit diffusion with respect to oxidation of transition series metals and alloys where diffusion measurements are available to correlate the oxidation kinetics with short-circuit diffusion in the single-phase scales formed. The oxidation properties of two pure metals, zirconium and nickel, are first considered since the scales are formed by either oxygen or metal diffusion, respectively. These considerations are then extended to oxidation of alloys under conditions in which the diffusion properties of scales formed can be compared to those of the oxide.

A diffusion model (Wallwork, Smeltzer and Rosa, 1964) to account for parabolic growth of a ZrO_2 scale and oxygen diffusion in α-Zr is depicted in Fig.5. ZrO_2 supports a very small degree of nonstoichiometry and oxygen migration by anion

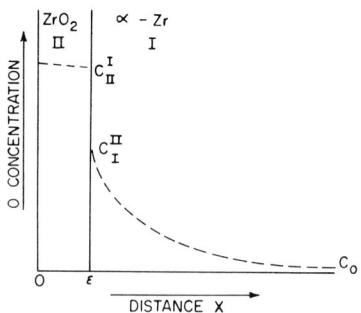

Fig.5. Diffusion model for growth of a ZrO_2 scale on α-Zr with concurrent dissolution of oxygen into the metal by parabolic reaction kinetics.

vacancies and free electrons. If a linear gradient and an error function represent the oxygen profiles in ZrO_2 and α-Zr, respectively, the oxidation kinetics are determined by parabolic rate constants for oxide layer growth and oxygen solution into the metal. An interpretation of parabolic oxidation in the range 400°-850°C (Madeyski, Poulton and Smeltzer, 1969) by this model and oxygen short-circuit diffusion is shown in Fig.6.

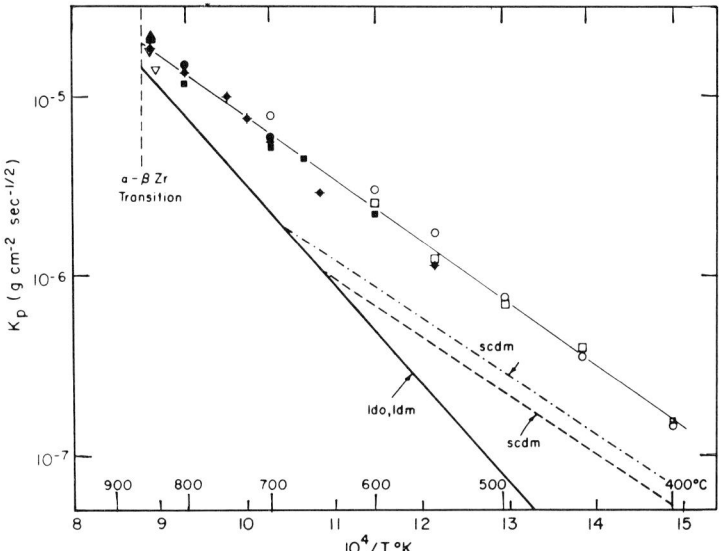

Fig.6. Experimental determination and calculation of the parabolic oxidation constant for α-Zr in the range 400-862°C. The curve ldo, ldm was calculated using values for lattice diffusion of oxygen in the scale and metal. The curves scdm were calculated using values determined for oxygen short-circuit in α-Zr.

The rate of oxygen solution into the metal is approximately of the same magnitude as the scaling rate at 862°C. These plots illustrate a characteristic oxidation feature associated with short-circuit diffusion: the total oxygen uptake using values for oxygen lattice diffusion in the oxide and metal is much smaller than the experimental value. These conclusions were substantiated by oxygen tracer diffusion measurements. In Fig.7, values are given for oxygen self-diffusion in ZrO_2 crystals and in scales exhibiting a microstructure as shown in Fig.8. A comparison is also given of these values to those calculated for oxygen diffusion in the ZrO_2 scales from the results of the oxidation kinetics analyzed by means of the oxidation model. The measured and calculated oxygen diffusivities for the scales are 10^4 larger than the coefficient for oxygen diffusion in ZrO_2 crystals and the ratio of the activation energies for the two types of oxides is 0.5. The polycrystalline ZrO_2 scale, therefore, supports long intervals of parabolic oxidation determined by oxygen short-circuit diffusion due to the extreme microstructural stability of the columnar scale.

As illustrated by the Arrhenius plots of the parabolic oxidation rate constants over the range 500°-1400°C in Fig.1, short-circuit diffusion is the active mechanism of NiO scale growth on nickel below 1000°C. The experimental determinations which show approximate agreement at lower temperatures to the plot of the parabolic rate constant obtained using the Wagner theory (eqn.[4]) are those obtained under special circumstances: values down to 800°C for nickel subjected to pre-oxidation at 1200°C to preform a thin NiO layer relatively free of easy diffusion paths and two evaluations at 700°C of the smallest thickness of large NiO grains on specific metal grains of polycrystalline nickel specimens. Short-circuit diffusion does determine the rate of oxide growth under normal conditions, its relative magnitude becoming larger with decreasing temperature and as large as 10^5 at 500°C. A cross-section of a scale formed at 800°C is

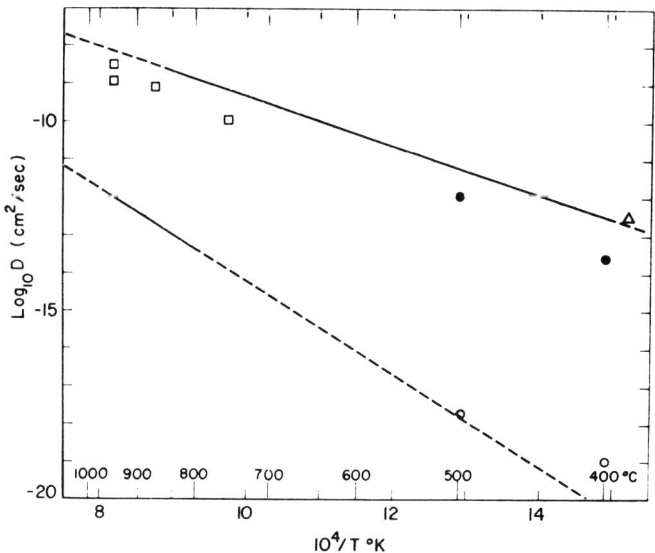

Fig.7. Arrhenius plots of oxygen diffusivities in monoclinic ZrO_2. Upper plot is for calculated oxygen diffusivities in scales; lower plot represents experimental values for lattice diffusion in ZrO_2. Upper and lower sets of independent experimental results represent short-circuit and lattice diffusion of oxygen, respectively.

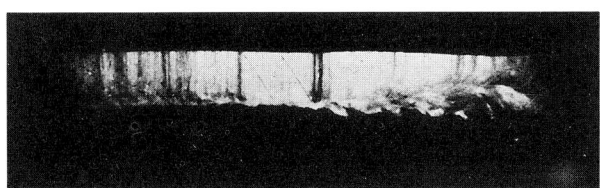

Fig.8. Microstructure of columnar ZrO_2 scale formed on α-Zr.

shown in Fig.9; NiO columnar grains overlay smaller equiaxed grains adjacent to the metal interface, the grains increasing in size at longer exposures.

Fig.9. Cross-section of NiO formed on a (100) face of nickel at 800°C showing growth of an inner equiaxed and outer columnar grained scale (Herchl and co-workers, 1972).

Since the oxidation kinetics conform more closely to lattice diffusion control
with increasing oxide grain-size, the kinetics for (100) nickel face at several
temperatures below 1000°C are shown in Fig.10 fitted to eqn.[10] describing decay
of boundary diffusion sites by grain growth. These curves exhibit decreasing

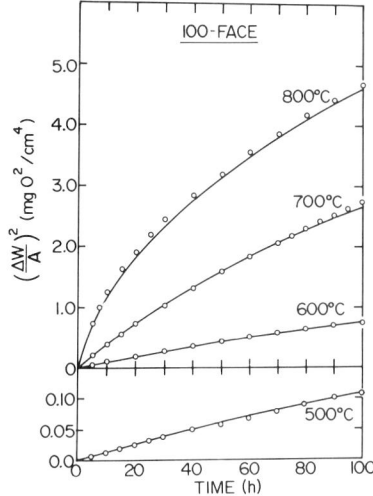

Fig.10. The experimental results and the curves
calculated by eqn.[10] for the growth of NiO
layers on the (100) nickel face (Herchl and co-
workers, 1972).

slope at longer time and it is limiting tangents to such curves which are used as
evaluations of the effective parabolic oxidation rate constants in Fig.1. If the
term for scale growth by lattice diffusion is taken into account, the activation
energy determined for nickel diffusion along NiO boundaries is 38 kcal/mole which
is much lower than the value of 60 kcal/mole for the activation energy of nickel
self-diffusion in NiO. Isotope tracer measurements have been completed to obtain
the various diffusivities of NiO (Atkinson, 1985). Nickel diffusion in boundaries of polycrystalline NiO is more rapid than in dislocations which, in turn,
is more rapid than lattice diffusion, the activation energies being 42, 47 and 60
kcal/mole, respectively. These diffusion measurements, therefore, substantiate
the above conclusions for growth of the scales by short-circuit diffusion.

Some remarks are now directed to growth by short circuit diffusion of single-
phase scales on alloys by selective oxidation of an alloying element. As
examples, growth of Al_2O_3 and Cr_2O_3 scales ensure useful lifetimes for several
iron and nickel alloys. These oxides contain minute point defect concentrations;
consequently, scaling kinetics are influenced by scale microstructures and by
reactant short circuit diffusion at temperatures up to 1000°C. Although diffusion in these oxides is governed by the influence of impurities on point defect
concentrations and oxide growth is complicated by scale adhesion, the oxidation
kinetics of nickel and iron alloys of chromium and/or aluminum correlate well
with short-circuit diffusion in the scales (Hindam and Whittle, 1982).

Growth of Cr_2O_3 scales on Fe-Cr and Ni-Cr alloys corresponds closely to pure
chromium. At 1000°C, chromium diffusion in Cr_2O_3 by low angle boundaries and
dislocations is 10^4 more rapid than lattice diffusion (Atkinson and Taylor,
1985). As illustrated in Fig. 11, a comparison of experimental parabolic oxidation rates to those calculated by the Wagner expression for lattice diffusion,
eqn. [3], demonstrates that chromium short-circuit diffusion by oxide grain
boundaries determines the growth rate of scales.

A comparison of lattice and grain boundary diffusivities for aluminum and oxygen
diffusion in single-crystal and polycrystalline Al_2O_3 as a function of temperature from different investigations is shown in Fig. 12. There is also shown

for comparison purpose, the value of the effective diffusivity calculated from the parabolic growth kinetics of Al_2O_3 on several iron and nickel alloys using eqn. [10]. These calculated values are equal to or larger than the values for boundary diffusion of oxygen which is the predominant diffusion species.

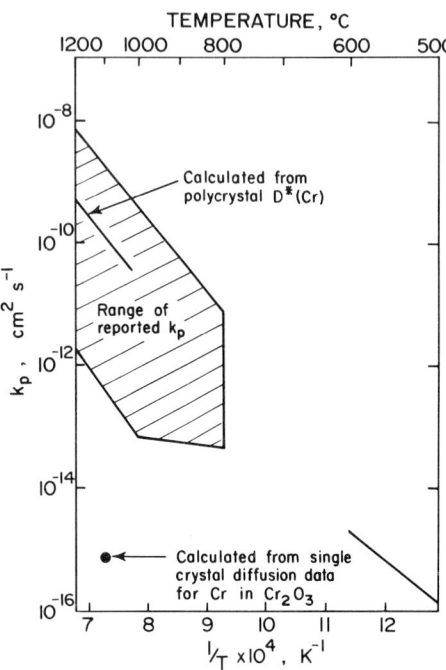

Fig.11. Arrhenius plot of the parabolic rate constant for Cr_2O_3 growth on chromium comparing experimental and predicted values by lattice and boundary diffusion in Cr_2O_3 (Atkinson and Taylor, 1985).

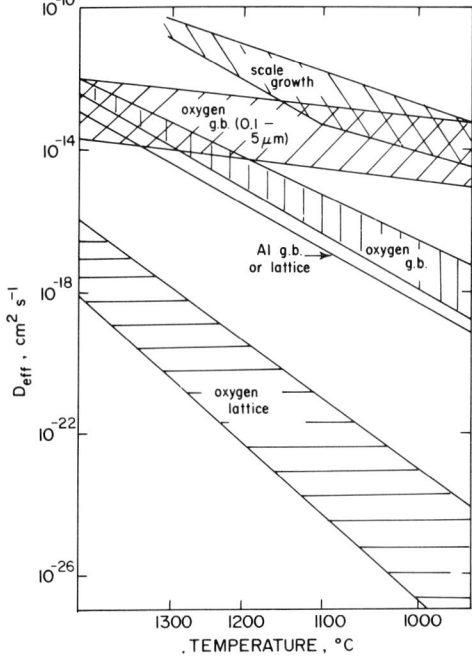

Fig.12. Lattice and grain boundary diffusivities for aluminum and oxygen in single-crystal and polycrystalline Al_2O_3. The calculated values of D(eff) and parabolic kinetics for Al_2O_3 growth on alloys are within the band or higher than the values for oxygen grain boundary diffusivities (Hindam and Whittle, 1982).

SCALE GROWTH AND INTERNAL OXIDATION OF BINARY ALLOYS

Superficial scale growth on an alloy can be accompanied by internal oxide precipitation in the metal substrate (designated as an internal oxidation zone or subscale), the external and internal layers often growing at high temperatures by parabolic kinetics. A brief summary is presented here of modelling these oxidation phenomena for a binary alloy by considering the problem by ternary diffusion and thermodynamics and, consequently, described by a diffusion path on the phase diagram. This approach was introduced in the early 1940's (Rhines, 1940) and its development to the present time is described in a recently published textbook on multicomponent diffusion in the condensed state (Kirkaldy and Young, 1988). This type of analysis has elucidated internal oxidation when it occurs by either general or cooperative precipitation and to a more complete understanding of a transition from internal oxidation to exclusive growth of an external scale with increasing content of the reactive alloying element.

Several instances are shown in Fig. 13 of oxide morphologies produced on binary alloys and their relationship to diffusion paths placed on either the ternary phase diagram or the alloy - P_{O_2} diagram. The alloying element is regarded as being selectively oxidized to form the oxide BO which exhibits only a small solubility of A. It is only in the case (a) of Fig. 13 that the diffusion path

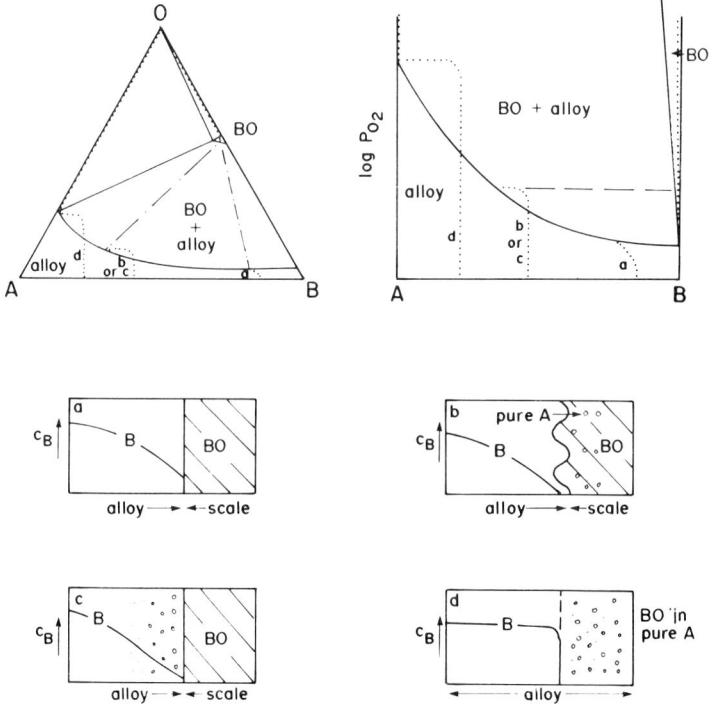

Fig.13. Diffusion paths and concentration profiles on a ternary phase diagram and on an equilibrium O_2 pressure diagram for a binary alloy undergoing parabolic oxidation kinetics: (a) BO scale growth; (b) or (c) scale interfacial instability or internal oxidation; (d) internal oxidation (Whittle, 1981).

is stable following equilibrium tie lies in two-phase regions and thus described by the diffusion model shown in Fig. 3. Paths b-d are virtual (unstable) paths as they cut into two-phase fields, their decay in these fields leading to interfacial instability and/or internal oxidation.

Two limiting cases of subscale formation involving general precipitation of small oxide particles in the metal matrix have been developed (Laflamme and Morral, 1978). One of these is the classical case with deep penetration of the diffusion path into the two-phase region (case d in Fig. 13) giving rise to an approximately constant amount of precipitate across the internal oxidation zone with precipitation confined to the subscale-matrix interface. It may be shown (Wagner, 1959) that the penetration depth, ε, of the internal oxidation zone is given by a parabolic relationship,

$$\varepsilon = (2N_O^{(s)} D_O t / \gamma N_B^O)^{1/2} \quad [11]$$

where N_O^S is the surface oxygen solubility, D_O is the oxygen diffusivity, N_B^O is the bulk alloy composition and γ represents the number of oxygen atoms per metal atom in the oxide. In the other limiting case which corresponds to low penetration into the two-phase region (case c in Fig. 13), precipitation occurs continuously through the subscale as it forms. Both of these limiting cases are approximations to the general equation for simultaneous diffusion and precipitation in ternary systems.

As the concentration of the alloying element which is selectively oxidized is increased to larger amounts, oxide precipitation by one of several cooperative modes may occur (Smeltzer, 1987). One common cellular type-precipitation is represented by decomposition of the oxygen saturated metal (α) to the solute-depleted but structurally identical phase (α') and an oxide precipitate phase (β). That is, $\alpha \rightarrow \alpha' + \beta$ by the growth of $\alpha' + \beta$ lamellae into α. This behavior is encountered upon internal oxidation or sulfidation and an example of this latter type is now discussed.

Fe-Aℓ alloys containing from 6 to 18 a/o Aℓ when sulfidized in H_2S-H_2 atmospheres at 900°C form rectangular plates of FeAℓ_2S$_4$ or FeAℓ_2S$_4$-Aℓ_2S$_3$ which extend inward completely across the internal sulfidation zone beneath the superficial scale. An example of this lamellar type sulfide growth is shown in Fig. 14. A model

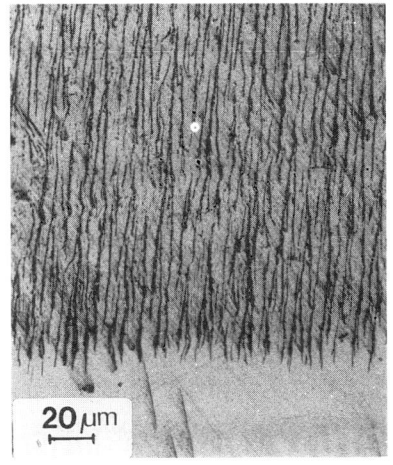 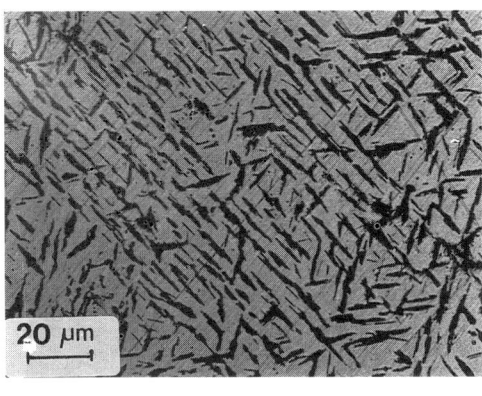

Fig.14. FeAℓ_2S$_4$-Aℓ_2S$_3$ plates extending across the internal sulfidation zone in an Fe-6Aℓ alloy sulfidized at 1173K: (a) cross-section; (b) lateral surface of internal sulfidation zone (Patnaik and Smeltzer, 1985).

appropriate for describing growth of this type of lamellar internal sulfidation zone must account for growth of sulphide lamellae by lattice and boundary diffusion of the oxidant. In this model, Fig. 15, the effective sulfur flux through the internal sulfidation zone is treated as proceeding through the alloy lattice, the internal sulfide/metal interfaces and the internal sulfide plates. One obtains the following expression for the ratio of the effective to

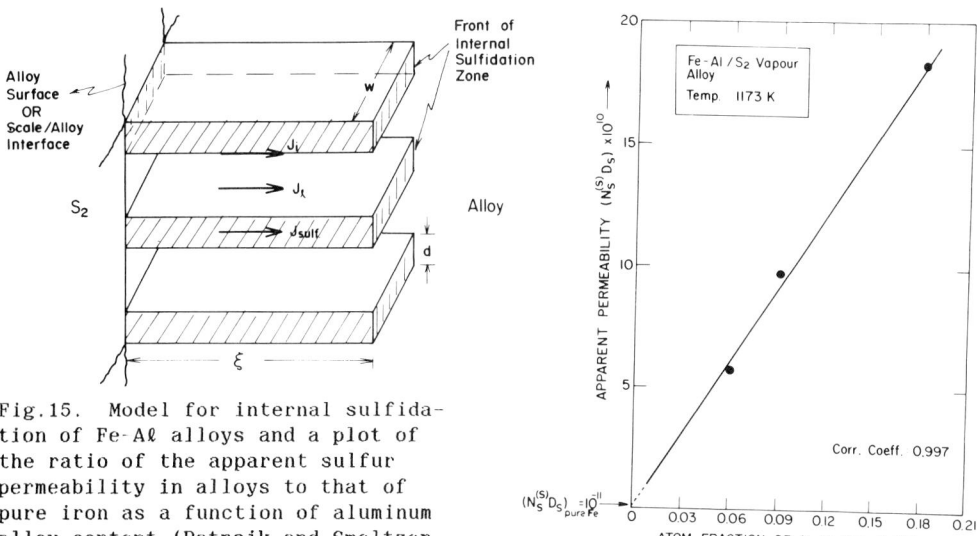

Fig.15. Model for internal sulfidation of Fe-Al alloys and a plot of the ratio of the apparent sulfur permeability in alloys to that of pure iron as a function of aluminum alloy content (Patnaik and Smeltzer, 1985).

lattice sulfur permeabilities if it is assumed that sulfur diffuses down a common sulfur concentration gradient to cause complete depletion of the alloying element by sulfidation at the internal front:

$$\frac{N_s D_s (\text{eff})}{N_s D_s (\text{lattice})} = 1 + [2\frac{b}{d} \frac{D_{s,B}}{D_{s,L}} - 1] \frac{V_{\text{sulf}}}{V_{\text{alloy}}} \alpha\, N_{Al}^o \qquad [12]$$

In agreement with this relation, the effective sulfur diffusion coefficient for the internal sulfidation zone is found to be a linear function of the original atom fraction of aluminum in the alloy as shown in Fig. 15. The value calculated for the ratio of the rates of sulfur boundary to lattice diffusivities of 10^4 from the sulfidation experimental results at 1173K agrees with this ratio from independently obtained diffusivity measurements.

Two methodolgies have been advanced to account for transition from internal oxidation to exclusive growth of a superficial scale. In the first method, the volume fraction of oxide particles at a critical alloy composition predicted by diffusion and precipitation as in eqn. [11] blocks inward oxygen migration. It has been found that oxide volume fractions in the range 0.3 to 0.5 lead to this transition for several noble-base alloys and to lateral growth of an Al_2O_3 layer at the cellular reaction fronts in Ni-Al and Fe-Al alloys. In the second method, internal oxidation occurs if diffusion causes the metal and oxygen concentrations to exceed the solubility product (path c and d in Fig. 13). Thus, internal oxidation does not occur in an alloy of solute concentration sufficiently large for the diffusion path to simply contact or tangent the alloy oxygen solubility curve (path a in Fig. 13). The critical alloy concentration is therefore related to the magnitude of the parabolic oxidation rate constant, diffusivities and alloy thermodynamic solution behavior (Smeltzer and Whittle, 1978). Moreover, in multicomponent alloys synergistic effects between elements may increase the flux of the reactive solute to the alloy surface (Morral, Thompson and Devereaux, 1987).

SUMMARY

An attempt has been made in this contribution to survey specifically selected topics of metal and alloy oxidation in terms of chemical diffusion and thermodynamics. Insofar as possible, models have been advanced based upon growth of compound layers, metal solid solution phases and diffusion determined composition profiles. These models account for growth of superficial scales by lattice and short-circuit diffusion on metals and alloys and for internal oxidation by general and cooperative modes of precipitation.

ACKNOWLEDGEMENT

The author wishes to express his appreciation to Professor Jack Kirkaldy for his friendship, many stimulating discussions and opportunities to actively collaborate on research contributions to this field.

REFERENCES

Atkinson, A. (1985) Rev. Mod. Phys. 57, 437.
Atkinson, A. and R.I. Taylor (1985). Transport in Nonstoichiometric Compounds, ed. G. Simkovich and V.S. Stubican, NATO AISI Series B: Physics 129, 385.
Bastow, B.D., G.C. Wood and D.P. Whittle (1985). Corros. Sci. 25, 253.
Elrefaie, F.A., A. Manolescu and W.W. Smeltzer (1985). J. Electrochem. Soc. 132, 2489.
Fryt, E.M., V.S. Bhide, W.W. Smeltzer and J.S. Kirkaldy (1979). J. Electrochem. Soc. 126, 683.
Herchl, R., N.N. Khoi, T. Homma, and W.W. Smeltzer (1972). Oxid. Metals 4, 35.
Hindam, H.M. and D.P. Whittle (1982). Oxid. Metals 18, 245.
Kirkaldy, J.S. and D.J. Young (1988). Diffusion in the Condensed State, Institute of Metals, London.
Laflamme, G.R. and J.E. Morral (1978). Acta Metall. 26, 1791.
Madeyski, A., D.J. Poulton, and W.W. Smeltzer (1969) Acta Metall. 17, 579.
Morral, J.E., M.S. Thompson, and O.F. Devereux (1987). Proc. Norman L. Peterson Memorial Symposium, Oxidation of Metals and Associated Mass Transport, AIME Publ. 315.
Myrowec, S., and K. Przybylski (1984). High Temp. Mater. and Process. 6, 1.
Narita, T., K. Nishida and W.W. Smeltzer (1982). J. Electrochem. Soc. 129, 209.
Patnaik, P.C. and W.W. Smeltzer (1985). J. Electrochem.Soc. 132, 1226.
Rhines, F.N. (1940). Trans. AIME 137, 246.
Smeltzer, W.W. and D.J. Young (1975). Prog. Solid State Chem. 10, 17.
Smeltzer, W.W. (1987). Oxidation Mechanisms of Metals and Alloys, Proc. Normal L. Peterson Memorial Symposium, Oxidation of Metals and Associated Mass Transport, AIME Publ. 109.
Smeltzer, W.W. and D.P. Whittle (1978). J. Electrochem. Soc. 125, 1116.
Wagner, C. (1933). Z. Physik Chem. B21, 25.
Wagner, C. (1951). Atom Movements ASM Cleveland, 153.
Wagner, C. (1959). Z. Elektrochem. 63, 772.
Wagner, C. (1969). Corros. Sc. 9, 91.
Wallwork, G.R., W.W. Smeltzer and C.J. Rosa (1964). Acta Metall. 12, 409.
Whittle, D.P. (1981). High Temperature Corrosion, ed. R.A. Rapp, NACE 6, 171.

Wetting Transition of Ethylene Adsorbed on Graphite

ICHIRO ARAKAWA*[1], YOSHIKATA KOGA**[2],
AND THE LATE JAMES A. MORRISON*

*Institute for Materials Reserch, and Department of Chemistry, McMaster University,
Hamilton, Ontario, Canada, L8S 4M1
**Department of Chemistry, The University of British Columbia, Vancouver,
British Columbia, Canada, V6T 1Y6

ABSTRACT

Adsorption isotherms and isosteres of C_2H_4 on graphite (Grafoil MAT) were determined volumetrically near the bulk saturation pressure, p_o (above 0.95 p_o), around the wetting temperature, T_w, which in this case is the same as the bulk triple point, T_t. The results indicate that the exponent x is 1/2, in $n_a^{sat} \sim (T_w-T)^{-x}$, where n_a^{sat} is the amount adsorbed at $p_o = p$ below T_w. This is in agreement with the exponent for O_2/graphite (Bartosch and Gregory, 1985). The values determined for the exponents y and z in $n_a \sim (p_o-p)^{-y}$ for $T>T_w$, and $n_a \sim (p_o'-p)^{-z}$ for $T<T_w$ were not conclusive but not inconsistent with those for O_2/graphite mentioned above. Within the range of about 4 to 8 monolayers, the heats of adsorption were 15.8 ± 0.2 kJ·mol^{-1} for $T>T_w$, slightly higher than the heat of bulk liquefication, and 17.8 ± 0.4 kJ·mol^{-1} for $T<T_w$, lower than the heat of bulk freezing. This indicates that two distinct adsorbed phases are present, bounded at about the bulk triple point.

KEYWORDS

C_2H_4/graphite; wetting transition; critical exponents; isoteric heats of adsorption near bulk condensation.

INTRODUCTION

The work presented here concerns the way in which two-dimensional adsorbed films on a smooth substrate grow to three-dimensional bulk phases. Two ways are known. In the so-called complete wetting scheme, the amount adsorbed increases asymptotically to infinity as the pressure increases isothermally to that of bulk condensation, p_o. In the other, the incomplete wetting scheme, the amount adsorbed remains finite, n_a^{sat}, at $p = p_o$, and the adsorbed films coexist with the

---

[1] Permanent address: Department of Physics, Gakushuin University, Tokyo, Japan.
[2] Correspondence should be directed to Y. Koga.

bulk phase. Menaucourt, Thomy and Duval (1977) discovered that the C_2H_4/graphite system undergoes a wetting transition, a changeover from the incomplete to the complete wetting scheme, at the wetting temperature, T_w. It turned out that T_w is equal in this case to the bulk triple point. Since then, a number of such transitions were reported, as summarized, not exhaustively, in Table 1.

TABLE 1 Summary of Wetting Transitions

Adsorbate	Substrate	$T_w(K)$	T_w/T_t	T_w/T_c	Exponent x	Ref.
Kr	graphite	83.0	0.72	0.396	1/3	(a)
Xe	graphite	116.3	0.72	0.401	1/3	(a)
CH_4	graphite	75.5	0.84	0.397	1/3	(b)
CF_4	graphite	37.0	0.41	0.16	1/3	(c)
O_2	graphite	54.4	1.0	0.31	1/2	(d)
C_2H_4	graphite	104.0	1.0			(e)
Kr, Xe Ar, N_2 O_2 CH_4	Au(111)		1.0		1/3	(f)
C_2H_6 C_2H_4	PbI_2	~104.0	1.0			(g)

Ref. (a) Inaba, Morrison and Telfer (1988).
     (b) Inaba and Morrison (1986).
     (c) Suzanne and co-workers (1984).
     (d) Bartosch and Gregory (1985).
     (e) Menaucourt, Thomy and Duval (1977).
     (f) Krim, Dash and Suzanne (1984).
     (g) Bassignana and Larher (1984).

Most of these wetting transitions are known to be continuous, and the amount adsorbed at saturation, n_a^{sat}, increases critically as $n_a^{sat} \sim (T_w-T)^{-x}$.

In terms of the wetting temperature, there appear to be at least two kinds of transition as seen in Table 1. In one, the wetting temperature is equal to the bulk triple point of the adsorbate. In this case, it is generally understood that even below $T_w(=T_t)$, the adsorption starts off as the complete wetting scheme with the virtual saturation pressure, p_o', which lies on the metastable extension of the bulk liquid-vapour coexistence line. The actual bulk condensation occurs when the pressure reaches the bulk solid-vapour coexistence line, p_o, before p_o'. (Bartosch and Gregory, 1985; Drir and Hess, 1986; Drir, Nham and Hess, 1986; Krim, Dash and Suzanne, 1984; Krim, Coulomb and Bouzidi, 1987). Bartosch and Gregory (1985) showed that for O_2/graphite the exponent x defined above as well as the exponent z in $n_a \sim (p_o'-p)^{-z}$ is about 1/2 below T_w, while the exponent y in $n_a \sim (p_o-p)^{-y}$ above T_w is about 1/3. If the substrate-adsorbate interaction is of the van der Waals type, in which the potential varies inversely as the third power of the distance, then both x and z should also be 1/3.

In the second kind of wetting transition, T_w is lower than the triple point. It is particularly noteworthy that for CH_4/graphite (Inaba and Morrison, 1986) and Kr and Xe/graphite (Inaba, Morrison and Telfer, 1988) the ratios T_w/T_c are equal to 0.399 ± 0.003, T_c being the bulk critical temperature of the adsorbate. Some preliminary results from neutron diffraction measurements on a thin film (∼5.5 monolayers) of CD_4 on Grafoil MAT indicate the occurrence of a time and temperature dependent wetting transition throughout the region from T = 75 K (≈T_w for CH_4/graphite) to the bulk melting point (Morrison, Shapton and Smith, 1988). This suggests a possible involvement of nucleation in the transition phenomenon.

To investigate the wetting transition further, particularly the latter kind, we decided to determine the exponents y and z in the very close proximity of p_o. We also allowed a very long equilibration time (10∼20 hrs) to eliminate any possible kinetic effect such as nucleation. However, we encountered experimental difficulties. While the amount adsorbed, n_a, was determined very accurately, the error in p was very sensitive to the presence of noncondensable impurities in the gas. Moreover, the uncertainties in p_o in the literature became very significant in comparison with the value (p_o-p). While some of the results thus obtained have been presented (Arakawa, Koga and Morrison, 1987), we decided to circumvent these difficulties by constructing a double cell in which the difference (p_o-p) is directly measured. What is described below are interim results on C_2H_4/graphite obtained by using the newly constructed double cell as well as the conventional single cell.

The effect of capillary condensation is known to be inevitable for exfoliated graphite in the pressure range >0.95 p_o. However, to what extent such effects disturb the observation of the intrinsic phenomena of interest has never been clearly understood. Bartosch and Gregory (1985) used a graphite-fibre oscillator to avoid a possible capillary condensation effect in studying the O_2/graphite system. As will be shown below, the present results on C_2H_4/graphite using Grafoil MAT are consistent with their conclusions. This indirectly provides some confidence in the conclusions of both their work and ours.

EXPERIMENTAL

The apparatus with a conventional single cell was described in detail elsewhere (Inaba, Koga and Morrison, 1986). The double cell, consisting of two compartments, was made of copper. One compartment was packed with Grafoil MAT pretreated in the same way (Inaba, Koga and Morrison, 1986) and the other was empty. The latter compartment was to contain bulk phase during the experiment. The cell was thermally connected to an aluminum shield, the temperature of which was regulated to ±4 mK or better. The temperature difference between the two compartments was estimated to be less than 1 mK at equilibrium. Each compartment was connected to the external gas handling system and the pressure difference, (p_o-p), was directly measured by a calibrated MKS Baratoron differential gauge with the sensitivity of 0.0002 torr. Ethylene of Research Purity (99.99%) from Matheson Gas was further purified by several freeze-pump-thaw operations. The residual pressure over the solid C_2H_4 at liquid nitrogen temperature was used to monitor the results of purification.

RESULTS AND DISCUSSION

Several isotherms were obtained by using the single cell. The recent study of C_2H_4 adsorption on a cleaved surface of pyrolytic graphite using an elipsometric technique (Drir, Nham and Hess, 1986) resolved up to 8 layers of the layer-by-layer growth. The present isotherms show up to 4 steps, and for p>0.95 p_o the isotherms were structureless, similar to those for O_2 adsorption on a graphite-fibre oscillator (Bartosch and Gregory, 1985). This is probably due to lack of energetic uniformity of the surfaces. Thus, the present isotherms provide an averaged-out behaviour, which still can be subjected to an analysis of the type $n_a \sim (\ln p/p_o)^{-y}$, or equivalently $n_a \sim (p_o-p)^{-y}$ for T>Tw, or $n_a \sim (p_o'-p)^{-z}$ for T<Tw. The iostherms obtained by means of the single cell were analysed by plotting $1/n_a^{(1/y)}$ (or $1/n_a^{(1/z)}$) against p with a trial value of y (or z). The value of y (or z) was thus decided as the value which provided the best straight line. For certain cases, this method of determining y (or z) was not sensitive. For example, as shown in Fig. 1 for 101.98 K, both plots of $1/n_a$ and $1/n_a^3$ against p/p_o appear to be straight lines for the range $0.96 < p/p_o < 1.00$, indicating that the exponent z is between 1/3 and 1. When the isotherms are obtained by means of the double cell, whereby (p_o-p) is determined directly, then more accurate evaluation of y is possible. That will be the subject of a future publication. Isosteric heats of adsorption were calculated in the ranges T>T_w and T<T_w from two isotherms. These results obtained by means of the single cell are summarized in Table 2. Note that the isosteric heats of adsorption are different below and above T_w.

Table 2 Results from the Single Cell

	T>T_w = T_t	T<T_w = T_t
Isosteric heats of adsorption	15.7 ± 0.3 kJ·mol^{-1} (106∿107 K, ∿9 monolayers)	17.2 ± 0.3 (102∿103 K, ∿5 monolayers)
Heats of bulk condensation (Meneaucourt, 1982)	15.74 kJ·mol^{-1}	19.73 kJ·mol^{-1}
Exponent x in $n_a^{sat} \sim (T_w-T)^{-x}$	—	1/2 (see Fig. 3)
Exponent y in $n_a \sim (p_o-p)^{-y}$	1/3 ?	—
Exponent z in $n_a \sim (p_o'-p)^{-z}$	—	1/3 ∿ 1 ? (See Fig. 1)

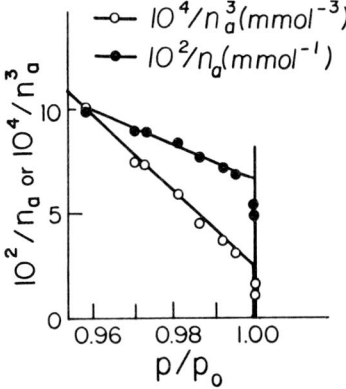

Fig. 1. Isotherms at 101.98 K

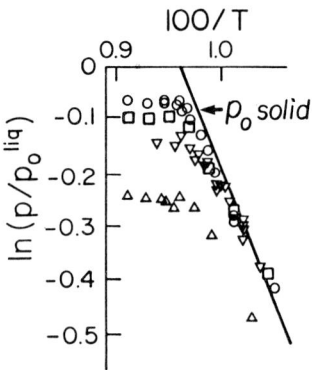

Fig. 2. Isosteres
Δ - 2 monolayers
∇ - 4 monolayers
□ - 5 monolayers
○ - 8 monolayers

At present, only isosteres have been measured using the double cell. The results are shown in Fig. 2. The pressure was normalized to p_o (liquid) calculated by the Clapeyron-Clausius equation for the liquid phase data obtained by Menaucourt (1982). For each isostere, the amount adsorbed was kept constant within 0.05%. The slopes change dramatically at about T_t, in contrast to the results of Drir, Nham and Hess (1986). This difference may be caused by the fact that they normalized the pressure to the value extrapolated to infinite thickness by the Frenkel-Halsey-Hill equation,

$$RT \ln (p/p_o) = \alpha \, n_a^{-3} \, .$$

This is equivalent to forcing the exponent z as well as x to be 1/3 below $T_w = (T_t)$. On the other hand, our results show that x = 1/2 (see Fig. 3 below) and z = 1/3 ~ 1, below $T_w = (T_t)$. Isosteric heats of adsorption for 4 to 8 monolayers calculated from slopes of Fig. 2 are 15.8 ± 0.2 kJ·mol^{-1} for T>T_w and 17.8 ± 0.4 kJ·mol^{-1} for T<T_w. They are independent of coverage and consistent with the results obtained by means of the single cell. These results are not consistent with the general understandings on the triple point wetting transition mentioned in the introduction. Such an interpretation calls for the isosteres to extend smoothly into the regin below T_t, until they hit the bulk solid vapour pressure line. Bartosch and Gregory (1985), however, suggested that while the surface liquid phase persists below the triple point for O_2/graphite, the structure may differ from that above it. Indeed, the present results show an additional phase boundary at about T_t, by which two regions with different isosteric heats of adsorption are separated.

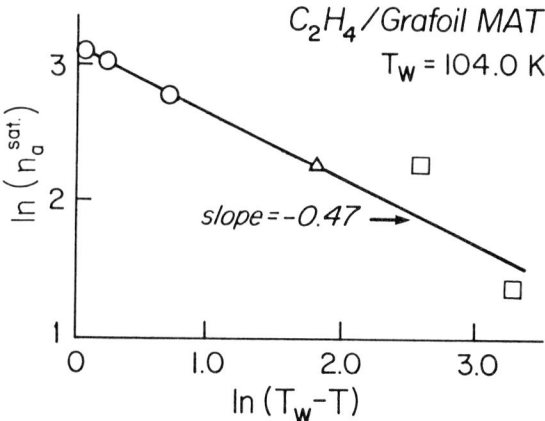

Fig. 3. $\ln(n_a^{sat})$ vs. $\ln(T_w-T)$
○ - from isotherms
△ - from isosteres
□ - Meanaucourt, Thomy and Duval (1977)

Fig. 3 shows the plots of $\ln(n_a^{sat})$ against $\ln(T_w-T)$ obtained mainly from isotherms measured by the single cell. From isosteres, only the case for 4 monolayers shows clearly the merging point into the bulk solid vapour pressure line, which is included in the figure after applying an appropriate conversion factor. The values by Menaucourt, Thomy and Duval (1977) are also treated in the same way. Fig. 3 indicates that the exponent x is 1/2. Thus, the exponents x, y and z for C_2H_4 on Grafoil MAT are consistent with those of O_2 on a graphite-fibre oscillator (Bartosch and Gregory, 1985). This provides some confidence in the genuine nature of the present observations. While the influences of capillary condensation can not be totally ruled out, it appears that such effects do not entirely obscure the intrinsic phenomena in question.

ACKNOWLEDGEMENT

We should like to thank Dr. T. Meichel for assistance with some measurements, and Dr. M.L. Klein and Dr. D.E. Sullivan for helpful suggestions and discussions. The work was supported by the Natural Science and Engineering Research Council of Canada through an International Scientific Exchange Award to I.A. and an operating grant to J.A.M. One of us (Y.K.) is grateful to Dr. B.A. Dunell for his help in improving the manuscript.

REFERENCES

Arakawa, I., Y. Koga and J.A. Morrison (1987). CAP Meeting, Paper #DF7, Toronto, June 1987.
Bartosch, C.E. and S. Gregory (1985). Phys. Rev. Lett., 54, 2513-2516.
Bassignana, I.C. and Y. Larher (1984). Surf. Sci., 147, 48-64.
Drir, M. and G.B. Hess (1986). Phys. Rev., B33, 4758-4761.
Drir, M., H.S. Nham and G.B. Hess (1986). Phys. Rev., B33, 5145-5148.
Inaba, A., Y. Koga and J.A. Morrison (1986). J. Chem. Soc., Faraday Trans. II, 82, 1635-1646.
Inaba, A. and J.A. Morrison (1986). Chem. Phys. Lett, 124, 361-364.
Inaba, A., J.A. Morrisona nd J.M. Telfer (1988). Mol. Phys., in press.
Krim, J., J.G. Dash and J. Suzanne (1984). Phys. Rev. Lett., 52, 640-643.
Krim, J., J.P. Coulomb and J. Bouzidi (1987). Phys. Rev. Lett., 58, 583-586.
Menaucourt, J. (1982). J. Chim. Phys., 79, 531-535.
Menaucourt, J., A. Thomy and X. Duval (1977). J. de Phys., 38, C4-195-200.
Morrison, J.A., R.A. Shapton and R.K. Thomas (1988). To be published.
Suzanne, J., J.L. Seguin, M. Bienfait and E. Lerner (1984). Phys. Rev. Lett., 52, 637-639.

Metastability/Unstability: Kinetics of Polymorphic Transition from Phase II to Phase III of C_2Cl_6

YOSHIKATA KOGA

Department of Chemistry, The University of British Columbia, Vancouver, British Columbia, Canada, V6T 1Y6

ABSTRACT

What is believed to be the changeover from metastability to unstability was observed in kinetics of the polymorphic solid-solid transition from phase II to phase III of C_2Cl_6 (the transition temperature: 43.64°C). Both the nucleation rates and the growth rates showed two distinct behaviours in the temperature regions bounded at about 41°C.

KEYWORDS

Changeover from metastability to unstability; transition from phase II to phase III of C_2Cl_6; growth rates; nucleation rates.

The recent development of theoretical studies on the dynamics of first order transitions including the possible distinction between metastability and unstability has been reviewed (Gunton, San Miguel and Sahni, 1983). Experimentally, the metastable states have been recognized in a variety of situations; hysteresis loops, superheated or supercooled states etc. However, there has been no experimental example which clearly shows the changeover from metastability to unstability. What is described below is the observation of the behaviour in the kinetics of polymorphic transition from phase II to phase III of C_2Cl_6 that may be interpreted as the changeover from metastability to unstability.

The low temperature phase III of C_2Cl_6 is known to be orthorhombic and there is no rotational disorder, while the high temperature phase II appears to have considerable disorder (Hohlwein, Nägele and Prandl, 1979). The polymorphic solid-solid transition between phases II and III of C_2Cl_6 belongs to a class called a single crystal-single crystal transformation. Namely, there is only one nucleus in a crystal and it grows to cover the entire crystal without fracture resulting in a single crystal of the new phase (Mnyukh, Petropavlov and Musaev, 1966). Structural investigations on both phases of this class of transitions generally suggest a very small change in the position or the orientation of the molecule through the transition on the essentially unchanged lattice site.

Kinetics of the transition were studied by means of a dilatometer. For measurements of the nucleation rates (Koga, 1987) and the growth rates (Koga, 1984; Koga and Miura, 1978), the methods were devised to minimize secondary effects caused by impurities and defects. Hence the reproducibility of the experimetal results was good.

The growth rates of the previous work (Koga and Miura, 1978) were replotted in Fig. 1 against temperature. The growth rate is zero at 43.64°C, i.e. the transition temperatures, T_{tr} is 43.64°C. Below 41.5°C, the growth rate, v, for the transition from phase II to phase III appears to increase rapidly to infinity. Assume that such increase depends on the following equation with a "critical exponent", n,

$$v \sim (T_\theta - T)^{-n},$$

where T_θ is the temperature at which the growth rate becomes infinity. With a trial values of 2, 3 and 5 for n, the values of $1/\sqrt[n]{|v|}$ were plotted against temperature in Fig. 2. With the given data points, n cannot be determined unequivocally. If n = 3, however, T_θ = 40.8°C. In any case, it may be reasonable to conclude that T_θ is about 41°C within half a degree of uncertainty. Thus, for the temperature range, $T_\theta<T<T_{tr}$, the growth rates are finite and have some functional dependence on temperature, which can be qualitatively explained in a similar way as a kinetic Ising model (Koga, 1984). For the range, $T<T_\theta$, on the other hand, the growth rates are infinite. Namely, $T_\theta \approx 41°C$ marks a boundary, in both sides of which the growth processes behave differently.

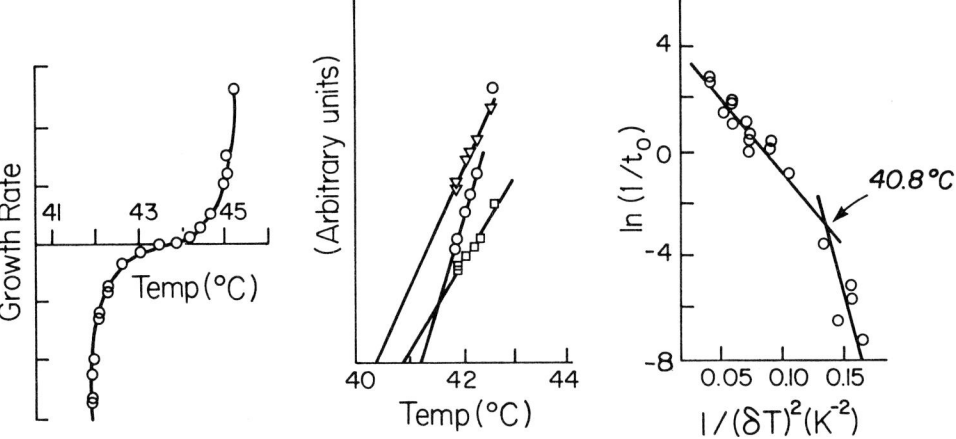

Fig. 1. Growth Rates

Fig. 2. Growth Rates
$\nabla - 1/\sqrt[5]{v}$
$\square - 1/\sqrt[3]{v}$
$\bigcirc - 1/\sqrt{v}$

Fig. 3. Nucleation Rates

The nucleation rates, or the inverses of the induction periods were also measured using the same specimen (Koga, 1987). All the existing theories, classical or modern, predict the nucleation rate, J, or the inverse of the induction period, t_o, to be expressed in the form

$$J \text{ (or } 1/t_o) \sim \exp\{-C_2/(\delta T)^2\}, \tag{1}$$

where C_2 is a constant specific to the theory and $\delta T = T_{tr} - T$. Fig. 3 shows the plots of $\ln(1/t_o)$ against $(\delta T)^2$. If the expression (1) is to be forced, there are two straight lines in Fig. 3 bounded at 40.8°C. The classical theory shows that C_2 is expressed for a spherical nucleus as, $C_2 = 16\pi\sigma^3/3RT\Delta S^2$, where σ is the interfacial tension and ΔS is the entropy of the transition per unit volume. Therefore, in terms of the classical nucleation theory, Fig. 3 indicates that $\sigma = 1.4$ erg cm^{-2} for 40.8°C<T<43.64°C and $\sigma = 0.4$ erg cm^{-2} for T<40.8°C. Note that how meaningful these values of σ are is not clear at present. The validity of equation (1) above for the entire temperature range is still questionable also. However, it is clear that 40.8°C marks a boundary in both sides of which the nucleation processes behave differently.

In conclusion, both the nucleation and the growth rates show two distinct behaviours in the temperature regions bounded at about 41°C. This 41°C marks a type of transition, which may correspond to the changeover from metastability to unstability.

REFERENCES

Gunton, J.D., M. San Miguel and P.S. Sahni (1983). In C. Domb and J.L. Lebowitz (Ed.) Phase Transitions and Critical Phenomena, Vol. 8. Academic, London. pp.
Hohlwein, D., W. Nägele and W. Prandl (1979). Acta. Cryst., B35, 2975-2978.
Koga, Y. (1984) J. Cryst. Growth, 66, 35-44.
Koga, Y. (1987). Physica B, 146B, 408-415.
Koga, Y. and R.M. Miura (1978). J. Chem. Soc., Faraday Trans. I, 74, 1913-1921.
Mnyukh, Yu, V., N.N. Petropavlov and N.I. Musaev (1966). Acta. Cryst., A21, 202-203.

Freezing in Transitions

ALFRED R. COOPER

*Department of Materials Science & Engineering,
Case Western Reserve University, Cleveland, Ohio, 44106*

ABSTRACT

A freezing in transition is a non equilibrium transformation from an equilibrium state to a state where some aspect of the configuration is fixed, changing neither with time nor with small changes in the external variables. The long studied glass transition is a prime example of a freezing in transition. Its phenomenology is reviewed with special emphasis on the significance of the Lillie number. Non equilibrium transitions in both crystalline and glassy systems are shown to display behavior characteristic of a glass transition, and it is argued that the methodology of studying the glass transition can be usefully applied to a wider class of materials.

KEYWORDS

Non equilibrium; glass transition; Lillie number; short range ordering; ferroelectrics; ferrites.

INTRODUCTION

Many materials at low temperatures (room temperature or below), have properties which are constant with time, provided the external conditions are fixed, but which depend not only on the current external variables, but also on the "history" the material has experienced. Such materials are said to be in a <u>frozen state</u>. Thus, "frozen" has a different definition from "solidified" or "crystallized." Frozen applies to a material with a non-equilibrium configuration which is not perceptibly changing toward equilibrium.

Typically, the external variables are pressure, P, temperature, T, and the chemical composition of the atmosphere, $\bar{C}$. We may combine these as the external variable vector, $\bar{E} = P, T, \bar{C}$. Then the history, function $\bar{\Phi}$ ($t_e < t < 0$), is described by the variation with time of $\bar{E}$ from the most recent occurrence of equilibrium ($t = t_e$) to the current instant ($t = 0$). It is usual that P and $\bar{C}$ remain constant, in which case $\bar{E} = T$ and $\bar{\Phi} = T$ ($t_e < t < t_0$).

While a particular frozen state can be obtained from another frozen state, here attention will be confined to "freezing in transitions," the conversion of a material from an equilibrium state to a frozen state. To examine such transitions, we will review and characterize the various pertinent states, examine the phenomenology of the glass transition, and show that freezing in transitions in crystalline solids are widespread and have characteristics of the glass transition. This will lead to the conclusion that the long studied glass transition may serve as a model for a broad class of freezing in transitions.

EQUILIBRIUM AND NON EQUILIBRIUM STATES

There are two types of equilibrium states of interest, stable and metastable. Both[1] require a minimum in the free energy, and in both cases the properties depend only on the external variables. Stable equilibrium refers to the global minimum of free energy, while metastable equilibrium is but a local minimum. At constant, $\bar{E}$, the properties of a material in a stable equilibrium state are constant. Frequently the same is true for a metastable equilibrium state, although a metastable equilibrium state may transform continuously to a stable equilibrium state. Usual examples of metastable states are: super cooled liquids and crystalline polymorphs. With $\bar{E}$ fixed, there is but one equilibrium state and only a few metastable states.

It has already been pointed out that, unlike an equilibrium state, a material in a frozen state has properties which depend on history as well as on $\bar{E}$. Its properties are independent of time, and therefore the frozen state is often called a false equilibrium state (Prigogine and Defay, 1954). Since variations in history are continuous, there are a continuum of frozen states at a fixed $\bar{E}$.

A material in a non equilibrium unstable state has similar characteristics to those of a frozen state, with the exception that in the unstable state properties depend on time, with $\bar{E}$ fixed. A summary of the characteristics of the various states is given in Table 1. In this table and henceforth, the symbols for external variable vector and history function have been simplified to E and ϕ.

TABLE 1. Characteristics of Different Classes of States

State	Properties Depend On	Minimum in G	Number of States at (E_0)
Stable Equil.	E	Global	1
Metas. Equil.	E, i*	Local	Few
Unstable	E, ϕ, t	None	∞
Frozen	E, ϕ	None	∞

*i indicates that i^{th} metastable phase is considered.

Since, with E fixed, frozen states and unstable states have properties which are different from each other and from the equilibrium state, there must also be a difference in their configuration. This configurational difference may be in: the progress of a chemical reaction, the atomic arrangement, the defect density, the microstructure, or the macrostructure.

---

1. When it is known which metastable state is being considered.

Though the configuration necessarily departs from equilibrium during a freezing in transition, vibrational spectra maintain equilibrium with E.

THE GLASS TRANSITION

From studies beginning toward the end of World War I motivated by a need for the manufacture of fine optical glass, it became clear that the index of refraction of glass depended, not only on the current temperature and glass composition, but also on the thermal history it had experienced. Continuing efforts almost up to the present have allowed the phenomenology of the glass transition, the conversion of material from a metastable[2] equilibrium liquid state to a frozen "glassy" state, to be clarified.

Glasses can be as rigid as crystalline materials. Thus, there are two paths by which a liquid can solidify: crystallization, or a glass transition. Figure 1a, a plot of $V(T)$, specific volume vs. temperature, shows the features of a glass transition observed at constant P and $\bar{C}$, when a glass forming liquid is cooled from the metastable equilibrium state to a frozen state at a constant rate $Q \equiv -\partial T/\partial t$, and then reheated, again at a constant rate, to the original state. The large hysteresis confirms the non equilibrium nature of the glass transition. Equilibrium volume, V_e, is represented by a dashed line. Its slope, the equilibrium thermal expansion,[3] $\alpha_e \equiv \partial V_e/\partial T$, is due to changes with temperature of both the atomic configuration and the amplitude of atomic vibrations. The upper glass transition temperature, T_u, is defined as the temperature at which differences between the actual volume and the equilibrium volume are first perceived during cooling. At the lower glass transition temperature, T_L, volume changes associated with atomic configuration are no longer detectable. Below T_L the configuration is effectively frozen, the material is in a glassy state, and $V_g(T)$ follows the dotted line. Only atomic vibrations contribute to the thermal expansion, $\alpha_g \equiv \partial V_g/\partial T$, of the glassy state. (For simplicity, both α_e and α_g are assumed to be constant functions of temperature.) The temperature interval (T_L, T_u) between the frozen and the equilibrium states is termed the glass transition range. In this region the system is unstable, and perceptible but incomplete relaxation toward the equilibrium state occurs. The volume, $V(T)$, in the unstable glass transition range is represented on Fig. 1a by the curved solid line.

Extrapolation of the dashed line into the glass transition range implies cooling and heating at a rate sufficiently low that deviations from equilibrium are imperceptible. Likewise, extrapolation of the dotted line above T_L implies heating and cooling so quickly that no perceptible relaxation of the configuration occurs.

Ceasing cooling at the point A and allowing the system to relax isothermally results in a monotonic decrease in volume with time. Likewise, stopping heating at the same temperature, i.e. at point B, causes an isothermal volume increase. That the long time results of both experiments converge to the same volume confirms the existence of a metastable equilibrium volume.

For constant rate cooling from above T_u to a frozen state, the history, ϕ, is completely described by the cooling rate, Q. To assure that freezing is complete, cooling must continue to a temperature significantly below T_L because, were the system held at T_L for long periods, it would relax to a lower volume.

---

2. Metastability is with respect to the crystalline phase.
3. Note that thermal expansion $\partial V/\partial T$ and not thermal expansion coefficient $(1/V)\partial V/\partial T$ is used.

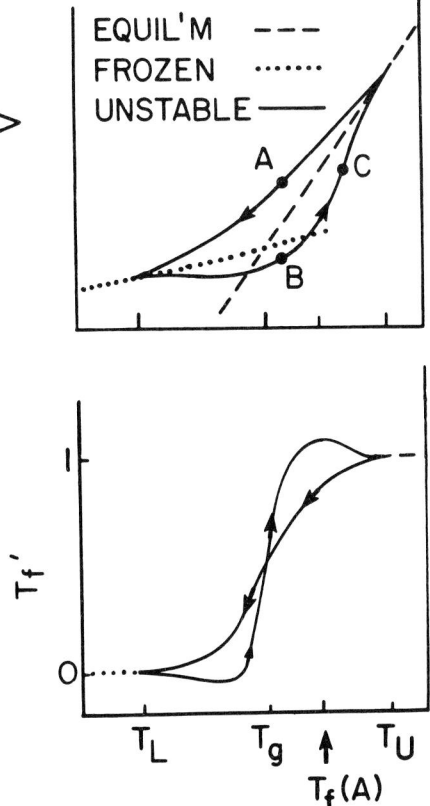

Fig. 1. Schematic showing behavior on cooling through a glass transition followed by reheating to the equilibrium state. Specific volume V(T) is shown in the upper curve (Fig. 1a). The derivative of fictive temperature with temperature, $T_f(T)$, is shown on the lower curve, (Fig. 1b). Characteristic temperatures shown on lower abscissa apply to both Fig. 1a and Fig. 1b.

If, for simplicity, it is assumed that for every point in (V, T) space corresponding to an unstable or a frozen state, there exists a point on the dashed equilibrium line with an equivalent atomic configuration, then the temperature at which a line parallel to the dotted line intersects the equilibrium line characterizes the configuration of all points along the line. Tool (1946) first introduced this temperature, and called it the fictive temperature, T_f. Points A and C on Fig. 1a have the same fictive temperature indicated by $T_f(A)$. The fictive temperature of a glass is called its "glass temperature," Tg. According to the stated assumption, the atomic configuration of a glass is fully characterized by Tg.

Tool postulated that the rate of such isothermal relaxation is given by:

$$T_f' \equiv dT_f/dt = -(T_f - T)/\tau \tag{1}$$

where τ is the configurational relaxation time. Often it is found that τ is approximately equal to the shear stress relaxation time, η/G, where η is the viscosity and G the shear modulus. Equation 1 with constant, τ, predicts an exponential decay of T_f with time, which usually only approximately describes careful isothermal experiments (Rekhson and Mazurin, 1974).

According to the definitions of α_g and α_e, it follows from Fig. 1a that

$$T_f - T = (V-V_e)/(\alpha_e - \alpha_g) \qquad (2)$$

Differentiating eqn. 2 with respect to temperature gives:

$$T_f' \equiv dT_f/dT = (\alpha-\alpha_g)/(\alpha_e - \alpha_g) \qquad (3a)$$

Eqns. 3 show that T_f' is the difference between the actual thermal expansion and the thermal expansion of the frozen state normalized by the thermal expansion difference between the equilibrium and the frozen state. An equivalent relation can be written for specific heat, i.e.

$$T_f' = (Cp - Cp_g)/(Cp_e - Cp_g) \qquad (3b)$$

According to eqns. 3, at equilibrium $T_f' = 1$, and for a glass $T_f' = 0$. Notice that during heating $T_f' = 1$ need not imply equilibrium, nor must $T_f' = 0$ imply a glassy state. Figure 1b shows a schematic of $T_f'(T)$ corresponding to the cooling and heating rates utilized for Fig. 1a. Because the volume in the unstable region becomes asymptotic to the equilibrium volume as $T \to T_u$ from below, and to the frozen volume as $T \to T_L$ from above, there is a degree of arbitrariness about the definition of T_u and T_L during cooling, which is clarified by defining T_u and T_L as follows (Cooper and Gupta, 1982):

$$T = T_u \text{ when } T_f' = (1-\delta) \qquad T = T_L \text{ when } T_f' = \delta$$

where δ is a small number (say 0.01), which depends on the sensitivity with which α or Cp can be determined.

Substitution of the definition of Q into eqn. 1 results in:

$$T_f' = (T_f - T)/Q\tau \qquad (4)$$

Kovacs and Hutchinson (1979) demonstrated that eqn. 4 can be solved analytically for both heating and cooling (at a constant rate) under the assumption that there is a single structural relaxation time with a temperature dependence:

$$\tau = \tau_R \exp - \theta (T-T_R) \qquad (5)$$

where τ_R, θ, and T_R are constants, τ_R being the configurational relaxation time at $T = T_R$. With this temperature dependence, T_f' during cooling from above T_u is a function only of $-Q\tau' \equiv d\tau/dt$, which is called the Lillie number, $\mathcal{Li}$ (Cooper, 1983) At the glass transition temperature: $\mathcal{Li} = 2$. At $T = T_u$: $\mathcal{Li} \approx \delta$, and at $T = T_L$: $\mathcal{Li} \approx 5/\delta$. Numerical solutions (Mazurin and Cooper, 1985), using more realistic conditions do not significantly alter the value of $\mathcal{Li}$ at $T = T_g$, but as would be expected, a distribution of relaxation times increases $T_u - T_L$.

The freezing in transition during cooling is nearly centered about $d\tau/dt = 1$. When $d\tau/dt <<< 1$, deviation from equilibrium during cooling is imperceptible. When $d\tau/dt <<< 1$, relaxation is so slow that the configuration is imperceptibly different from that of the frozen state.

Figure 1a and 1b apply to a certain cooling rate. A faster rate causes freezing to occur at a higher temperature. Thus, at any temperature in the frozen state, the faster the glass has been cooled, the greater is the specific volume, and the higher is Tg, consistent with the recognition that the configuration in the frozen state depends on history.

EXPERIMENTAL RESULTS WITH TRADITIONAL GLASSES

Oxide, halide, nitrate, sulfate, polymeric, chalcogenide, and metallic glasses all show the behavior indicated schematically by Figs. 1a and 1b. As an example, Fig. 2 shows Moynihan et al's (1976) data on specific heat of B_2O_3 glass on reheating after cooling at a rate 8 times the heating rate. Tg of the glass before reheating is indicated on the plot. The close similarity of Fig. 2 and Fig. 1b is obvious. In fact, the authors' conversion of $C_p(T)$ to $T_f'(T)$ gives a result almost identical to Fig. 1b.

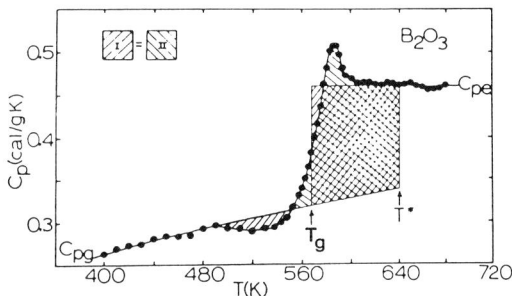

Fig. 2. Specific heat versus temperature for B_2O_3 glass. See text.

FREEZING IN OF CRYSTALLINE DEFECTS

Truly perfect crystals never exist. At sufficiently low temperatures neither do crystals with their defect population at equilibrium. Ionic conductivity in a crystal is proportional to point defect density. Hence, a glass-like transition resulting from the freezing in of a non equilibrium population of vacancies or interstitials can be observed by measurement of ionic conductivity as a function of temperature. Kirk and Pratt's (1967) results on Na Cl displaying such behavior are shown as Fig. 3. At the higher temperatures defects are maintained at equilibrium, while at lower temperatures the defect population is essentially frozen, causing the temperature dependence of the conductivity to be solely due to the mobility of the defect. Additional evidence for the quenching in of a non equilibrium vacancy population has been pointed out by Kinoshita et al. (1978).

SITE OCCUPANCY ORDERING

Crystalline binary alloys often exhibit at low temperature, short range ordering, a preference for unlike species as nearest neighbors. As temperature is increased, deviations from random site occupancy are diminished because of the greater influence of entropy. Such a binary alloy cooled from a temperature at which the atom species distribution is at equilibrium, orders at a decreasing rate as temperature is decreased, causing deviations from equilibrium and ultimately the freezing in

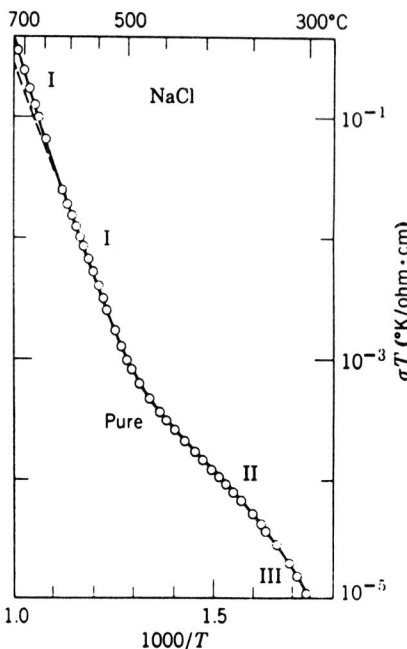

Fig. 3. Electrical conductivity of NaCl vs. reciprocal temperature. Note "glass transition" at about 450°C.

of an incompletely ordered non equilibrium state.

The site occupancy freezing in transition phenomenology is most clearly demonstrated, then, by the work of Kinoshita et al. (1978). By careful measurements of specific heat during reheating of a $Cu_{.155}Al_{.845}$ alloy that was furnace cooled, they found the behavior shown in Fig. 4. Its similarity to the behavior obtained by reheating of a glass, is seen by comparing figures 1a, 2, and 4. While the authors did not mention the analogy of their results to glass transition behavior, they did develop a comprehensive model based on the decreasing density and mobility of lattice defects during cooling. The model's success is indicated by the close agreement between its prediction (solid line) and the experimental points on Fig. 4. The authors measured electrical resistivity during reheating, which also revealed similar behavior. Further, Eguchi et al. showed (1978) that thermal history affected room temperature resistivity in this system, and pointed out that Lang and Schuele (1970) had previously found a similar dependence of the resistivity of CuZn on thermal history. Such dependence is the characteristic of the frozen state.

To compare Kinoshita et al's results with the prediction that $\mathcal{L}\iota(Tg) \approx 2$, we use the method of Moynihan et al. (1976) to determine from Fig. 4, that after cooling in the furnace, the alloy had a "glass temperature," $\overset{*}{T}g$, of ~ 250°C, 523K, and note that in the vicinity of the transition $-Q = 4 \times 10^{-2}$ K/s. (Actually, Newtonian cooling occurred, but in the temperature range of interest the deviation from a constant cooling rate was slight.)

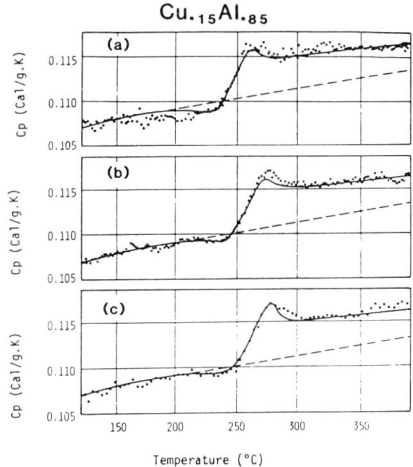

Fig. 4. Specific heat Cp(T) during reheating of $Cu_{.155}Al_{.845}$ at various heating rates.
Upper: 0.013K/s. Middle: 0.033K/s. Lower: 0.067K/s.

Kinoshita et al. cite Li and Nowick's (1956) expression for τ_z, the Zener relaxation time, and Welch's (1969) relation, $\tau = \tau_z/3$ for τ, the relaxation time for short range ordering. Together they give:

$$\tau = (0.4 \times 10^{-15}) \exp(21,000 \pm 900/T).$$

Combining these values yields: $0.3 < \mathcal{L}\tau(\tilde{T}_g) < 2$, which is considered to be satisfactory agreement with the prediction noted above for a glass transition.

Spinel ferrites also exhibit site occupancy ordering. Mason and Sujata (private communication) measured electrical resistivity versus temperature upon reheating of a rapidly quenched magnesium ferrite. The results are shown in Fig. 5. They interpret the near discontinuity at about 250°C as resulting from the relaxation of a frozen state toward an equilibrium state. The step in resistivity which is partially masked by the strong temperature dependence of the equilibrium resistivity, is deemed to be analogous to the step in volume on reheating shown in Fig. 1. Evidence suggesting that freezing in transitions are common in spinel ferrites is the isothermal decay of resistivity with time, found by the same workers on a different ferrite, and shown in Fig. 6. The behavior follows that predicted by eqn. 1 for a single relaxation time.

Perovskite ferroelectrics, e.g. $Pb(Sc_{1/2}Ta_{1/2})O_3$ display disorder of the B site cations Sc and Ta at high temperatures ~1400°C, and ordering at 1000°C., according to Setter and Cross (1980). Fig. 7 and Fig. 8 show their results on the effect of B site disorder on the temperature dependence of the spontaneous polarization and the dielectric constant of this ferroelectric, displaying the marked effect of quenched in disorder (and hence thermal history) on dielectric properties.

Depression of glass thermometers refers to the isothermal change with time of the bore of a thermometer causing a change in the reading at the ice point. As pointed out by Charles (1971), this phenomena has been recognized for more than a century. Its distinguishing features are:

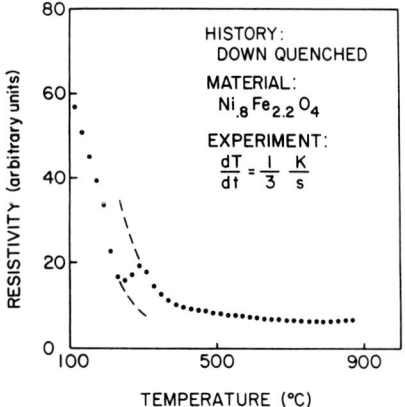

Fig. 5. Resistivity versus temperature during reheating a nickel ferrite.

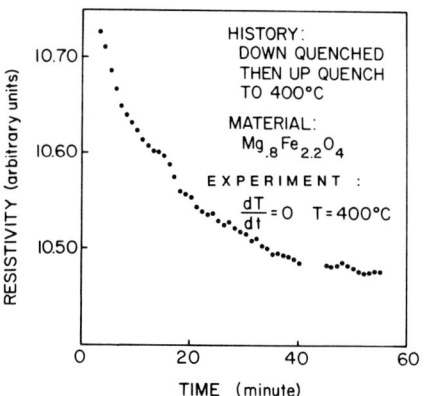

Fig. 6. Isothermal change of resistivity with time of a magnesium ferrite.

1. "Secular Rise": after a thermometer is manufactured and calibrated at the ice point, the mercury rises slowly with time.

2. "Depression": if a thermometer is heated to above 100°C and allowed to cool, the mercury will first equilibrate below the calibrated ice point, and then slowly rise. Equilibrium is obtained only if a very long time elapses between temperature cycles.

3. If a thermometer is raised to temperatures above 300°C, then cooled and placed in ice water, the mercury level will stand above the original ice point. The amount of elevation is determined by the time and temperature of the $T > 300°C$ soak. Such treatment generally reduces the extent of the ice point depression and the secular rise.

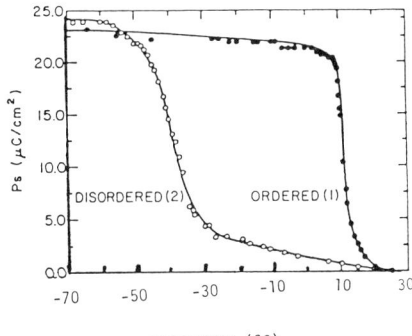

Fig. 7. Spontaneous polarization vs. temperature of $Pb(Sc_{1/2}Ta_{1/2})O_3$ showing effect of B site order.

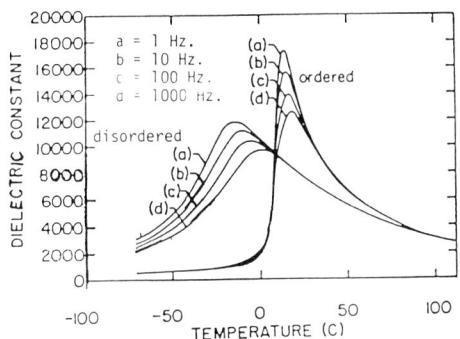

Fig. 8. Dielectric constant vs. temperature of $Pb(Sc_{1/2}Ta_{1/2})O_3$ showing effect of B site order. A = 1Hz, b = 10Hz, c = 100Hz, d = 1kHz.

4. Large depression constants are obtained only if the glass contains more than one alkali metal oxide.

Charles was able to rationalize this behavior by assuming that residual and thermal stresses caused interdiffusion of alkali ions which were the source of the volume change.

Another explanation follows from the recognition near neighbor alkali ordering can occur in glasses as well as in crystals and that such ordering may affect the molecular volume of the glass enough to be perceptible by this high sensitivity test.

At high temperature, alkali neighbors are essentially randomly distributed. As temperature is lowered, there is presumably a favoritism for unlike alkali near neighbors, resulting in a denser structure. The secular rise is explained by the isothermal densification of the glass at ~ 0°C as its alkali ordering equilibrates. Heating to 100°C decreases the site occupancy order and results in a less dense glass on cooling. Hence, the so-called "depression." Heating to 300°C allows slow densification due to atomic configuration changes associated with the primary glass transition, and hence an increase in the density at the ice point. The decreased sensitivity may result from lower alkali mobility caused by this densification. The necessity for more than one alkali ion type is obviously required for site occupancy ordering.

SPIN GLASSES AND MULTI POLAR GLASSES

Binder and Young (1986) have recently reviewed the theoretical and experimental development of spin glasses. They suggest that such materials may have their spins frozen into a non equilibrium configuration near to 0K. Similarly, pseudo spin glasses may have dipole motion frozen into a non equilibrium configuration (Samara, 1985).

SUMMARY AND CONCLUSIONS

There are a broad range of materials whose properties are affected by freezing in phenomena. Measurements of properties during cooling, heating, and isothermal soaking, which have proven invaluable in clarifying the phenomenology of the traditional glass transition, would seem to offer the same benefits to the systematic study of the effect of thermal history on the properties of ferrites, ferroelectrics, as well as high T_c ceramic superconductors. "History" should be easy to control or alter. The potential for increased uniformity as well as intrinsic enhancement of properties should justify the effort.

ACKNOWLEDGEMENT

The assistance of J. Cerny and A. Sayir is acknowledged with thanks.

REFERENCES

Binder, K., and A.P. Young. (1986). Spin glasses: experimental facts, theoretical concepts, and open questions. Revs. Mod. Phys., 58, 801-976.

Charles, R.J. (1971). The origin of depression of glass thermometers. Glass Tech., 12, 24-26.

Cooper, A.R., and P.K. Gupta. (1982). A dimensionless parameter to characterise the glass transition. Phys. Chem. Glasses, 23, 44-45.

Cooper, A.R. (1983). A proposal to name a dimensionless number pertinent to the glass transition for Howard R. Lillie. Glastechn. Berichte, 56, 1160-1164.

Eguchi, T., C. Kinoshita, and Y. Tomokiyo. (1978). Kinetics of short range ordering and behavior of vacancies in binary substitutional alloys. Trans. JIM, 19, 198-202.

Kinoshita, C., Y. Tomokiyo, and T. Eguchi. (1978). Kinetic behavior of vacancies and atoms in binary alloys. Phil. Mag., B38, 221-240.

Kirk, D.L., and P.L. Pratt. (1967). Ionic conductivity in pure-doped crystals and magnesium sodium chloride. Proceedings of Brit. Ceram. Soc., 9, 215-232.

Kovacs, A.J., and J.M. Hutchinson. (1979). Isobaric thermal behavior of glasses during uniform cooling and heating: dependence of the characteristic temperatures on the relative contributions of temperature and structure to the rate of recovery. II. A one-parameter model approach. J. Polym. Sci. Polym. Phys., 17, 2031-2058.

Lang, E., and W. Schuele. (1970). Z. Metallk, 61, 866.

Li, C.Y., and A.S. Nowick. (1956). Atomic mobility in a Cu-Al alloy after quenching and neutron irradiation. Phys. Rev., 103, 294-303.

Mason, T. (1987). Northwestern University, Chicago, IL, USA (priv. comm.)

Mazurin, O.V., and A.R. Cooper. (1985). An analysis of the suitability of the Lillie number to characterize the glass transition of real glass-forming substances. J. of Non-Crystalline Solids, 72, 65-71.

Moynihan, C.T., A.J. Easteal, M.A. DeBolt, and J. Tucker. (1976). Dependence of the fictive temperature of glass on cooling rate. J. Am. Ceram. Soc., 59, 12-16.

Prigogine, I., and R. Defay. (1954). Chemical Thermodynamics, Longman's, London, p. 41.

Rekhson, S.M., and O.V. Mazurin. (1974). Stress and structural relaxation in Na_2O-CaO-SiO_2. J. Am. Ceram. Soc., 57, 327-328.

Samara, G.A. (1985). Nature of the phase transition in $KTaO_3$ with random site impurities. Jap. J. of App. Phys., 24, 80-84.

Setter, N., and L.E. Cross. (1980). The role of B-site cation disorder in diffuse phase transition behavior or perovskite ferroelectrics. J. Appl. Phys., 51, 4356-4360.

Tool, A.Q. (1946). Relation between inelastic deformability and thermal expansion of glass in its annealing range. J. Am. Ceram. Soc., 29, 240-253.

Welch, D.O. (1969). Kinetics of short-range order and zener relaxation in substitutional solid solutions. Mater. Sci. Engng., 4, 9-21.

Recent Advances in the Synthesis of Amorphous Metallic Materials*

RICARDO B. SCHWARZ

Center for Materials Science, Los Alamos National Laboratory, Mail Stop K-765, Los Alamos, NM 87544, USA

ABSTRACT

Amorphous metallic alloys have been traditionally prepared by techniques based on the rapid solidification of molten alloys. New methods of synthesis have been discovered recently that are based on isothermal solid-state reactions and transformations. It has further been shown that the products of these reactions are predicted by free-energy diagrams that treat the amorphous alloy as an undercooled liquid. These discoveries have enhanced our understanding of metastable alloys, both crystalline and amorphous, and has opened new windows for the synthesis of novel materials.

KEYWORDS

Amorphous alloys; glassy alloys; solid state reactions; mechanical alloying; mechanical grinding; glass forming range

INTRODUCTION

Ionic and covalent glasses often form when the corresponding melt fails to crystallize during relatively slow cooling. In these, the directional nature of the inter-atomic bonds limits the rate at which the atoms or molecules can rearrange to maintain thermodynamic equilibrium during cooling, and thus the melt solidifies into a glass even at cooling rates as low as 10^{-2} K s^{-1}. Metallic melts, in contrast, have non-directional bonding, and thus exhibit far less resistance to crystallize and do not form glasses unless cooled at much higher rates.

New methods of synthesis of amorphous metallic alloys have been developed over the last 10 years which are based on isothermal solid state reactions. For example, Malik and Wallace (1977) and Yeh et al. (1983) reported that the reaction of certain crystalline intermetallics with hydrogen leads to the formation on an amorphous hydride. In 1983, Schwarz and Johnson demonstrated that amorphous La-Au alloy films can be formed by a low-temperature interdiffusion reaction between crystalline thin films of pure lanthanum and gold. White (1979) and Koch and co-workers (1983) showed that amorphous alloy powders can be prepared by mechanically grinding a mixture of powders of two pure metals. In the three examples, the

---

(*) Work supported by the U. S. Department of Energy, Basic Energy Sciences.

amorphous metallic phase forms as the result of a chemical reaction that involves mass transport by diffusion. Amorphous alloys can be also prepared by polymorphous isothermal solid state reactions. For example, Yermakov, Yurchikov, and Barinov (1981, 1982), and Schwarz and Koch (1983), reported the formation of amorphous alloy powders by simply mechanically grinding powders of crystalline intermetallics of the same composition. In 1987, Von Allmen and Blatter reported the formation of amorphous alloys by massive transformations from a metastable crystalline phase to the amorphous phase.

In general, we can classify the methods of synthesis of amorphous alloys in terms of the process involved as follows: (a) rapid solidification from the melt or vapor phase, (b) solid state reactions between (crystalline) pure elements, (c) atomic disordering of crystalline lattices, (d) polymorphic transformations from metastable crystalline states to the amorphous state, and (e) deposition from electrolytes. In the following sections we summarize the various methods of synthesis of amorphous alloys. We describe in grater detail the synthesis of amorphous thin films by interdiffusion reactions and the synthesis of amorphous metallic powders by mechanical alloying, which we have been studying. Method (e), which has been sometimes classified as a variant of the rapid-solidification technique, will not be discussed here.

AMORPHOUS ALLOYS PREPARED BY THE RAPID SOLIDIFICATION OF MELTS

The preparation of amorphous alloys by the rapid solidification (RS) of melts requires bypassing the formation of crystals while cooling the melt from the melting temperature T_m to the glass transition temperature T_g. Crystallization is a nucleation and growth process. If special precautions are not taken to remove heterogeneous nucleants (oxide inclusion, etc) from metallic melts, these crystallize at small values of undercooling. However, researchers have found that in the absence of heterogeneous nucleants, most pure metal melts can be largely undercooled without inducing crystallization (Turnbull and Cech, 1950; Turnbull, 1952; Perepezko and Paik, 1982). Turnbull and co-workers (1952, 1961) measured the isothermal kinetics of nucleation in undercooled droplets of Hg and showed that the nucleation was stoichastic in time and space and could be described by simple *classical* nucleation theory. In this theory the transformation is driven by the difference in free energy between the liquid and crystal phases and is opposed by the reversible barrier to form the crystal-melt interface. The observation that the barrier to homogeneous nucleation is high suggested researchers that crystal nucleation requires significant reconstruction of the short-range order of the liquid. A mechanism for such reconstruction was proposed by Frank (1952).

Although early experiments showed that pure metal melts of small dimensions could be easily undercooled to $T/T_m = 0.8$ without inducing crystallization (Turnbull, 1950), researchers found it difficult to trap these melts in the glassy state by further cooling to below T_g. Buckel and Hilsch (1954) demonstrated that amorphous thin films of pure metals could be formed by condensing the metal vapor onto a cryogenically cooled substrate, a process that gives an equivalent cooling rate of the order of 10^{12} K s^{-1}. However, it is likely that the amorphous structure in these films was stabilized by impurities trapped in the film during deposition.

The field of amorphous alloys grew at a fast rate following the experiments of Pol Duwez and co-workers (1960) who developed a variety of techniques to rapidly solidify alloy melts at cooling rates of 10^6 to 10^8 K s^{-1}. In the course of their effort to extend the solubility limit of binary solid solutions, the authors discovered that certain metallic alloys could be melt-quenched to amorphous solids. It soon became apparent that the glass-forming range (GFR) of metallic alloys is strongly composition dependent, being easier to form metallic glasses near the compositions of *deep* eutectics in the phase diagram. At these compositions the temperature interval between the liquidus, T_ℓ, and the glass transition temperature, T_g, is reduced so that the probability of being able to cool the melt through the interval without inducing crystallization is enhanced. Thus, there is a tendency for the GFR to vary inversely with the reduced glass-transition temperature, $T_{rg} = T_g/T_\ell$. For alloys with low T_{rg}, the required cooling rates to avoid crystallization are large and thus one dimension of the amorphous product must be small (50 µm or less for cooling rates of the order of 10^6 K s^{-1}). For alloys with large T_{rg}, such as $Pd_{40}Ni_{40}P_{20}$ ($T_{rg} = 0.67$), the cooling rate requirement is relaxed allowing the preparation of cm-size amorphous samples at cooling rates of few K s^{-1} (Drehman, Greer, and Turnbull, 1982).

The prediction of the GFR of metallic alloy systems has been the subject of numerous studies over the last 20 years (Davies, 1983). Experience has shown that under conditions of fast cooling, polymorphic crystallization is kinetically favored over crystallization involving solute partitioning. Thus, if partitionless crystallization is avoided while cooling the melt through the regime $T_\ell - T_g$, the molten alloy will be trapped in the glassy state. For binary alloys the thermodynamic transformation temperature for partitionless crystallization can be described by a T_o curve which is the temperature-composition locus at which the driving force for the transformation is zero. The dash-dot curves in Fig. 1 are the T_o curves for the terminal solid solutions in the nickel-titanium system. Below T_o the thermodynamic force for crystallization increases continuously. Recently Nash and Schwarz (1988) used *classic* nucleation theory to calculate the volume fraction of crystallized material formed by partitionless crystallization during the continuous cooling of Ni-Ti melts as a function of cooling rate and composition. The thermodynamic force for the transformation was deduced from a thermodynamic model for the Ni-Ti system (Murray, 1988). The liquid-crystal interfacial energy was assumed proportional to the enthalpy of fusion (Turnbull, 1950). The authors integrated the transformation kinetics equation to calculate continuous-cooling-transformation (CCT) curves as a function of alloy composition, x. They then used these results to calculate the temperature $T'(x,\varsigma,dT/dt)$ for the formation of a volume fraction ς of crystalline material during cooling at the constant rate dT/dt. Figure 1 shows T' curves for $\varsigma = 10^{-6}$ (this value of ς is used as a criterium for defining the retention of the glassy structure) and $dT/dt = 10^6$ K s^{-1}. The solid curves in Fig. 1 are the $T'(x)$ curves for the terminal solutions. Because the thermodynamic model describes the crystalline intermetallic compounds Ni_3Ti, $NiTi$, and $NiTi_2$ as composition invariant, the T' values for partitionless crystallization into the crystalline compounds were only calculated at the stoichiometric compositions and are shown as solid triangles. The dashed curves are estimated T' curves for the compounds. In this model the GFR for the Ni-Ti system is determined by the condition $T'(x) < T_g$. The solid circles in Fig. 1 show the crystallization temperatures of amorphous Ni-Ti alloys prepared by the RS of melts. In agreement with observation, the model predicts the difficulty in preparing glasses at compositions near Ni_3Ti and $NiTi$. It also predicts (marginally) a gap in the GFR at the composition of $NiTi_2$ which is not observed.

AMORPHOUS ALLOYS PREPARED BY SOLID-STATE REACTIONS

Hauser (1981) and later Herd et al. (1983) reported that metals can diffuse at low temperatures into amorphous semiconductors, such as tellurium, selenium, and silicon, without causing the amorphous semiconductor to crystallize. The first example of two pure crystalline metals reacting to form single-phase amorphous alloy was reported by Schwarz and Johnson (1983). In this experiment, thin films of pure gold and lanthanum, a few tenths of a nm in thickness, were fully reacted at 70°C within a few hours. At higher reaction temperatures (T > 125°C), intermetallic compounds of gold and lanthanum were formed.

The solid state amorphization reaction (SSAR) method requires clean (oxide free) interfaces between two reacting metals. Such interfaces can be obtained by depositing crystalline thin films of metals A and B in a high vacuum or by rolling an alternating stack of metallic foils (Atzmon et al., 1985; Schultz, 1987). The chosen reaction temperature for the SSAR must be a few tens to a few hundred Kelvin lower than T_x. Three requirements have been proposed for the SSAR (Schwarz and Johnson, 1983; Schwarz, 1986): (1) metals A and B must have a large negative heat of mixing in the amorphous phase, (2) metals A and B must have vastly different diffusivities in each other and in the amorphous alloy to be formed, and (3) the binary system AB should have no stable crystalline phase in which the sub-lattice of the larger atom can be derived by a diffusionless transformation from the crystalline lattice of the larger atom. The first condition ensures that a thermodynamic driving force for the reaction exists. The negative heat of mixing also favors the reaction kinetics by increasing the chemical diffusivities (Darken, 1948). The second condition ensures that the SSAR will be kinetically favored over the formation of crystalline intermetallic compounds. This favoring occurs because one species diffusing in the other and in the amorphous alloy is sufficient for the SSAR; generally, the formation of intermetallics, which have crystalline structures quite different from those of the two starting metals, requires the atomic motion of both species. Thus, a *temperature window* opens for the SSAR by choosing a pair of elements with vastly different diffusivities from each other while in the amorphous phase. The third condition ensures that the binary system does not have a crystalline intermetallic that can be formed when the smaller atom diffuses through the crystalline lattice of the larger one, i.e., by the same mechanism responsible for the SSAR. If this mechanism was possible, the interdiffusion reaction could lead to forming stable crystalline intermetallics that have a free energy lower than that of the metastable amorphous alloy.

The SSAR has been observed in a large number of thin film couples. Some are: Au-La, Au-Ti, Au-Y, Au-Zr, Co-Sn, Co-Zr, Cr-Ti, Fe-Zr, Ni-Ti, and Ni-Zr. These systems clearly satisfy the first two requirements for the SSAR listed above - they all have a large negative heat of mixing (Niessen and co-workers, 1976) and the first element in each pair is an anomalously fast diffuser in the crystalline lattice of the second element (Le Claire, 1978). The requirement that the two elements have vastly different diffusivities in the amorphous alloy has only been tested for the Ni-Zr system (Cheng, Johnson, and Nicolet, 1985; Hahn, Averback and Rothman, 1986; Hahn, Averback, and Shyu, 1988), where the diffusivity of Ni in amorphous NiZr alloy is some three orders of magnitude larger than that of Zr. A violations of the third condition for SSARs has been reported in the Ni-Sn system (Tiainen and Schwarz, 1988).

Schwarz and Johnson (1983) discussed the thermodynamic aspects of the SSAR in terms of a metastable free energy diagram for the La-Au system and showed that the diagram predicts the reaction products and their homogeneity ranges. Figure 2a shows a schematic phase diagram for a binary system A-B that has the main characteristics necessary for a SSAR. Phases α and β are crystalline primary solid solutions and phase γ is a crystalline intermetallic. Figure 2b shows the free energies of phases α, β, and γ and of the amorphous phase, λ, evaluated at the reaction temperature T_r. If the interdiffusion reaction at the A/B interfaces in Fig. 1 reached a state of thermodynamic equilibrium, the reaction products would be determined by the common tangents between α, β, and γ in Fig. 2b (these tangents are not shown). However, if the reaction temperature is low, insufficient thermal activation exists to promote the nucleation and growth of the crystalline phase γ. In the absence of phase γ, the SSAR products are determined by the common tangents between phases α, β, and λ (shown in Fig. 2b). These tangents predict five reaction products: a crystalline solid solution, α, for $0 < x < x_1$; a two-phase product of $\alpha(x_1)$ and $\lambda(x_2)$; a single-phase amorphous alloy, λ, for $x_2 < x < x_3$; a two-phase product of $\lambda(x_3)$ and $\beta(x_4)$; and the crystalline solid solution, β, for

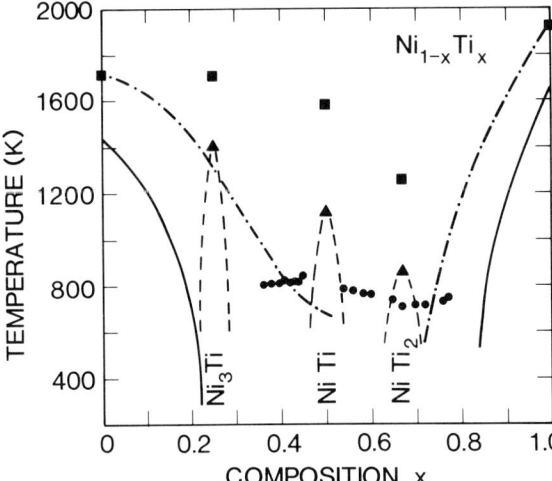

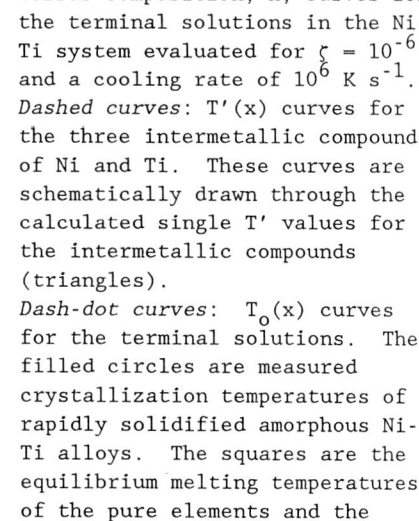

Fig. 1. *Solid curves:* T' versus composition, x, curves for the terminal solutions in the Ni-Ti system evaluated for $\zeta = 10^{-6}$ and a cooling rate of 10^6 K s^{-1}. *Dashed curves:* T'(x) curves for the three intermetallic compounds of Ni and Ti. These curves are schematically drawn through the calculated single T' values for the intermetallic compounds (triangles).
Dash-dot curves: $T_o(x)$ curves for the terminal solutions. The filled circles are measured crystallization temperatures of rapidly solidified amorphous Ni-Ti alloys. The squares are the equilibrium melting temperatures of the pure elements and the compounds.

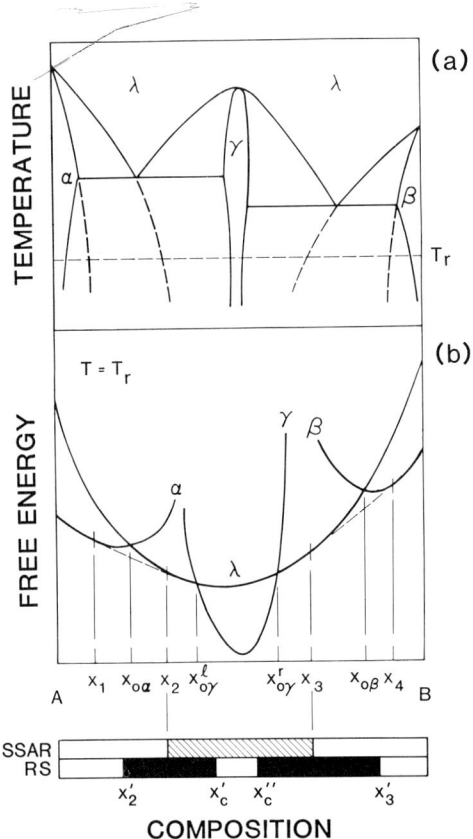

Fig. 2. Schematic phase diagram (a) for a binary system, AB, with a negative heat of mixing in the liquid state, and corresponding free-energy diagram at the temperature T_r (b). The bars at the bottom of the figure give the GFR predicted for the SSAR and the RS synthesis methods.

$x_4 < x < 1$. We note that the GFR for alloys prepared by SSARs is wide and continuous and is located near the center of the composition range. For the La-Au system and $T_r = 340$ K, $x_2 \approx 0.2$ and $x_3 \approx 0.65$ (Schwarz and Johnson, 1983).

Figure 2b allows us to compare the GFR for the SSAR and the RS methods. The GFR for theSSAR method equals the regime (x_2, x_3) defined by the common tangents in the free energy diagram in Fig. 2b. The GFR for the RS method consists of two regimes, (x_2', x_c') and (x_c'', x_3'), shown at the bottom of Fig. 2b. We have estimated these regimes qualitatively based on the results shown in Fig. 1. These calculations predict that the values of x_2', x_c', x_c'', and x_3' are determined by the intersections of the T' curves for the three crystalline phases, α, β, and γ, and the T_g-versus composition curve. The $T'(x)$ and $T_g(x)$ curves are not shown in Fig. 2a. However, T_g usually has a weak composition dependence and is only 50 to 100 Kelvin higher than the temperature T_r chosen for the SSAR. Furthermore, the approximate location of the T' points in Fig. 2b can be inferred from the location of the four T_o-points, $x_{o\alpha}$, $x_{o\gamma}^\ell$, $x_{o\gamma}^r$, and $x_{o\beta}$, shown in Fig. 2b. Because crystallization must overcome a nucleation barrier, and because $T_r < T_g$, for a finite cooling rate we must have $x_2' < x_{o\alpha}$, $x_c' > x_{o\gamma}^\ell$, $x_c'' < x_{o\gamma}^r$, and $x_3' > x_{o\beta}$. These considerations allow us to estimate the location of x_2', x_c', x_c'', and x_3'. We notice that the compositions x_2' and x_3' lie *outside* the corresponding T_o-compositions, $x_{o\alpha}$ and $x_{o\beta}$, for the terminal solid solutions at $T = T_r$, whereas the extreme compositions of the amorphous phase formed by the SSAR, x_2 and x_3, lie *inside* these T_o-points.

Although we have discussed the thermodynamic aspects of the SSAR in terms of a schematic free-energy diagram, the predicted differences between the GFRs for the SSAR and the RS synthesis methods, summarized by the bars at the bottom of the Fig. 2b, should be quite general. Most of these predictions are indeed observed. For example, Hellstern et al. (1988) compared the GFRs for the SSAR[*] and the RS techniques in alloys of zirconium with nickel, cobalt, and iron. For the three alloy systems studied they observed, in the notation of Fig. 2b, that $x_2' < x_2$ and $x_3' > x_3$. In addition, the homogeneity range of the amorphous phase obtained by MA is continuous whereas that for the RS method is fragmented into zirconium-rich and zirconium-poor regimes. Figure 2 further predicts that with increasing cooling rates the GFRs for the RS method, (x_2', x_c') and (x_c'', x_3'), should become wider and the gap (x_c', x_c'') may eventually close. These effects have been observed by Lin and Spaepen (1986), who compared the GFR in nickel-niobium alloys prepared by slat quenching, R.F. Sputtering, and quenching of surface melt formed by picosecond-duration laser pulses. Whilst the GFR for the splat quenched samples has gaps near the compositions of NiNb and Ni_3Nb, the GFR for the laser-quench experiments, which have an estimated cooling rate of 10^{12} K s^{-1}, is continuous from approximately $Ni_{20}Nb_{80}$ to $Ni_{85}Nb_{15}$. Finally, in contrast to the results for the RS method, the GFR for the SSAR method, (x_2, x_3), is independent of kinetics (provided the reaction is carried on to completion) and has a weak dependence on T_r (Zöltzer and Bormann, 1987).

AMORPHOUS ALLOY POWDERS PREPARED BY MECHANICAL ALLOYING

Amorphous metallic alloys in powder form can be produced by mechanical alloying (MA) (White 1979; Koch and co-workers, 1983; Schwarz, Petrich and Saw, 1985;

---

[*] Helstern and co workers (1988) prepared the Ni-based amorphous alloys by mechanical alloying. We discuss in a later section that mechanical alloying involves a SSAR and thus the GFR for this technique is described by a free-energy diagram evaluated at the ball milling temperature.

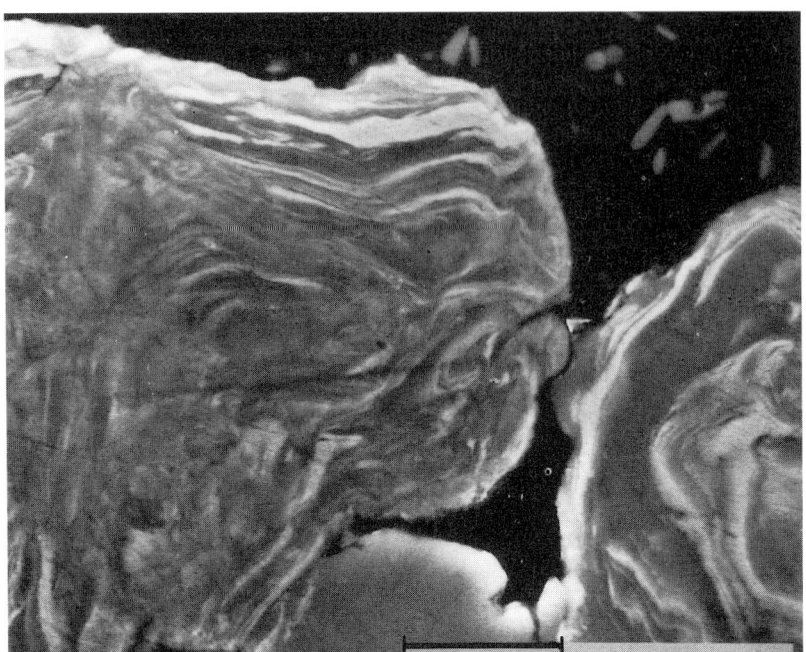

Fig. 3. Scanning electron micrograph of the cross section of a powder particle obtained after mechanically alloying a mixture of Ni and Ti powders for 0.5 h. The black bar corresponds to 10 μm.

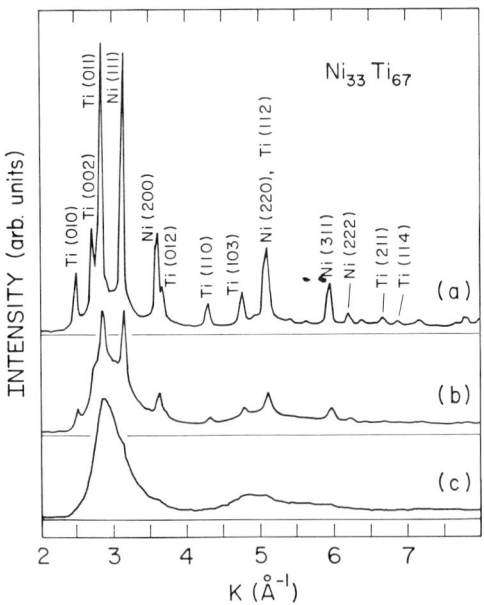

Fig. 4. X-ray diffraction intensity as a function of wavenumber K for a mixture of Ni and Ti powders mechanically alloyed for 2, 5, and 25 h.

Hellstern and Schultz, 1986). MA is a high-energy ball milling process that was developed for synthesizing dispersion-strengthened alloy powders (Benjamin and Volin, 1974; Benjamin, 1976). For the synthesis of amorphous alloy powders by MA, a mixture of elemental crystalline powders are placed into a sealed, hardened-steel or tungsten carbide container together with balls of the same material. The strong agitation of the balls in the container repeatedly deforms, breaks, and cold welds the powder particles trapped between colliding balls.

To date, the majority of the amorphous powders that have been prepared by MA are binary alloys of the early-transition metals titanium and zirconium and various 3-d late-transition metals: Ti-(Mn, Fe, Co, Ni, Cu) and Zr-(Mn, Fe, Co, Ni, Cu). Other systems in which amorphous alloy powder has been prepared by MA are Ti-Pd, Nb-Ni, Hf-Al, Fe-Al, Cu-Nb-Sn, Sn-Ni and Ti-Pd-Cu. Although the synthesis of amorphous metal-metal alloy powders by MA is quite straight forward, the synthesis of amorphous metal-metalloid alloy powders, such as $Fe_{80}B_{20}$, have been found more difficult. For the metal-metal systems, the powder morphology changes characteristically with increasing MA time. Figure 3 shows the internal structure of a particle obtained after ball-milling a mixture of nickel and titanium powders for 0.5 h. The particle consist of alternating layers of nickel and titanium. With increasing MA time, the average thickness of the layers decreases (Benjamin, 1976). We believe that a SSAR reaction takes place at the clean Ni-Ti interfaces created by the mechanical deformation of the particles. We discuss next the evidence supporting this hypothesis.

As for the reaction products of SSARs, the reaction products from MA powder mixtures of an early- and a late-transition metal have been found in good agreement with the predictions from free-energy diagrams evaluated for the average ball-milling temperature. (See, for example, Fig. 1 in Schwarz, Petrich and Saw, 1985). This agreement has been considered supporting evidence for the claim that the amorphization process during MA involves a solid-state reaction similar to that observed in the thin film experiments. (Schwarz, Petrich and Saw, 1985; Hellstern and Schultz, 1986). The chemical diffusion necessary for the SSAR at the metal-metal interfaces is most likely assisted by the excess point and lattice defects generated by the severe plastic deformation of the particles and by the momentary temperature increase in the powder particles trapped between colliding balls. Although there are no direct measurements of this increase, calculations (Schwarz, Petrich and Saw, 1985) suggest that the temperature increase is at most a few hundred Kelvin.

Figure 4 shows x-ray diffraction patterns for a MA mixture of nickel and titanium powders in molar ratio 1:2 taken after 2 h (a), 5 h (b), and 25 h of MA (c). (Schwarz, Petrich and Saw, 1985) With increasing MA time, the integrated intensity of the Bragg peaks for the starting pure metals decreases and is replaced by the broad diffraction bands characteristic of the amorphous alloy phase. Schwarz and Petrich (1988) used differential scanning calorimetry to study the reactions that take place during MA. From these measurements the authors conclude that during the first sages of MA the rate of amorphization is controlled by the SSAR at the Ni-Ti interfaces. For the later stages of MA, the remaining crystalline nickel and titanium phases become separated by the amorphous alloy already formed. The rate of amorphization becomes then controlled by the rate of attrition which must bring the yet-unalloyed nickel and titanium into contact. This slows down the rate of amorphization considerably.

All present measurements indicate that amorphous NiTi alloys prepared by MA and by the RS of melt have very similar structures. The thermal stability (crystallization temperature, enthalpy of crystallization, and apparent activation energy for crystallization) of amorphous $Ni_{33}Ti_{67}$ prepared by MA and by RS are the same (Schwarz, Petrich and Saw, 1985). Also, for these alloys, their total-reduced atomic pair distribution functions are the same. In addition, the crystallization temperatures of all the amorphous $Ni_{1-x}Ti_x$ alloys that have been prepared by the two techniques can be superimposed on the same set of T_x versus composition curves (Schwarz, 1988).

AMORPHOUS ALLOY PREPARED BY THE MECHANICAL DEFORMATION OF INTERMETALLICS

The formation of amorphous alloy powders by the mechanical deformation of powders of intermetallics was first reported by Yermakov, Yurchikov and Barinov (1981, 1982). These authors suggested that the amorphous alloy results from the RS of melt pools that are formed when the the powder particles are deformed heterogeneously during milling. Schwarz and Koch (1983) prepared amorphous alloy powders of $Ni_{32}Ti_{68}$ and $Ni_{45}Nb_{55}$ by ball milling a mixture of crystalline elemental powders (the MA technique discussed earlier) and by ball milling powders of the intermetallics $NiTi_2$ and $NiNb$, respectively. The authors showed that for each alloy system, the amorphous powders obtained by the two techniques were practically identical. In addition, the authors estimated the temperature increase in the particles trapped between colliding balls and concluded that this increase was too low to explain the amorphization phenomenon by a melting and RS sequence.

The observation that the same amorphous alloy powder can be synthesized by ball milling a mixture of elemental powders or by ball milling powders of an intermetallic compound is somewhat surprising. In the first case the synthesis involves a chemical interdiffusion reaction that is aided by the negative heat of mixing of the starting elements. In the second case, the mechanical attrition must raise the free energy of the intermetallic to a value equal or larger than that of the amorphous phase. This is thought to occur by the chemical disordering of the crystalline lattice and by the accumulation of point and lattice defects. In order to differentiate the two processes it is convenient to denote by MA the amorphization when starting from a mixture of elemental powders an to denote by MG the amorphization by the mechanical grinding of powders of intermetallics. Schwarz and Petrich (1988) used differential scanning calorimetry (DSC) to study the kinetics of both amorphization processes. The MA and MG were performed under identical conditions (ball mill, powder to balls mass ratio, attrition temperature, etc). At periodic time intervals, small aliquots of the MA and MG powders were removed from the ball mill and were analyzed by DSC. The DSC measurements show that as soon as the MG starts, there is a rapid increase in the total energy stored in the crystalline powder. However, this increase soon reaches a saturation while most of the powder is still crystalline. Extensive additional MG is necessary to make the powder fully amorphous. Based on these observations, the authors suggested that the rate of crystal-to-amorphous transformation is limited by the dynamic recovery of the lattice. The authors propose that MG causes the lattice of the intermetallic compound to dilate (the result of chemically disordering the lattice and introducing point and lattice defects) and that the dilation controls both the rate of dynamic recovery and the rate of amorphization. If this mechanism is valid, then amorphization by MG will be difficult to achieve in crystalline compounds where dynamic recovery sets in at a value of dilation lower than that necessary for the crystal-to-amorphous transformation. These observations should also apply to crystal-to-amorphous transformations induced by the irradiation with energetic particles (ions, neutrons, electrons).

Because many amorphous alloy powders can be produced by either MA or MG (Lee, Jang and Koch, 1988), it is logical to ask whether during the MA of a mixture of powder of metals A and B, the amorphous alloy $(AB)_{am}$ forms by the direct reaction $A + B \rightarrow (AB)_{am}$ or through the indirect reaction $A + B \rightarrow (AB)_{cr} \rightarrow (AB)_{am}$, which involves the intermediate crystalline phase $(AB)_{cr}$. The literature has examples of both reactions. For example, in the MA of nickel and titanium powders to form amorphous $Ni_{1-x}Ti_x$ with $(0.3 < x < 0.7)$, Schwarz, Petrich and Saw (1985) found no evidence of the formation of intermetallic compounds. However, Kim and Koch (1987) reported that during the MA of niobium and tin powders in molar ratio 3:1, the Nb_3Sn intermetallic (A15 structure) forms first which, with continued ball milling, transforms to the amorphous $Nb_{75}Sn_{25}$ alloy. An indirect amorphization reaction was also observed by Tiainen and Schwarz (1988) during the ball milling of pure nickel and tin powders, which we will discuss in more detail. In the composition range studied (0.2 to 0.4 molar percent tin), after only 2 h of MA the powder was a mixture of crystalline α-nickel and Ni_3Sn_2. With increasing MA time, one of the crystalline phases was progressively consumed with the concurrent formation of amorphous $Ni_{75}Sn_{25}$. Which of the two crystalline phases was consumed depended on the average alloy composition: for $Ni_{1-x}Sn_x$ with $x < 0.25$, the crystalline Ni_3Sn_2 disappeared with increasing MA as it was replaced by amorphous $Ni_{75}Sn_{25}$. For $x > 0.25$, α-Ni was consumed as the amorphous $Ni_{75}Sn_{25}$ formed. For an initial mixture of nickel and tin powders with average composition $Ni_{75}Sn_{25}$, both crystalline compounds were consumed simultaneously during MA leading to the formation of amorphous $Ni_{75}Sn_{25}$. The authors notice that the *sub-structure* of tin in Ni_3Sn_2 can be derived from a diffusionless transformation applied to the lattice of pure tin (see Fig. 8 in Tiainen and Schwarz, 1988), and thus the nickel-tin system violates the third condition for the synthesis of amorphous alloys by SSAR, discussed earlier. The authors propose that the easy formation of Ni_3Sn_2 during the early stages of MA results from the diffusion of nickel into pure tin, without the need for the long-range diffusion of tin, i.e., by the same mechanism that operates during SSARs. Thus, it would be interesting to investigate whether this mechanism is general and explains all cases where intermediate crystalline phases form during MA.

AMORPHOUS ALLOYS PREPARED BY MASSIVE TRANSFORMATIONS

von Allmen and Blatter (1987) reported the formation of cm-size amorphous $Cr_{0.40}Ti_{0.60}$ alloys by long-term anneals of a metastable bcc phase of the same composition at a temperature of 600°C. The starting bcc alloy was obtained by solidifying the molten alloy at moderate cooling rates (≈ 100 K s^{-1}). The same effect was reported in several Ti- and Nb-based alloys with Cr, Mn, Fe, Co, and Cu (Blatter et al., 1987). The transformation starts at grain boundaries and at the sample surface, suggesting a similarity to melting. The transformation requires that the free energy versus temperature curves of the liquid and the bcc alloy cross twice. The two curves certainly cross at the melting temperature, T_m. A second crossing is possible at lower temperatures if the free energy of the (undercooled) liquid decreases more rapidly than the free energy of the crystalline phase as the temperature is lowered from T_m. Greer (1988) has proposed that, in principle, the re-entrant melting behavior is reversible an occurs because the liquid has a large degree of chemical order. In Ti-Cr the second crossing apparently occurs above 600°C. Thus at 600 °C a thermodynamic driving force exists for the amorphization reaction. This transformation temperature is apparently too low to promote the chemical partitioning necessary for the nucleation and growth of the equilibrium crystalline phases, Ti and $TiCr_2$.

DISCUSSION

The number of papers dealing with amorphization by reactions in the solid state by now exceeds 100. It has become clear that these phenomena are rather ubiquitous (Johnson, 1986). The cases of amorphization induced by chemical interdiffusion in gas/solid or solid/solid couples, by MA, by MG, and by massive transformations are all examples of solid-state transformations from the crystalline to the amorphous state. The products of these transformations are simply *amorphous* alloys, whereas the amorphous phases formed by rapid solidification of a parent liquid phase are termed *glasses*. By and large, the properties (radial distribution functions, thermal stability, magnetic properties) of *amorphous* and *glassy* alloys of the same composition are the same suggesting that these alloys are structurally similar. Investigators often presume that this result is general and that amorphous alloys prepared by different techniques relax to a more or less common structure following thermal annealing at temperatures near the glass transition temperature. If this theory is true, then we can associate well-defined thermodynamic and physical properties with an amorphous phase of specified composition. This assumption is, in fact, necessary if we are to understand solid-state amorphizing transformations in terms of conventional thermodynamic potentials and driving forces.

REFERENCES

Atzmon, M., K. M. Unruh, and W. L. Johnson (1985). *J. Appl. Phys.* **58**, 3865-70.
Benjamin, J. S. (1976). *Sci. Am.* **234**, 40-8.
Benjamin, J. S. and T. E. Volin (1974). *Metall. Trans.* **5**, 1929-34.
Blater, A., N. Baltzer and M. von Allmen (1987). *J. Appl. Phys.* **62**, 276.
Buckel, W. and R. Hilsch (1954). *Z. Phys.* **138**, 109-20.
Cheng, Y. T., W. L. Johnson, and M-A. Nicolet (1985). *Appl. Phys. Lett.* **47**, 800-2.
Davies, H. A. (1983). In F. E. Luborsky (Ed.), *Amorphous Metallic Alloys*, Butterworths, London, pp. 8-25.
Darken, L. S. (1948). *Trans. Met. Soc. AIME*, **175**, 184-201.
Drehman, A. J., A. L. Greer, and D. Turnbull (1982). *Appl. Phys. Lett.* **41**, 716-7.
Duwez, P., R. H. Willens, and W. Clement Jr. (1960. *J. Appl. Phys.* **31**, 1136-7.
Frank, F. C. (1952). *Proc. Roy. Soc.* **215A**, 43.
Greer, A. L. (1988). The thermodynamics of inverse melting. In Schwarz, R. B. and W. L. Johnson (eds.), *Solid State Amorphizing Transformations*, J. Less-Common Met. (in press).
Hahn, H, R. S. Averback, and H. -M. Shyu (1988). Diffusion studies in amorphous NiZr alloys and their relevance for solid state amorphizing reactions. In Schwarz, R. B. and W. L. Johnson (eds.), *Solid State Amorphizing Transformations*, J. Less-Common Met. (in press).
Hahn, H., R. S. Averback, and S. J. Rothman (1986). *Phys. Rev. B.* **33**, 8825-8.
Hauser, J. J. (1981). *J. de Phys. Colloq.* **42**, C4-943.
Hellstern, E. and L. Schultz (1986). *Appl. Phys. Lett.* **49**, 1163-5.
Hellstern, E., L. Schultz, and J. Eckert (1988). Glass-forming ranges of mechanically alloyed powders. In Schwarz, R. B. and W. L. Johnson (eds.), *Solid State Amorphizing Transformations*, J. Less-Common Met. (in press).
Herd, S. R., K. N. Tu and K. Y. Ahn (1983). *Appl. Phys Lett.* **42**, 597; *This Solid Films* **104**, 197.
Johnson, W. L. (1986). *Prog. in Mater. Science* **30**, 81-134.
Kim, M. S. and C. C. Koch (1987). *J. Appl. Phys.* **62**, 3450-3.
Koch, C. C., O. B. Cavin, C. G. McKamey, and J. O. Scarbrough (1983). *Appl. Phys. Lett.*, **43**, 1017-1019.
Le Claire, A. D. (1978). *J. Nuclear Mater.* **69** & **70**, 70-96.
Lee, P. Y.,J. Jang, and C. C. Koch (1988). Amorphization by mechanical alloying: the role of mixtures of intermetallics. In Schwarz, R. B. and W. L. Johnson (eds.), *Solid State Amorphizing Transformations*, J. Less-Common Met. (in press)
Lin, C. -J. and F. Spaepen, (1986). *Acta Metall.* **34**, 1367-75.
Malik, S. K., and W. E. Wallace (1977). *Solid State Commun.* **24**, 283-285.
Murray, J. (1988). *Bull. Alloy Phase Diagrams*, in press.
Nash, P. and R. B. Schwarz (1988). Calculation of the glass forming range in binary metallic systems using thermodynamic models. *Acta Metall.*, in press.
Niessen A. K., F. R. de Boer, R. Boom, P. F. de Châtel, W. C. M. Mattens, and A. R Miedema (1983). *CALPHAD* **7**, 51-70.
Perepezko, J. H. and J. S. Paik, in B. H. Kear, B. C. Giessen and M. Cohen (Eds.), *Rapidly Solidified Amorphous and Crystalline Alloys*, North Holland, Amsterdam, pp. 49.

Schultz, L (1987). In M. Tenhover, L. E. Tanner, and W. L. Johnson (Eds.), *Science and Technology of Rapidly Quenched Alloys*, Mater. Res. Soc. Symp. Proc. 80, 97-104.

Schwarz, R. B. and W. L. Johnson (1983). *Phys. Rev. Lett.* 51, 415-418.

Schwarz, R. B., R. R. Petrich, ans C. K. Saw (1985). *J. Non-Cryst. Solids* 76, 281-302.

Schwarz, R. B. and C. C. Koch (1985). *Appl. Phys Lett.* 49, 146-8.

Schwarz, R. B. (1986). *Mater. Res. Bull.*, 11, 55-58.

Schwarz, R. B. (1988). Formation of amorphous alloys by solid state reactions. *J. Mater. Sci. and Eng.*, in press.

Schwarz, R. B. and R. R. Petrich (1988). Calorimetry study of the synthesis of amorphous Ni-Ti alloys by mechanical alloying. In Schwarz, R. B. and W. L. Johnson (eds.), *Solid State Amorphizing Transformations*, *J. Less-Common Met.* (in press).

Tiainen T. J. and R. B. Schwarz (1988). Synthesis and characterization of mechanically alloyed Ni-Sn powders. In Schwarz, R. B. and W. L. Johnson (eds.), *Solid State Amorphizing Transformations*, *J. Less-Common Met.* (in press).

Turnbull, D (1950). *J. Appl. Phys.* 21, 1022

Turnbull, D. and R. E. Cech (1950). *J. Appl. Phys.* 21, 804.

Turnbull D. and R. L. Cormia (1961). *J. Chem. Phys.* 34, 820.

Turnbull, D (1952). *J. Chem. Phys.* 20, 411.

Von Allmen, M. and A. Blatter (1987). *Appl. Phys. Lett.* 50, 1873-75.

White, R. L. (1979). Ph.D. Dissertation, Stanford University.

Yeh, X. L., K. Samwer, and W. L. Johnson (1983). *Appl. Phys. Lett.* 42, 242-44.

Yermakov, A. Ye., V. A. Barinov, and Ye. Ye. Yurchikov (1982). *Phys. Met. Metall.* 54, 935-41.

Yermakov, A. Ye., Ye. Ye. Yurchikov, and V. A. Barinov (1981). *Phys. Met. Metall.* 52, 50-8.

Zöltzer K. and R. Bormann (1988). Thermodynamics of stable and metastable phases in the NiTi system and its application to amorphous phase formation. In Schwarz, R. B. and W. L. Johnson (eds.), *Solid State Amorphizing Transformations*, *J. Less-Common Met.* (in press).

Structural Relaxation and Phase Trans-formations in Silicon/Transition Metal Multilayers

BY S. RADELAAR

*Centre for Submicron Technology, Department of Applied Physics,
Delft University of Technology, Lorentzweg 1, 2600 GA Delft*

ABSTRACT

In this paper recent work on stress measurements and phase transformations in amorphous molybdenum/silicon and titanium/silicon multilayers is summarized. Stresses in as-deposited material depend on the period of the multilayer structure. The dependence of the stress on period can be explained by the existence of a thin intermixed layer between transition metal and silicon. The multilayers remain amorphous during annealing at low temperatures. Crystallization takes place after homogenization of the multilayer composite. The homogeneous amorphous silicide crystallizes in a metastable structure which transforms to the stable equilibrium structure during high-temperature annealing. The mechanical stress in crystallized layers is mainly due to the difference in thermal expansion between silicide and substrate.

INTRODUCTION

Materials problems abound in the microelectronic industry. The need for high reliability and the ever-shrinking feature size impose heavy constraints on materials utilization. In the early days of integrated circuit technology the solution of materials problems could be left to process developers and electrical engineers; now the problems have become so complex that a solid background in materials science is required. Since many processing steps take place at elevated temperatures diffusion problems abound; it is fortunate that the work by Jack Kirkaldy and his co-workers, summarized in the recent book by Kirkaldy and Young, has provided the materials scientists with a solid base from which these problems can be attacked.

With the growing complexity of integrated circuits an ever larger fraction of the chip area is used for interconnect. Also the resistance of the interconnections becomes a limiting factor in the speed of integrated circuits. For this reason a very large effort has been made to find a suitable replacement for doped polycrystalline silicon, a material which was widely used up to a few years ago. An increasing number of IC-manufacturers nowadays use transition metal silicides since the disilicides of these elements have a reasonably low resistivity and are compatible with integrated circuit fabrication technology.

No attempt is made to review the existing literature on silicides, since several good reviews exist already, see e.g. the reviews by d'Heurle and by Nicolet and Lau. Instead I will address a few problems that still hamper the application of silicides in practical circuits, despite the enormous amount of work that has been done so far. These problems are, among others, the occurrence of large mechanical stresses (on the order of 1GPa) and the adherence to the substrate. It will be shown that the stresses in well-annealed silicides can be largely accounted for by the difference in thermal expansion between silicide and substrate. Another problem concerns the control of the argon content of sputtered silicide layers. I will also briefly discuss the phase transformations occurring during annealing of crystallized silicide layers.

The work described in this paper is the result of a cooperation between the Centre for Submicron Technology and the Department of Materials Science, both of the Delft University of Technology, and Philips Research Laboratories, Eindhoven, The Netherlands.

EXPERIMENTAL TECHNIQUES

A variety of techniques is used for the deposition of silicides and the subsequent processing. The simplest possible way would be to deposit a transition metal on top of an amorphous or polycrystalline silicon layer and anneal the combination. Unfortunately this process can result in very large stresses, which are predominantly due to the large reduction in volume which accompanies the formation of the silicide from the pure elements.

To minimize the stresses during processing one tries to premix the components of the layer. In the case of sputter deposition one can either sputter from a compound target, or sputter alternately from a silicon and a metal target by moving the substrate under the two targets. Sputtering from a compound target has certain drawbacks, e.g. no independent control of the composition of the deposited layer is possible, and often results in a relatively high impurity content. A disadvantage of sputtering of multilayers is the possible incorporation of the sputtering gas, argon in our case, in the deposited silicon layers. During the annealing step this incorporated argon agglomerates in bubbles which have a detrimental effect on the integrity and the adhesion of the silicide layer. We prefer sputtering from different targets, the period is varied by changing the rotation speed of the substrate table. Details of the deposition techniques are given in the original papers. The layers are subsequently processed either by means of rapid thermal processing or by conventional furnace annealing. Heat treatments at steadily increasing temperatures result in the following sequence of transformations: At low temperatures (to about 400°C) homogenization of the multilayer occurs. The amplitude of the concentration variation decreases slowly but the layers remain amorphous. Subsequent heating at higher temperatures induces crystallization of the multilayer. The first crystalline phase to appear is often metastable. Heating at still higher temperatures finally yields the equilibrium structure.

INCORPORATION OF SPUTTERING GAS

As was mentioned above sputtering gas can be incorporated in the growing layer. The largest concentrations of argon are found in the amorphous silicon layers. This can be expected since amorphous silicon, like crystalline silicon, is covalently bonded and has a rather open structure. The incorporation of argon in the multilayer can be reduced by making the period of multilayer structure smaller. Measurements of the argon content of titanium-silicon multilayers as a function of the repeat distance by Janssen and Wessels (1988) are shown in Fig.1. This figure shows that the incorporation of argon can be almost completely avoided if sufficiently thin silicon layers (< 0.5 nm) are deposited. This can be accomplished by a sufficiently high rotational speed of the substrate table. Since the argon content goes to zero for a finite thickness of the repeat distance Janssen and Wessels concluded that argon is not readily incorporated in the thin intermixed layer which results when the deposition changes from silicon to metal and vice versa. From these measurements one can estimate the thickness of the intermixed layer; one finds 0.4 nm for deposition at room temperature.

STRESSES IN DEPOSITED LAYERS

Stresses in the as-deposited material can be determined from measurements of the radius of curvature of the wafers and their elastic properties. The radius of curvature is measured with a Fizeau interferometer. Using the Stoney equation (Hoffmann, 1983),

$$\sigma_{//} = \frac{E_s}{6(1-\nu_s)} \frac{t_s^2}{t_l R} \qquad (1),$$

the weighted average stress parallel to the surface of the wafer can be determined. In this equation E_s and ν_s denote Young's modulus and Poisson's ratio of the substrate and t_s and t_l the thickness of the substrate and the layer respectively. R is the radius of curvature of the substrate.

In the case of polycrystalline layers strains in the deposited layer can be measured directly by means of the so-called $\sin^2\Psi$-method, see Hauk and Macherauch (1981). In this method the spacing of a given set of crystallographic planes in the film is measured as a function of orientation of these planes in the specimen. From these measurements the difference in strain perpendicular and parallel to the film, ($\varepsilon_{//} - \varepsilon_{\perp}$)is determined.

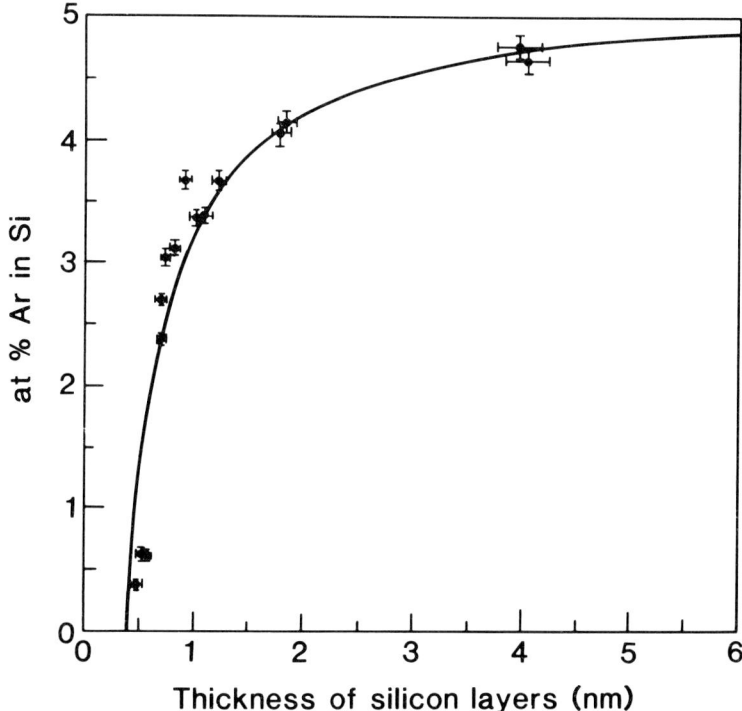

Fig.1 Measurements of the amount of argon gas trapped in amorphous multilayers as a function of the thickness of the individual silicon layers

The obvious advantage of the latter method above measurements of the curvature is that the deformation of the substrate is not necessarily elastic and therefore can lead to erroneous results. If however the deformation of the substrate is elastic the combination of the $\sin^2\Psi$-method with measurements of the curvature of the substrate makes it possible to determine one unknown elastic constant of the layer if the other is known (assuming isotropic elasticity).
In Fig.2 a model for the distribution of stress in deposited multilayers is given. Basically there exist three different layers each with a different intrinsic stress: amorphous silicon, amorphous

or microcrystalline transition metal and a relatively thin layer in which extensive intermixing has occurred. The transitions between these three layers are assumed to be abrupt. The intrinsic stress in the amorphous silicon and metal layers can be determined independently from measurements of the bending of wafers with thick layers of pure amorphous silicon or metal.

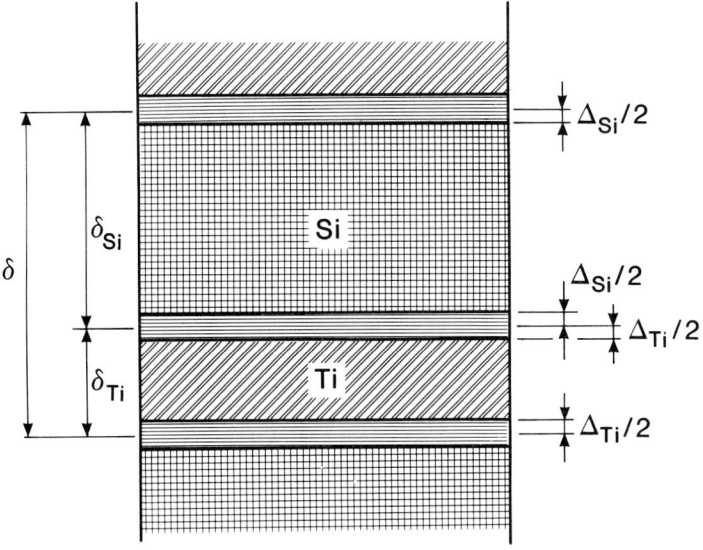

Fig. 2 Schematic structure of deposited multilayers.

The overall stress in the multilayer structure is the weighted average of the stresses in the interfacial layers and the pure titanium and silicon layers. Since a difference in the thickness of the Ti-Si and Si-Ti interfaces does not influence the results of our model, the thicknesses of the interfaces are assumed to be equal.

We denote the actual thickness of the layers in the multilayer structure by d_{Si}, d_{Ti} and d_{int} for silicon, titanium and the interfacial layer respectively. $\Delta_{Si}/2$ and $\Delta_{Ti}/2$ are defined as constants which represent the thickness of silicon and titanium consumed in the formation of the interfacial layer. These values are assumed to be independent of the deposited Ti/Si ratio, provided of course that enough Ti and Si is available. The thickness of the silicon, titanium and interfacial layers can be expressed in terms of the nominal thickness the titanium (δ_{Ti}) and silicon (δ_{Si}) layers, i.e. the thickness of silicon and titanium layers if no mixing at the interface would have occurred.

$d_{Si} = \delta_{Si} - \Delta_{Si}$,

$d_{Ti} = \delta_{Ti} - \Delta_{Ti}$ and:

$d_{Int} = (\Delta_{Si} + \Delta_{Ti})/2$.

For simplicity it is assumed that the formation of an interfacial layer has no effect on the total thickness of the multilayer structure.
The mean stress $\sigma_{//}$ in the multilayer structure can now be expressed as the weighted average of the stresses σ_{Ti}, σ_{Si} and σ_{int} in the Ti, Si and interfacial layers respectively:

$$\sigma_{//} = \{d_{Si}\, \sigma_{Si} + d_{Ti}\, \sigma_{Ti} + d_{Int}\, \sigma_{int}\} / \delta \qquad (2)$$

where δ is the period of the multilayer.
Obviously the nominal thickness of the individual layers of silicon and metal should be larger than the total thickness of the intermixed layers: $\delta_{Si} > \Delta_{Si}$ and $\delta_{Ti} > \Delta_{Ti}$. If that is not the case the average stress will of course be equal to the the stress in the mixed layer: $\sigma_{//} = \sigma_{int}$. Equation (2) can be rewritten as

$$\sigma_{//}\, \delta = C_{Ti,Si}\, \delta + K_{Int} \qquad (3)$$

with

$$K_{Int} = \Delta_{Si}(\sigma_{int} - \sigma_{Si}) + \Delta_{Ti}(\sigma_{int} - \sigma_{Ti}), \text{ and} \qquad (4)$$

$$C_{Ti,Si} = (\delta_{Si}/\delta)\, \sigma_{Si} + (\delta_{Ti}/\delta)\, \sigma_{Ti} \qquad (5)$$

The relation (3) between stress and period is convenient, since in our experiments the nominal silicon to metal ratio was kept constant, hence (δ_{Si}/δ) and (δ_{Ti}/δ) are constant. With the assumption that the stresses in the individual layers are independent of the thickness it follows that $C_{Ti,Si}$ and K_{int} are constants.
In Fig. 3 measurents of the stress $(\sigma_{//})$ as a function of the period (δ) of the multilayer are presented. The data can also be presented by plotting the product of stress and spacing versus the spacing (Fig.4). The values of K_{int} and $C_{Ti,Si}$ can be determined from a linear least squares fit to the data. This fit is represented in Fig.4 by the line with negative slope. From the slope and the intercept of the line values for K_{int} and $C_{Ti,Si}$ are determined:

$K_{int} = 1.05 \pm 0.16$ N.m^{-1} \qquad (6a)

$C_{Ti,Si} = -0.55 \pm 0.07$ GPa \qquad (6b)

The datapoint at d=1.05 nm must be excluded since the condition for the application of eq. 2 is not fulfilled. $C_{Ti,Si}$ can also determined from the ratio of the thickness of the titanium and silicon layers and from the independently measured stresses in pure silicon and titanium layers:

$\sigma_{Si} = -1.03 \pm 0.03$ GPa and $\sigma_{Ti} = 0.23 \pm 0.03$ GPa.

From these values $C_{Ti,Si}$ can be calculated: -0.60 ± 0.03 GPa. This value is in good agreement with the results obtained from the intrinsic stress as a function of the period δ.

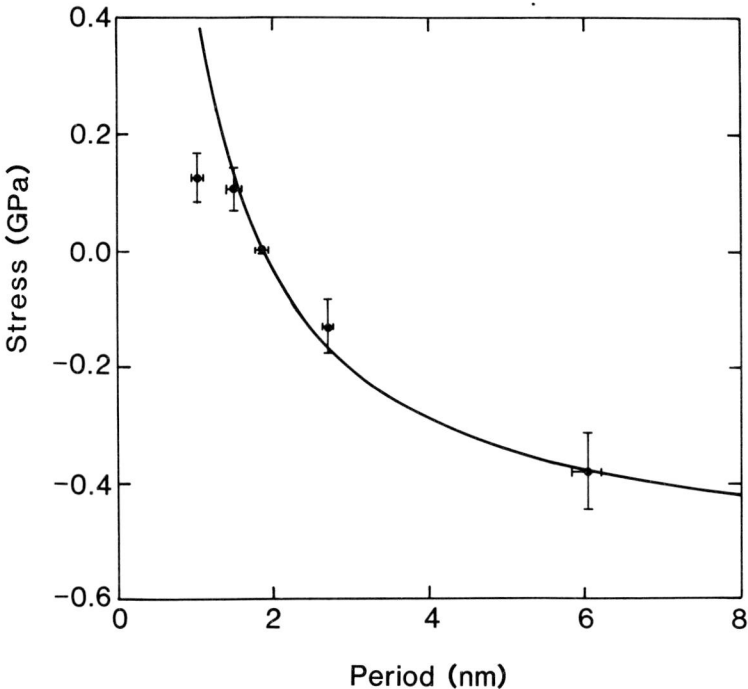

Fig.3. Stress in the deposited multilayer structure as a function of the period.

The fact that the datapoints in Fig.4 with exception of the point at d=1.05 nm can be described by a straight line shows that the assumption that σ_{Si}, σ_{Ti} and σ_{int} are independent of the thickness of the layers is correct. Deviations are of course expected to occur if the period of the multilayer is of the order of the thickness of the mixed layer. For a period smaller than 1.2 nm the stress will be equal to the stress in the interface layer σ_{int}. Discrepancies between data and the model are observed for periods (δ) below 1.2±0.2 nm. This yields a value of 1.2±0.2 nm for the thickness of the mixed layer: $\Delta_{Si}+\Delta_{Ti}$. The stress in the mixed layer is estimated as:

$\sigma_{int} = 0.11 \pm 0.04$ GPa.

If one assumes that the Si/Ti ratio in the interfacial layer ranges from 1 to 2 one can estimate the value of Δ_{Si}: 0.7 ± 0.2 nm. As described above the variation of argon concentration with period can also be described by the formation of an interfacial layer. From these experiments the thickness of the silicon layer consumed in the formation of an interfacial layer was found to be 0.4 nm. This value is approximately half the value for the thickness of the intermixed layer found from the evaluation of the stress measurements.

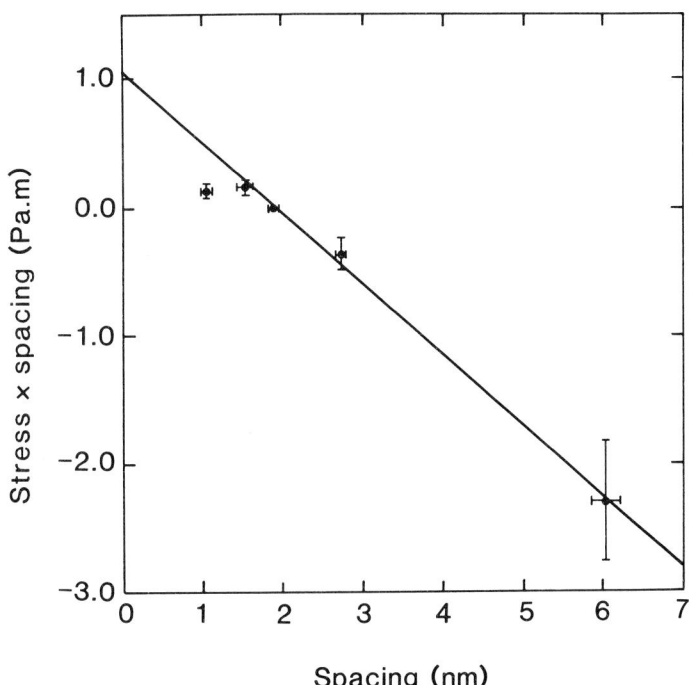

Fig.4. Plot of the data in Fig. 3 according to eq. (3).

STRESSES IN ANNEALED LAYERS

The stress in the plane of the film was measured again after rapid thermal processing (RTP) and furnace annealing. Annealing in the RTP resulted in $\sigma_{//} = 1.9 \pm 0.2$ GPa. Furnace annealing at 800 °C resulted in $\sigma_{//} = 2.3 \pm 0.3$ GPa. The period of the multilayers used in these experiments ranged from 1 to 6 nm. The resulting stress was independent of period.

The stress in a thin layer, formed at an elevated temperature, resulting from a difference in thermal expansion of layer and substrate is given by

$$\sigma_{//} = \frac{E_1}{1-\nu_1} (\alpha_1 - \alpha_{Si}) \Delta T \tag{10}$$

where the indices 1 and Si denote the properties of the silicide layer and the substrate resp. One of the assumptions underlying eq.10 is of course that the silicide layer is unstressed at the annealing temperature. This is a reasonable assumption since both the dislocation mobility and diffusivity are high at the annealing temperature.

Using the known elastic constants of titanium-disilicide one finds $\sigma_{//} = 2.0$ GPa, in agreement with the measured stress for furnace annealed samples. The agreement is less good for samples treated in the RTP (1.4 GPa) but it is conceivable that stress relaxation is incomplete at the somewhat lower annealing temperatures. One must also keep in mind that temperature measurement of the wafers in RTP-equipment is an art with many pitfalls. The conclusion is that at least for the wafers annealed in a tube furnace the stress can be completely accounted for by the difference in thermal expansion between layer and substrate. Similar results were obtained for $MoSi_2$ by Loopstra, Sloof and co-workers (1988) and for $TaSi_2$, see e.g. d'Heurle (1987).

HOMOGENIZATION OF AMORPHOUS MULTILAYERS

Multilayers with small periods are completely amorphous. If the period of the multilayer increases the transition metal layers become micro-crystalline. Obviously these two types of layers behave differently during annealing, but interestingly in both cases interdiffusion at relatively low temperatures results in completely amorphous, homogeneous layers. In our study we limited ourselves so far to alloys which were completely amorphous in the as-deposited condition. From the decay in intensity of the (000) first-order satellite reflection it is possible to determine the chemical diffusion coefficient of amorphous multilayers.

$$\frac{d}{dt} \left[\ln \frac{I(t)}{I(t=0)} \right] = - (8\pi^2/\delta^2) D(\delta)$$

where δ is the period of the composition modulation and $D(\delta)$ the chemical diffusion coefficient. Since the spacing of the silicon and transition metal layers is very small, δ is about 0.8 nm, very low values of the diffusion coefficient, on the order of 10^{-23} m^2.s^{-1}, can be measured. A problem for the interpretation of diffusion measurements in these amorphous layers is the fact that the diffusivity decreases rapidly during annealing due to a phenomenon known as structural relaxation. For a phenomenological model of structural relaxation see Van den Beukel and Radelaar (1983). One way to avoid these problems is to preanneal the specimen at a higher temperature such that no structural relaxation takes place during the subsequent diffusion anneals, so-called iso-configurational annealing. A disadvantage is of course that diffusion will also take place during the pre-anneal, thereby drastically diminishing the observable effects. Sloof, Loopstra and co-workers developed another approach. They incorporated the annealing behaviour explicitly into the model used to describe the diffusion process. For this purpose one can make use of the fact that during structural relaxation the viscosity of amorphous materials increases linearly with time. Assuming that the Stokes-Einstein relation, which says that the viscosity and diffusion coefficient are inversely proportional, is valid, one can easily show that the time dependence of the diffusion coefficients can be described by:

$$D^{-1}(t) = D^{-1}(t=0) + Bt$$

where B is a temperature dependent constant. From fits of this equation to the experimental data, see Fig.5, Sloof and co-workers found an activation energy for diffusion of 73 kJ.mole^{-1} and a pre-exponential constant of $\approx 10^{-11}$ cm^2.s^{-1}. These values deviate rather strongly from the values found for comparable crystalline alloys, where the pre-exponential factor is ≈ 1 and the activation energy approximately a factor 2 higher. Similar effects have been observed in other amorphous alloys. The cause of the vastly different behaviour of crystalline and amorphous materials could of course be sought in the different structure of the materials, but it is also possible that the validity of the Stokes-Einstein relation must be questioned. E.g.there are indications that the defects responsible for diffusion and viscous flow of amorphous materials are of different nature (van den Beukel). This is a rather complex subject which definitely needs further theoretical and experimental work.

CRYSTALLIZATION

Around 350°C the homogenized amorphous molybdenumsilicide crystallizes. Interestingly the first phase to appear is a metastable, hexagonal MoSi$_2$ phase (C40 Strukturbericht, hP9 Pearson notation). For material with less than 66 at% silicon no other phases with lower silicon content appear before 600°C. Apparently hexagonal MoSi$_2$ can accomodate a large fraction of excess metal atoms although the phase diagram indicates a very narrow solubility range for the equilibrium (tetragonal) phase.Rutherford Backscattering shows that the Mo$_5$Si$_3$-phase nucleates at or near the substrate interface and at the outer surface. A similar sequence is observed during the crystallization of amorphous titanium-silicon multilayers: the first, metastable crystalline phase that appears has a C49-structure.

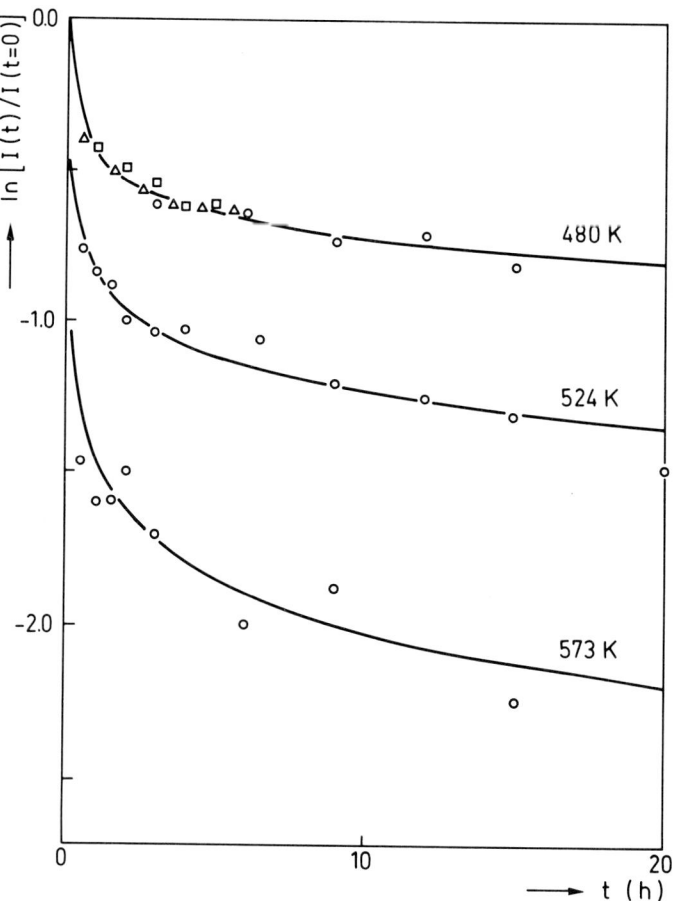

Fig. 5 Measurements of the decay of the first order satellite of the (000) reflection of amorphous molybdenum-silicon superlattices.

PHASE TRANSFORMATIONS

As was mentioned above the transition metal-silicides often crystallize into a metastable structure. When these materials are heated at higher temperatures the stable phase appears. The metastable phase that appears during crystallization of amorphous molybdenum-silicide is the hexagonal structure (C40 in the old Strukturbericht notation, hP9 in the Pearson notation) with symmetry $P6_222$. The metastable hexagonal material transforms to a tetragonal structure (C11b Strukturbericht notation, tI6 Pearson notation) with symmetry Fddd during annealing in the temperature range of ≈800°C. For material deposited on polycrystalline silicon the Mo_5Si_3 phase can be transformed to tetragonal $MoSi_2$ by supplying the additional silicon from the underlying silicon. It is interesting to note that the Mo_5Si_3 structure disappears in the same temperature range in which the stable tetragonal phase is formed. Apparently transformation only takes place if the atomic mobility is such that the distance over which silicon can move is on the order of the film thickness ($D \approx 10^{-13}$ cm^2.s^{-1}). This supports the idea that the hexagonal phase is stabilized by excess silicon. Fig.6 shows the resistivity as a function of the of temperature during isochronal annealing.

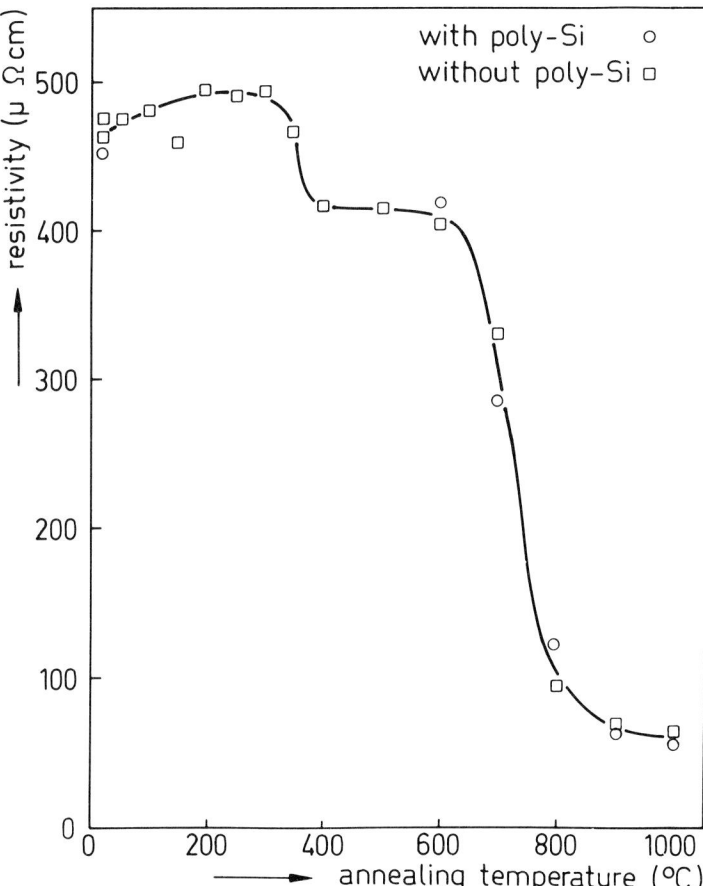

Fig. 6. Resistivity as a function of annealing temperature for isochronal annealing of MoSi multilayers.

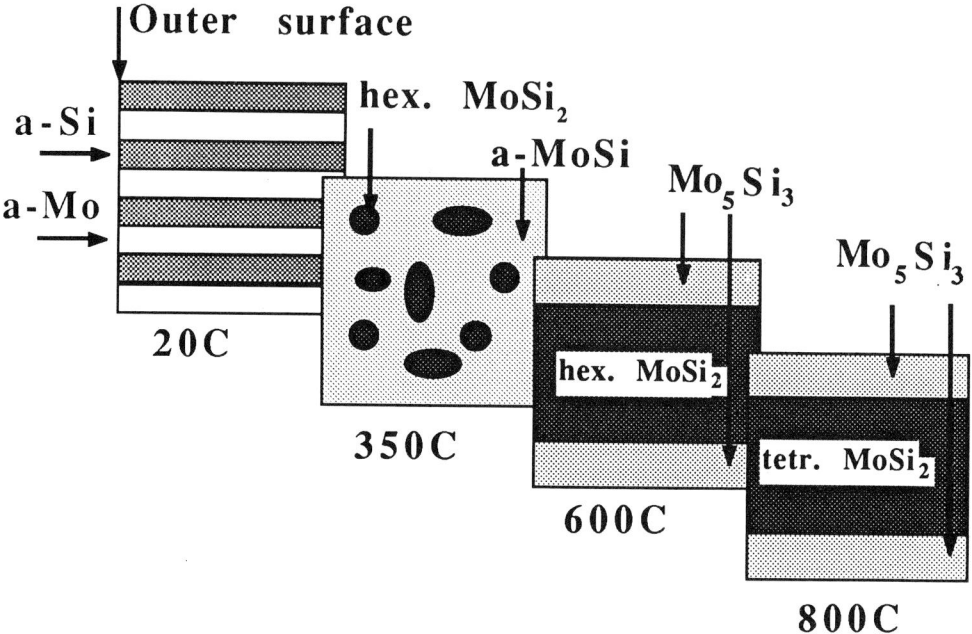

Fig. 7. Schematic picture of the sequence of phase transformations in MoSi-layers.

The sequence of transformations for material with a nominal silicon concentration below 66 at% deposited directly on oxide is summarized in Fig.7. Annealing of material deposited on polycrystalline silicon results in a single phase disilicide layer. As was mentioned above amorphous TiSi2 crystallizes in the C49 structure. X-ray diffraction (Fig.8) shows that transformation to the equilibrium C54 structure takes place at ≈600 °C. Prolonged annealing at low temperatures does not result in the reverse transformation, which is a strong indication that the C49 is really metastable. However, it is also possible that the reverse transformation does not occur because of low atomic mobility; most defects will have been annealed out after the high temperature heat treatment and the equilibrium diffusivity is low.

A review of the literature shows that the disilicides of transition metals show quite a variety of structures. It is therefore not surprising to find that most silicides crystallize in metastable structures before transformation to the final equilibrium structures. As was mentioned above both MoSi2 and TiSi2 crystallize in metastable structures before their stable structures appear. It is interesting to note that the recent study of the stability regions of AB2-compounds by Villars(1984) predicts two stable phases viz. C40 and C11b for MoSi2 and WSi2.

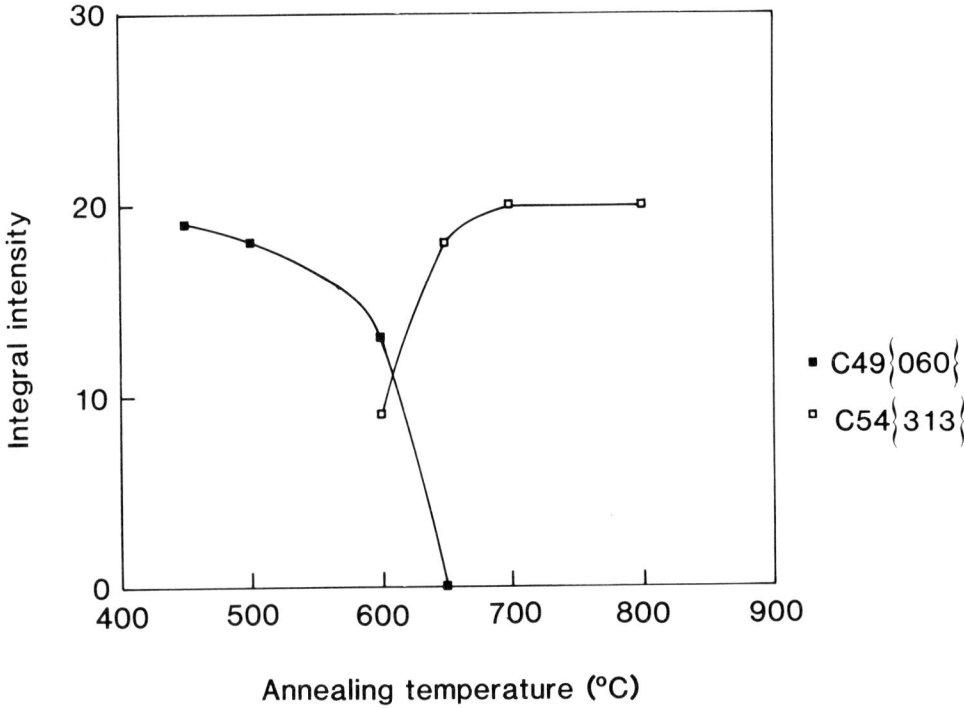

Fig.8. Integrated intensity of the {060} reflection of C49 and {313} reflection of C54 TiSi$_2$ as a function of annealing temperature.

CONCLUSIONS

It was demonstrated that the stress in deposited titanium-silicon multilayers is a function of the period of the multilayer. This variation can be modelled by assuming a mixed layer with constant thickness at each interface. From the stress-strain relationship in titanium–disilicide films Poisson's ratio is determined to be 0.22. The high stresses in annealed TiSi$_2$ and MoSi$_2$-layers ($\approx$2 GPa) can be totally accounted for by the difference in thermal expansion of silicon and titaniumdisilicide. Annealing of the amorphous multilayers produces a homogeneous amorphous alloy, subsequent crystallization at higher temperatures result in metastable phases. Equilibrium phases only appear at still higher temperatures.

ACKNOWLEDGEMENTS

The project was sponsored by the " Innovatiegericht Onderzoekprogramma IC-technologie". This work is part of the research program of the "Stichting voor Fundamenteel Onderzoek der Materie (FOM)", which is financially supported by the " Nederlandse Organisatie voor Zuiver Wetenschappelijk Onderzoek (ZWO)". Financial support of the "Innovatie gericht onderzoeksprogramma voor de IC-technologie (IOP-IC)" is gratefully acknowledged. The author wishes to thank his colleagues A.E.T.Kuiper and R.A.M.Wolters, both from Philips Research Labs, Th.de Keijser, E.J.Mittemeijer, G.C.A.M. Janssen and A.L.Mulder, all from the Technical University Delft and the (graduate) students O.B.Loopstra, W.G.Sloof, P.J.J.Wessels and J.F.Jongste for their contributions to this project and for stimulating discussions. Dr.G.Janssen is thanked for his critical reading of the manuscript.

REFERENCES

A.van den Beukel and S.Radelaar (1983) . Acta Metallurgica 31, 419.

A.van den Beukel, private communication

J.S.Kirkaldy and D.J.Young (1987) Diffusion in the Condensed State, The Institute of Metals, London.

F.M.d'Heurle (1986) . in Solid State Devices, P.Balk and O.G.Folberth eds. Elsevier Science Publishers V.M. Hauk and E.Macherauch (1981) .Adv. in X-ray Anal. 27, 81.

R.W. Hoffman (1966) in: Physics of Thin Films, eds. G. Hass and R. E. Thun, vol.3, Academic Press, New York, , p. 211.

G.C.A.M.Janssen, P.J.J.Wessels (1987) . Journal of Applied Physics 62, 3993.

G.C.A.M.Janssen, P.J.J.Wessels, J.F.Jongste (1988) . Vakuum-Technik 8 (to be published).

O.B.Loopstra, Th.H.de Keijser, E.J.Mittemeijer and S.Radelaar (1987) Le Vide, Les Couches Minces no 236, 71 .

O.B.Loopstra, W.G.Sloof, Th.de Keijser, E.J.Mittemeijer, S.Radelaar, A.E.T.Kuiper and R.A.M.Wolters (1988) Journal of Applied Physics, to be published.

M.-A.Nicolet, S.S.Lau in: VLSI electronics, Microstructure Science, N.G.Einspruch,

G.B.Larrabee editors (Ac.Press, New York. 1983) Vol.6, p.330.

W.G.Sloof, O.B.Loopstra, Th.H.de Keijser and E.J.Mittemeijer (1986) Scripta Metallurgica, 20 , 1683-1687.

Villars (1984) . J.of Less Common Metals 99, 33-43

P.J.J.Wessels, J.F.Jongste, O.B.Loopstra, G.C.A.M. Janssen, A.L.Mulder and S.Radelaar (1988) Journal of Applied Physics, (to be published)

The Role of Stresses on Phase Transformations

FRANCIS C. LARCHE

Université des Sciences et Techniques du Languedoc
34060 Montpellier Cedex, France.

ABSTRACT

The role of stress on equilibrium and diffusion is examined on specific examples. In special geometries, the classical descriptions are quantitatively modified: shift of phase diagrams lines, addition of a composition dependent term to the apparent diffusion coefficient. This behaviour is not generic, and stresses generally produce qualitative changes. Simple examples are presented where new phase diagrams features appear, and where diffusion becomes non-local. The use of these notions in complex phenomena associated with phase transformations is briefly outlined.

KEYWORDS

Thermodynamics; stress; equilibrium; phase-diagram; diffusion.

INTRODUCTION

Stresses are generated in a solid by applying forces on its external surface. But many situations can give rise to internal stresses, i.e. stresses that are present without externally applied forces. A non-uniform composition, a solid-state phase transformation accompanied by a volume change, the presence of a coherent interface are the most common ones, but dont exaust the possibilities. In view of such a list, one can conclude that, as a rule, stresses are generated in solids during phase transformations. Their effects depend on the available mechanisms to release them, and on the time scale of the process compared to the time scale of the phase transformation. In this review, we shall consider the extreme case where no mechanical release is possible. Several examples will be described where this hypothesis is indeed verified. In each practical case, this assumption should of course be checked.

The thermodynamics of stressed solids, developed over the years in collaboration with J.W.Cahn, provides the framework for this paper. The equilibrium equations (Larché and Cahn, 1985, 1986) contain only local quantities: the stress tensor, its divergence, and the composition. At this level, the origin of the stress (external or internal) is irrelevant. It will appear in the boundary conditions, and thus in the final solutions. But if the distinction is important in practice, it is not of great importance in the theoretical treatment. The role of stress will be classified according to whether it affects equilibrium or kinetics. The effects on a boundary condition of a kinetic equation obtained with the assumption of local equilibrium will be included in the kinetic section.

For simplicity, only constant temperature binary alloys are considered in this paper. Isotropic elasticity is assumed, except when explicitely stated. Einstein summation convention is used throughout.

THE ROLE OF STRESS ON EQUILIBRIUM

Epitaxial growth from a liquid is a good example to show that stresses could just shift phase diagrams boundaries, but also to see the limitations of this interpretation. Details of the calculations not included in the following section can be found in the original article (Larché and Cahn, 1987). When a solid and a liquid are at equilibrium under hydrostatic stress (i.e. under pressure) the chemical potentials are equal in each phase

$$\mu_1^L = \mu_1^P \quad ; \quad \mu_2^L = \mu_2^P \tag{1}$$

where the superscripts L and P refer respectively to the liquid and the solid. This system of equations can be rewritten

$$\mu_2^L - \mu_1^L = \mu_2^P - \mu_1^P \tag{2}$$

$$\mu_1^L = \mu_1^P \tag{3}$$

The phases are homogeneous, and these equations are valid at the interface as well as in the bulk of the phases. Since the system is under pressure, mechanical equilibrium simply implies that the pressure is the same throughout

$$P = \text{constant} \tag{4}$$

If the solid is under stress, the equilibrium equations are (Larché and Cahn, 1985)

$$M_{21} = \mu_2^L - \mu_1^L \tag{5}$$

and

$$f - c\,M_{21} = \mu_1^L \tag{6}$$

The quantity M_{21}, called diffusion potential, is related to the stress and the chemical potentials in the solid under negligible pressure μ_1^S and μ_2^S by

$$M_{21} = \mu_2^S - \mu_1^S - V_m \eta \sigma_{kk} \tag{7}$$

where V_m is the molar volume, σ_{kk} the trace of the stress tensor. The mole fraction of component 2 is c. The linear chemical expansion coefficient η is related to the change in lattice parameter a with composition (for cubic crystals) by

$$\eta = d\ln a/dc$$

f is the molar Helmholtz free energy of the solid[1]. Equation (5) is valid in the bulk and at the interface of the solid. Equation (6) applies only at the interface liquid-solid. Equations (5) and (6) contain the stress tensor, and an equation is needed for this quantity. This is the usual local force balance equation

$$\partial \sigma_{ij}/\partial x_j = 0 \qquad (8)$$

where the x_j are the spatial coordinates. This is a partial differential equation, and boundary conditions have to be specified. For an epitaxial layer they are as follows. a) the tractions acting on each side of the solid-liquid boundary are equal in absolute value and opposite in sign. Since we have assumed a negligible pressure in the liquid, one obtains

$$\sigma_{ij} n_j = 0$$

where n_j are the components of the normal to the interface. b) along the periphery of the layer, the traction is also zero. c) finally, the displacement in the plane of the interface layer-substrate suffers a discontinuity, directly related to the mismatch between the lattice parameters of the stress-free substrate and layer. An equation of state relating stress, strain and composition in the solid is also needed. A simple adaptation of the linear thermoelastic equations (Boley, 1960) gives an adequate formulation for most crystalline materials

$$\varepsilon_{ij} = \eta \Delta c\, \delta_{ij} - (\nu/E)\sigma_{ij}\delta_{ij} + [(1+\nu)/E]\sigma_{ij} \qquad (9)$$

In this expression, ν is Poisson's ratio, E is Young's modulus, and Δc the difference between the local composition and a reference composition. With such a model an expression for the Helmoltz free energy of the solid can be computed and used in equation (6). The final result is

$$\mu_1^S + \left[-\frac{\nu}{2E}(\sigma_{kk})^2 + \frac{1+\nu}{2E}\sigma_{ij}\sigma_{ij} + c\eta\sigma_{kk} \right] V_m = \mu_1^L \qquad (10)$$

In isotropic solids, only the trace of the stress tensor appears in equation (7), while all the components are present in the interface condition (10). As a consequence, if an isotropic phase is under pure shear, its composition is homogeneous. The stress affects its equilibrium only through the strain energy at the interface.

---

[1] In equations (6) and (7) the elastic constants are assumed independant of composition. A more complete treatment can be found in Larché and Cahn (1985).

The computation of the equilibrium stress was trivial for a system under pressure. Here there is no such simple solution. One possible method is to assume an arbitrary composition field, and find a solution of the elastic problem. This solution is in general a functional of the composition. For thin epitaxial films, the stress tensor has only two non-zero components, and they are equal

$$\sigma_{xx} = \sigma_{yy} \equiv \sigma = -Y\eta(c-c_0) \qquad (11)$$

where the axis x and y are parallel to the layer, and c_0 is the composition for which there is lattice matching of the layer and the substrate. The elastic coefficient Y is given by

$$Y = E/(1-\nu) \qquad (12)$$

This solution is easily extended to crystals with cubic symmetry. The coefficient Y has simply another expression. In equation (11), c is a function of z. To obtain an explicit value for this composition, the components of the stress tensor are replaced in equations (5) and (10). A system of algebraic equations is obtained, where z does not appear explicitly. As a consequence, the solution c cannot depend on z : the composition is uniform in the solid layer. This implies, through equation (11) that the stress is also uniform.

The presence of strain changes the equilibrium composition in the liquid and the solid. As a consequence, the lattice parameter is changed. This is easily measured, and the results can be compared with the theoretical predictions. It has been done for $Ga_x In_{1-x} P$ grown on GaAs at 785°C. The relative variation of the lattice parameter with equilibrium liquid composition $d(\ln a)/dc_{Ga}$ has a measured value of -1.5 and a calculated one of -1.6. There is no adjustable parameter, and taking into account the precision of the data used, this can be considered as very good agreement.

The equilibrium of a thin epitaxial layer with a melt is an example where the presence of stress shift the boundaries of the phase diagram. (for ternary III-V compounds, it also rotate the tie lines). In this particular instance the elastic energy per mole f_e

$$f_e = V_m \eta^2 Y(c-c_0)^2 \qquad (13)$$

behave as if it were an excess free energy. This property simplifies the resolution of equilibrium problems, where classical thermodynamics methods can be used. But the stress tensor with its specific properties is always hidden behind a formula like (13). The stability of the layer against decomposition, for instance, cannot be treated as usual (Larché, 1988), since an arbitrary fluctuation in composition would change the expression for the stress (equation 11). More important to note, the elastic energy does not generally behave as an excess free energy. Consider for instance the case of a thicker film (what is meant by "thin" and "thicker" is discussed at length in Larché and Cahn, 1987). Equations (5-6) are always valid. The difference with the thin film appears in the value of the stress tensor. It includes a term function of the position z, as well as the local composition $c(z)$

$$\sigma = -\eta(c-c_0)Y + K_1 Y + K_2 zY \qquad (14)$$

where K_1 and K_2 are two constants given by

$$K_1 = 2\eta\lambda Z \frac{2(<c>-c_0)(1+Z\lambda^3) + \lambda M(1-Z\lambda^2)}{1+4Z\lambda + 6Z\lambda^2 + 4Z\lambda^3 + Z^2\lambda^4} \qquad (15a)$$

$$K_2 = 6\eta\lambda Z \frac{(<c>-c_0)(1-Z\lambda^2) + 2\lambda M(1-Z\lambda)}{h_s(1+4Z\lambda+6Z\lambda^2+4Z\lambda^3+Z^2\lambda^4)} \qquad (15b)$$

M is a dimensionless moment defined by

$$M = \frac{1}{h_e^2} \int_0^{h_e} (c-c_0) z \, dz \qquad (16)$$

Z is the ratio of the quantity Y for the layer to that for the substrate. λ is the ratio of the thickness of the layer h_e to that of the substrate h_s. $<c>$ is the average composition in the layer. The last term of expression (14) is responsible for the bending sometimes observed (Nuese and co-workers, 1974; Hitchen and co-workers, 1974; Minagawa, Nakamura and Sano, 1985). It produces a variable composition at equilibrium. Such a possibility was also predicted by Li (1978) in its thermodynamic analysis of stressed solids. This result is a direct consequence of the long range nature of the elastic forces. In such case, the effects of the stress cannot be described as a shift of the phase diagram boundaries.

These examples illustrate the essential aspects of the equilibrium of stressed solids with liquids. Equations (5,6,8) provide the basic framework for all these equilibria. Only the boundary conditions are specific for each case. A simple inspection of equations (2-4) and equations (5-6) reveals their different mathematical structure. On one side, there is a set of algebraic equations, and Gibbs (1961) found an elegant graphical solution. On the other side, there is a set of coupled algebraic and partial differential equations. The properties of the solutions of this last system of equations is not yet fully known. As we have seen on the example, one can sometimes recover the classical results of equilibrium under hydrostatic stress. But as it appears on the thick layer case, the result is not generic. It is simply a consequence of the special relationship between stress and composition for a thin film.

For the equilibrium of stressed solid phases, the structure of the equations is similar (Larché and Cahn, 1978, 1985). Their expressions depend on the nature of the interface. The coherent interface is important, as it usually generates internal stresses. With the same equation of state for the solid as before, the following applies for an equilibrium between phases α and β.

$$M_{21}^\alpha = M_{21}^\beta \qquad (17)$$

$$\mu_1^\alpha + \left[-\frac{\nu^\alpha}{2E^\alpha}(\sigma_{kk}^\alpha)^2 + \frac{1+\nu^\alpha}{2E^\alpha}\sigma_{ij}^\alpha\sigma_{ij}^\alpha + c^\alpha \eta^\alpha \sigma_{kk}^\alpha \right] V_m + \epsilon_{ik}^\alpha \sigma_{kj}^\alpha n_i^\alpha n_j^\alpha V_m =$$

$$\mu_1^\beta + \left[-\frac{\nu^\beta}{2E^\beta}(\sigma_{kk}^\beta)^2 + \frac{1+\nu^\beta}{2E^\beta}\sigma_{ij}^\beta\sigma_{ij}^\beta + c^\beta \eta^\beta \sigma_{kk}^\beta \right] V_m + \epsilon_{ik}^\beta \sigma_{kj}^\beta n_i^\beta n_j^\beta V_m \quad (18)$$

as well as equation (8). The n_i's are the components of the normal to the interface, directed toward the exterior of the phase. Within the small strain approximation used here, one can use indifferently the molar volume of unstressed α or β for V_m. The main difficulty in solving this system of equations is usually found in the computation of the stress field. The plate geometry proved very interesting, because equation (8) can be solved analytically for a variety of boundary conditions and an arbitrary composition in the direction of the thickness of the plate. The case of two phases with such a geometry and self stressed because of the presence of the coherent interface has been recently explored (Cahn and Larché, 1984). The model used is very simple. The phases have a different, but constant, lattice parameter (i.e. η is zero) and they have identical elastic properties. Under such conditions, composition and stresses are homogeneous within each phase. The elastic energy is given by

$$f_e = W(1-W)\epsilon^2 Y \quad (19)$$

where W is the volume fraction of a phase, and ε the lattice mismatch. Obviously it does not have the properties of an excess free energy. As a result, the features of such equilibria differ markedly from those of liquid-liquid equilibria. In particular the phase rule does not apply, the composition of the phases varies with their volume fraction, and the extremities of the tie-lines do not coincide with the limit of the two-phase field. Similar conclusions have been reached for spherical precipitates by Johnson and Voorhees (1987). A more extensive discussion of these surprising results is given by Cahn and Larché (1984).

ROLE OF STRESS ON DIFFUSION

The basic approach is an extension of Einstein, Hartley and Darken's formalism (Darken and Gurry, 1953). To simplify the presentation, the case of a dilute interstitial solid solution is considered here. Only the interstitial species (called 2) can diffuse, and be exchanged with an external fluid. At equilibrium, if the solid is under stress, the potential M_2 is constant

$$M_2 = \mu_2^S - V_m \eta \sigma_{kk} \quad (20)$$

(An extended discussion on the differences between substitutional and interstitial solid solutions can be found in Larché and Cahn, 1985). The gradient of the potential M_2 can be considered as the driving force for the diffusion of species 2. A similar suggestion was made by Hillert (1957) for the diffusion flow of a component in non-isobaric systems. The flux equation can be written

$$J = - Bc \, \text{grad} \, M_2 \qquad (21)$$

where B is a mobility and c the concentration². B can be expressed in term of the tracer diffusion in dilute solution D

$$B = D/RT \qquad (22)$$

The conservation of mass equation reads as usual

$$\partial c/\partial t + \text{div} \, J = 0 \qquad (23)$$

As for the equilibrium problem, an equation is needed for the stress tensor. Because the relaxation time for elastic deformation is usually much smaller than the caracteristic time for chemical diffusion, elastic equilibrium can be assumed at all times. As a result, equation (8) is obeyed. The equation of state for the solid is the same as for equilibrium (equation 9). Equations (23) and (8) are both partial differential equations and require boundary conditions. Only the case of internal stresses has been explored (Larché and Cahn, 1982). In such instance, the tractions are zero along the external surface of the solid. Along the interface in contact with the fluid, several hypotheses can be used, just as in the classical diffusion problems. A very common one, which will be used here, is equilibrium at the interface. This is expressed by

$$M_2 = \mu_2^F \qquad (24)$$

where μ_2^F is the chemical potential of species 2 in the fluid. The set of equations (8,9,20,23,24) contains all the informations needed for the problem. Composition and stress appear in all of them, a result of the coupling between these two quantities. To present some properties of the solutions, we restrict ourselves to a one dimensional case. The solution is obtained via the route used for equilibrium. An arbitrary composition field c(z) is assumed, and the resulting stress field is found. It is a functional of c(z). The result is used to eliminate the stress tensor from equations (23) and (24). After this operation, the diffusion equation and its boundary conditions contain only the composition as variable. The final result depends on the thickness of the sample compared to its lateral dimensions.

In the case where the sample can be considered as semi-infinite (the diffusion distance is small compared to the thickness), the stress tensor is given by expression (11). The flux is

$$J_z = - D \left[1 + (\partial \ln \gamma / \partial \ln c) + (2cV_m \eta^2 Y/RT) \right] \partial c/\partial z \qquad (25)$$

It is proportional to the composition gradient, and can thus be considered

---

² This equation is valid with a frame of reference rigidly attached to the lattice. A correction for the displacement due to the strain has to be made to the diffusion profile obtained from this equation to get the profile in the laboratory frame of reference.

as Fickian. The effect of the stress is to add a composition dependent term to the apparent diffusion coefficient. As this term is always positive, the presence of self-stress always increases the apparent diffusion coefficient. If the stress tensor is also replaced in the boundary condition (24), the following is obtained

$$\mu_2^F = \mu_2^S + V_m \eta^2 Y \Delta c^* \qquad (26)$$

where Δc^* is the difference between the interface composition and the composition at time zero, assumed here homogeneous. As for the first case seen in the equilibrium section, the stress has produced a shift of the composition at the interface.

When the sample is thin compared to its lateral dimensions, the stress tensor has again two equal non-zero components, but the expression for σ_{kk} is now (Larché, 1988)

$$\sigma_{kk} = -2\eta V \left[\Delta c - \frac{1}{L} \int_0^L \Delta c \, dz - \frac{12(z-L/2)}{L^3} \int_0^L \Delta c(z-L/2) dz \right] \qquad (27)$$

where L is the thickness of the plate and Δc is the difference between the local composition and the composition at time zero. After replacement in (21), the flux given by

$$J_z = -D \left[1 + \frac{\partial \ln \gamma}{\partial \ln c} + \frac{2\eta^2 Y c V_m}{RT} \right] \frac{\partial c}{\partial z} - \frac{24 c \eta^2 Y V_m D}{L^3 RT} \int_0^L \Delta c(z-L/2) dz \qquad (28)$$

This expression contains a term proportional to the composition gradient, as in the previous case. In contrast the last term includes an integral of the composition profile which transforms the diffusion into a non-local phenomenon. It is proportional to the composition and affects the whole sample at all times. The boundary condition is obtained likewise by elimination of the stress from (24)

$$\mu_2^F = \mu_2^S + 2 V_m \eta^2 Y \left[\Delta c^* - \frac{1}{L} \int_0^L \Delta c \, dz + \frac{6}{L^2} \int_0^L \Delta c(z-L/2) dz \right] \qquad (29)$$

This expression contains the integral of the concentration profile. Since this profile evolves as diffusion proceeds, the composition at the interface c^* will vary with time. This result is at variance with the behaviour obtained for interface equilibration when the stress is negligible. (Note that since the sample is finite, a boundary condition at $z = L$ is needed to completely specify the problem).

These examples have a simple enough geometry to permit the elimination of the stress and lead to an apparent diffusion equation. This might prove difficult in other cases, but general conclusions can nevertheless be drawn. Larché and Cahn (1982) have shown on several other examples that the expression for the flux depends on the geometry. Furthermore, even when an apparent diffusion coefficient can be defined, its symmetry properties are not characteristic of a second rank tensor. They contain, in a somewhat hidden way, the symmetries of the fourth rank elastic coefficient tensor. In a cubic crystal, for instance, the apparent diffusion coefficient is not isotropic, as classically expected (Nye, 1957). The simple result, where diffusion is local and of Fickian character (i.e. the flux is proportional to the concentration gradient), is not generic. It has been found for one dimensional diffusion in simple geometries. In general, the flux is expected to be non-local. It will be a functional of the composition field. In other words the flux at a particular point will depends on the composition everywhere in the sample. Lewis and co-workers (1983, 1987) report experimental observations on diffusion of hydrogen in Pd-Pt alloys, where the unusual behaviour is attributed to this effect.

The case of a substitutional solution is more complex, since the solid can crystallize or dissolve at the interface in contact with the fluid. The diffusion and boundary conditions are simple extensions of the equations presented for interstitial solutions (Larché and Cahn, 1982). But the boundary may change shape. A detailed theoretical investigation of the solutions is, to our best knowledge, not yet available.

ROLE OF STRESS ON COMPLEX PHASE TRANSFORMATION PROCESSES

Equilibrium and kinetics are the basic ingredients in the modelisation of complex phase transformation phenomena. The notions covered in the preceeding sections are the basis when stresses are involved. We shall just briefly review cases where experimental observations point to internal or external stresses as important factors.

Diffusion induced grain boundary migration (DIGM) and liquid film migration are covered in an other communication (Purdy, 1988). The work of Song, Ahn and Yoon (1985) and of Baik and Yoon (1985) has shown conclusively that the effects are due to a shift of solubility produced by an internal stress resulting from the alloying of a surface layer. A similar treatment has been proposed for coherent resolidification after surface melting (Perepezko and Boettinger, 1985).

The effects of internal and applied stresses on precipitation of a coherent phase, or the coarsening of coherent precipitates has been the subject of much research. The $\gamma-\gamma'$ region of Ni alloys has been a favored field for experiments. At very low volume fraction, Miyasaki, Inamura and Kosakai (1981, 1982) have shown that the precipitates split rather than coarsen. This observation is thought to result from a reduction of elastic energy that is larger than the corresponding increase in surface free energy. At large volume fraction the situation is complex, and not yet completely understood (Tien and Copley, 1971; Cornet and Martin, 1987). Theoretical calculations (Johnson, 1983, 1984, 1987; Johnson and Cahn, 1984; Johnson Voorhees, 1984; Voorhees, Johnson and Laria, 1986; Berkenpas, Johnson and Laughlin, 1986) indicate that the results can be quite diverse. They depend on the misfit, the relative elastic constants of matrix and precipitate, as well as the applied stress. The coarsening process may be substancially changed by these elastic interactions.

SUMMARY AND CONCLUSION

In the preceeding sections, we have presented some aspects of the role of stress on thermodynamics and diffusion. The examples were chosen so as to show two important features. On one side, there exist special geometries where the classical properties are simply modified : shift of equilibrium lines in phase diagrams, extra term in the apparent diffusion coefficient.

On the other side, this simple behaviour is not general. Because of the long range nature of the elastic forces, the shape of the precipitates, their volume fraction, their spatial arrangement will interact with the phase transformation. We can also expect orientation effects, because elastic properties of crystalline solids are not isotropic.

The generation of stresses is practically unavoidable in phase transformations. But their importance upon the transformation itself depends critically on the presence of mechanisms for mechanical relief. The increased use of single crystals with low dislocation density, and of materials like ceramics where dislocation movements may be difficult, make these considerations more important at all stage of phase transformations. Comparison with experimental results have confirmed the validity of the thermodynamic theory. Recent observations seems also to confirm the validity of the predictions on the effects of self-stress on diffusion.

The basic equations are available for applications to phase transformations. The resolution of the equations remains a major problem, so that one has to resort to simplifying models in many cases. In this respect, the plate geometry is particularly interesting, because analytical solutions are usually possible. It seems to be particularly well suited for comparison between theory and experiment.

ACKNOWLEDGEMENTS

Enlightening discussions with G. Martin and W.C. Johnson are gratefully acknowledged.

REFERENCES

Baik, Y.-J. and D.N. Yoon (1985). Acta Metall. 33 1911-1917.
Boley, B.A. and J.H. Weiner (1960). Theory of Thermal Stresses. Wiley, New-York.
Berkenpas, M.B., W.C. Johnson and D.E. Laughlin (1986). J. Mater. Res.,1, 635-645
Cahn, J.W. and F.C. Larché (1984). Acta Metall., 32, 1915-1923.
Cornet, M. and G. Martin (1987). Scripta Metall.,21, 1091-1095.
Darken, L.S. and B.W. Gurry (1953). Physical Chemistry of Metals. McGraw-Hill, New-York.
Gibbs, J.W. (1961). Thermodynamics. Vol.1, Dover, New-York.
Hillert, M. (1957). Jernkontorets Ann., 141, 67-89.
Hitchens, W.R., N. Holonyak Jr, M.H. Lee and J.C. Campbell (1974). J. Cryst. Growth, 27, 154-165.
Johnson, W.C. (1983). Metall. Trans., 14A, 2219-2227.
Johnson, W.C. (1984). Acta Metall., 32, 465-475.
Johnson, W.C. (1987). Metall. Trans., 18A, 233-247.
Johnson, W.C. and J.W. Cahn (1984). Acta Metall., 32, 1925-1929.
Johnson, W.C. and P.W. Voorhees (1985). Metall. Trans., 16A, 337-347.
Johnson, W.C. and P.W. Voorhees (1987). Metall. Trans., A18, 1213-1224.
Larché, F.C. (1988). In G.Martin and L.Kubin (Edts). Non-linear Phenomena in Materials Science. In press.
Larché, F.C. and J.W. Cahn (1978). Acta Metall. 26, 1579-1589.
Larché, F.C. and J.W. Cahn (1982). Acta Metall., 30, 1835-1845.
Larché, F.C. and J.W. Cahn (1985). Acta Metall., 33, 331-357.
Larché, F.C. and J.W. Cahn (1986). In S.Saimoto, G.Purdy and G.Kidson (Eds) Solute-Defect Interaction, Theory and Experiment, Pergamon Press, Toronto pp. 1-27.
Larché, F.C. and J.W. Cahn (1987). J. Applied Phys., 62, 1232-1239.

Lewis, F.A., B. BARANOWSKI and K. Kandasamy (1987). J. Less-Commun Metals, 134, L27-L31.
Lewis, F.A., J.P. Magennis, S.G. McKee and P.J.M. Ssebuwufu (1983). Nature (London), 306, 673-675.
Li, J.C.-M. (1978). Metall. Trans., 9A, 1353-1380.
Minagawa, S., H. Nakamura and H. Sano (1985). J. Cryst. Growth, 71, 377-384.
Miyasaki, T., H. Inamura, H. Mori and T. Kozakai (1981). J. Mat. Sc., 16, 1197-1203.
Miyasaki, T., H. Inamura and T. Kozakai (1982). Mat. Sci. Eng., 54, 9-15.
Nuese, C.J., A.G. Sigai, J.J. Gannon and T. Zamerowski (1974). J. Electron Mater. 3, 51-78.
Nye, J.F. (1957). The physical properties of crystals. Clarendon Press, Oxford.
Perepezko, J.H. and W.J. Boettinger (1985). In L.E.Rehn, S.T.Picraux and H.Wiedersich (Edts), Surface Alloying by Ion, Electron, and Laser Beams. ASM Material Science Seminar, Toronto, pp. 51-89.
Purdy, G. (1988) This volume.
Song, Y.-D., S.-T. Ahn and D.N. Yoon (1985). Acta Metall., 33, 1907-1910.
Tien, J.K. and S.M. Copley (1971). Metall. Trans., 2, 543-553.
Voorhees, P.W., W.C. Johnson and V.J. Laria (1986). In S. Saimoto, G. Purdy and G. Kidson (Eds) Solute-Defect Interaction, Theory and Experiment, Pergamon Press, Toronto, pp. 409-425.

On the Role of Elastic Energy in Diffusional Phase Transformations

GARY R. PURDY AND JOHN R. DRYDEN

*Department of Materials Science and Engineering,
McMaster University, Hamilton, Ontario, Canada, L8S 4M1.*

ABSTRACT

Two sources of elastic strain fields are distinguished: In one, a coherent or semicoherent misfitting precipitate gives rise to an elastic strain field in an otherwise homogeneous matrix; in the other, a solute gradient in parent or daughter phases results in a variation of stress-free lattice parameters, which leads in turn to elastic strain if lattice continuity is maintained. The two sources are here termed "misfit" and "coherency" respectively.

Examples of the two sources of strain energy and of their influence on transformation morphology and kinetics are considered.

KEYWORDS

Elastic energy, coherency strains, misfit strains, discontinuous precipitation, chemically induced grain boundary migration, massive transformation

INTRODUCTION

The importance of elastic strain energy in phase transitions has long been recognized. The comprehensive analyses of Eshelby (1957) and Cahn and Larche (1986) come immediately to mind. It is not the purpose of this contribution to summarize an already well-reviewed field but to examine several cases of diffusional transformation in which elastic energy plays a major part, and where the transformations are rather far from equilibrium; for example, initial growth from solid solution, steady discontinuous precipitation.

It is at first surprising that strain energy is important in these examples; total strain energies are generally considerably smaller than the chemical free energy changes attending transformation. Strain energies are known to influence coarsening and other processes close to equilibrium quite profoundly. However, as will be demonstrated here, they also play a critical role in certain nucleation and initial growth processes.

At the beginning, a distinction will be drawn between strain energies due to misfitting precipitates, where the stress-free transformation strain is confined to the precipitate phase, and those cases where strain energies are due to solute gradients in the parent phase. (For these, the coherently connected parent crystal far from the precipitate acts to constrain a solute enriched or depleted layer of parent material adjacent to the transformation interface.) Thus, the "misfit" and "coherency" fields will be separately considered.

Eshelby (1957), following the tradition of Crum (1940) and Mott and Nabarro (1940) has considered the elastic field of a misfitting ellipsoidal inclusion. He shows that the elastic strain energy is proportional to the volume of the misfitting precipitate.

The coherency field caused by the solute concentration gradient was examined by Cahn (1961) in his treatment of spinodal decomposition. The elastic energy develops through the "self-stressing" of adjacent coherently connected regions of differing lattice parameter. For the case of a layer of composition C in coherent contact with a constraining matrix of composition C_0, the strain energy density w is given by

$$w = Y(C-C_0)^2 \eta^2 \qquad [1]$$

where Y is a collection of lattice vectors and elastic constants, which reduces to a function of Young's modulus, E, and Poisson's ratio, ν, for isotropic solids: $Y = E/(1 - \nu)$. η is a linear coefficient of expansion; $\eta = d(\ln(a))/dC$; a is the lattice parameter.

The distinction between the two types of strain field will be maintained here for convenience and clarity. Experimental cases exist where one, but not the other, type of strain is present. In other cases, the two fields are both present and the interaction of strain energies is of interest for the purpose of estimating local equilibria, as shown by Voorhees and Johnson (1986).

A number of examples of misfit and coherency strains - as they are thought to affect the early stages of phase transitions - will be considered in the following sections.

I. Misfit strains and nucleation

In textbooks, (e.g. Shewmon (1969)) precipitate misfit strains are often incorporated as a simple additive term in the volume free energy charge for nucleation. This can be justified on the basis of Crum's result (1940) as generalized by Eshelby (1957). The elastic energy in the precipitate and matrix is directly proportional to the volume of the precipitate.
For cases where the misfit is pure dilation, we would expect no nucleation autocatalysis, since the interaction of dilational strain fields is minimal.

Hydrostatic stress fields can act to aid or inhibit nucleation, to the extent that they diminish or abet the work of nucleation. Thus any applied stress, or internal stress, with a hydrostatic component is capable of catalyzing or suppressing the nucleation of a misfitting precipitate.

We next consider the influence of a solute field around a growing precipitate, and the possibility of the resultant elastic field influencing the nucleation of nearby misfitting precipitates. (Since the critical nucleus is supposed to be in equilibrium with an homogeneous supersaturated matrix, the nucleating particle will not possess a solute field). The solute concentration variation near the growing precipitate will correlate with a stress-free lattice parameter variation (indexed by the linear expansion coefficient, η). If the precipitate is a sphere, and solute is rejected to the solid matrix during growth, the annular region near the precipitate will tend to undergo a change in volume, but will be constrained by the surrounding matrix. At a point far from the perturbed region, this will appear as a simple dilation; the interaction between such coherency strains and the nucleating misfitting particles will be negligiblY small, provided the nucleating particle is not in the solute field of the growing one.

Points inside the solute field of a growing particle will experience superimposed misfit and coherency fields. It has recently been shown that there is no interaction between these fields; for an arbitrary, radially symmetric solute field $C(r)$ outside the precipitate, the component of displacement of the precipitate/matrix interface which can be attributed to the coherency field is negligible (Purdy and Dryden, 1988).

When the transformation strains are not simple dilations, many of the above simplifications are lost, and some interesting behaviour is observed. Even the definition of "misfit" require further consideration as in the case of tetragonal θ' in dilute aluminium-copper alloys. The broad faces are essentially immobile and fully coherent with the f.c.c. matrix; the plate edges, (and the (equivalent) edges of ledges) are mobile and non-coherent.

The elastic misfit in the habit plane differs from that normal to it. The latter quantity also varies with the precipitate thickness (Stobbs and Purdy, 1978), during initial growth from supersaturated solid solution. Integral numbers of unit cells (or equivalent structural units) of each phase meet at the plate-edge, and in general a residual elastic misfit results from the lack of perfect coincidence, as illustrated in Fig. 1. The residual term can be of either sign, as suggested by the figure.

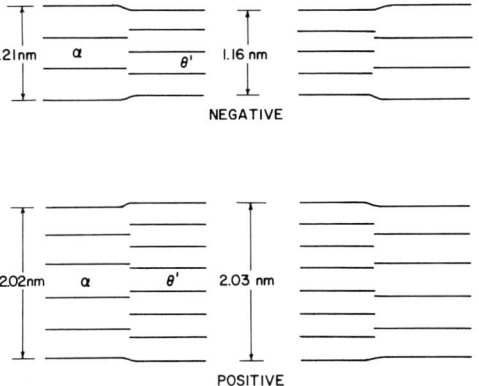

Fig. 1. Schematic cross-of θ' plates in α Al-Cu indicating the sense of the residual (elastic) misfit normal to the habit plane. The misfit in the habit plane is small.

The effect of the tetragonal transformation strains on nucleation of adjacent θ' particles was first noted by Lorimer (1968), who considered that the misfit field, and the solute field around a growing plate could interact to produce an optimum location and orientation for the "homogeneously" nucleating particle.

This type of autocatalysis, which, it is believed, leads to the linear arrays shown in Fig. 2, has been studied in detail by Perovic, Purdy and Brown (1981). Modelling of the elastic interaction between existing and nucleating particles, (for example, Fig.3) was used to demonstrate the probable origin of the autocatalytic event, and to rationalize the geometry of the observed precipitate arrays.

It is interesting to note that precipitates in the one-dimensional arrays so generated also appear to be strongly metastable with respect to further growth

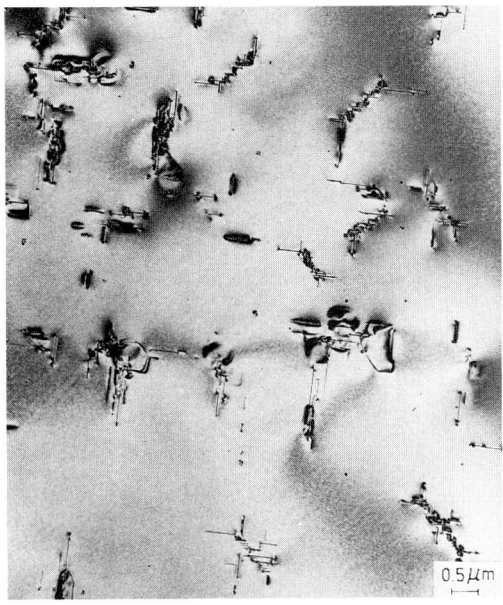

Fig. 2. Bright field T.E.M. image of [100] foil of Al-4% Cu, showing initial stages of linear array formation. Most of the θ' plates are imaged edge-on; the strain fields give rise to diffraction contrast. After Perovic (1980).

Fig. 3. Elastic interaction energy, E_{12}, between mutually perpendicular large and small (nucleating) plates as a function of separation $D(= l/r)$. The misfit is ε, the shear modulus μ, and the volume of the large plate V_o.

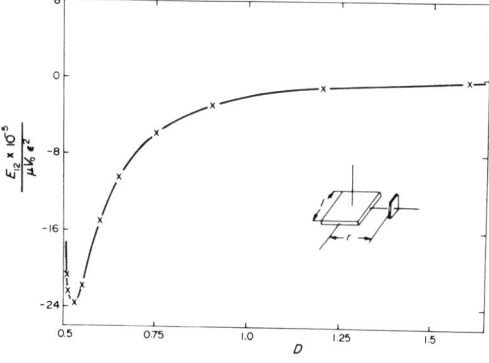

(lengthening) and with respect to coarsening (Fig. 4), as would be expected for precipitates in a strain energy-minimizing configuration (Perovic, Purdy and Brown, 1979).

These considerations apply, in slightly modified form, to the thickening of the θ' plates, for cases where the thickening is dependent on the supply of ledges through a "homogeneous" two-dimensional nucleation process. Then the interaction of the nucleating player with strain fields of the existing plate and any ledges on it will influence the location and frequency of "pillbox" nucleation (Perovic and Purdy, 1983).

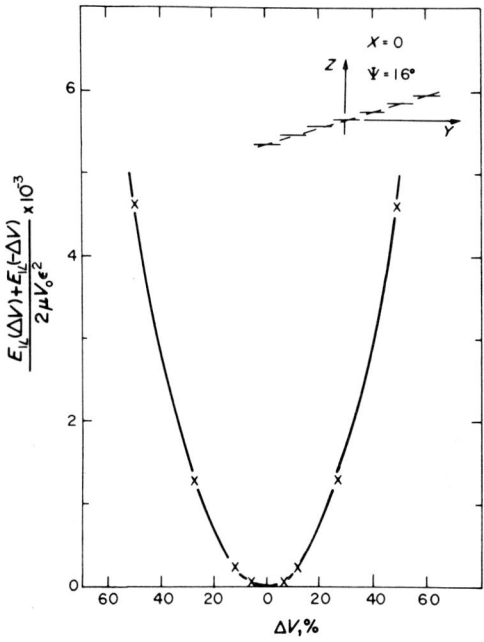

Fig. 4. The change of elastic interaction energy of a linear array of parallel plates due to volume exchange between a central plate and a near neighbour.

These studies, taken in conjunction with others (eg. Hosford and Agarwal 1975) suggest that comparatively small elastic strain energy terms can have a decisive influence on the nucleation event, which is generally supposed to involve an equilibrium between parent and daughter phases; indeed, from studies of autocatalysis, and of the effects of externally applied stress on precipitate nucleation, one can gain further and perhaps critical insight into the nature of the nucleus in diffusional solid-sold transitions.

II Coherency strains and interface migration

a) Liquid film migration. In many cases, misfit strains will coexist with coherency strains. For example, in the growth of a spherical particle from solid solution under conditions of local equilibrium, essentially independent misfit and coherency strain terms would each exert an influence on the equilibrium boundary conditions for precipitate growth, and hence on the kinetics of precipitate growth.

In the process of liquid film migration (Yoon and co-workers, 1986) the product is arguably strain-free; it forms as an homogeneous solid solution from a thin liquid layer. Thus the solute field in advance of the liquid layer is the main possible source of elastic strain energy, provided that the solute-enriched region of the parent crystal maintains coherent contact with the unaffected parent.

A series of recent experiments by Rhee and Yoon (1988), designed to test the idea that liquid film migration is driven by solute field coherency strain energy, has provided striking confirmation of the effect. The velocity of migrating liquid films in the sintering of Ni-Mo-Co-Sn alloys is proportional to the square of the misfit (see equation 1); the effect vanishes when the misfit is brought to zero.

The means by which the elastic term couples to the migration process is of interest. A solute concentration gradient across the liquid film results from the equilibration of the liquid with a stressed solid at the leading edge, and with an unstressed solid at the trailing edge, as illustrated in Fig. 5. The rate of diffusion through the liquid layer then determines the film velocity.

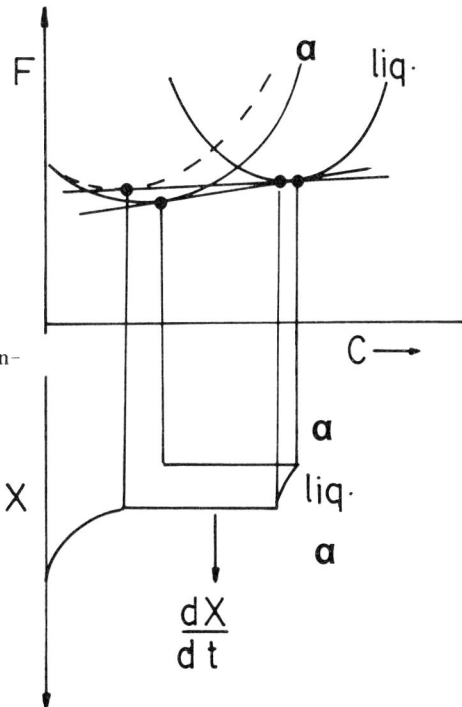

Fig. 5. Schematic free energy curves for a binary liquid-phase-sintering system. The concentration field in the parent α crystal raises its free energy, and results in a concentration gradient across the liquid film.

This model is now supported by quite conclusive experimental evidence; Yoon and co-workers (1986, 1987). We thus have an example of a phase transformation in which the dominant (only?) elastic term is the coherency strain associated with a solute field; the phenomenon depends on this term for its existence.

b) <u>Chemically Induced Grain Boundary Migration</u>. This process, the subject of at least two recent and comprehensive reviews (King, 1987; Handwerker, 1988) has certain characteristics in common with liquid film migration: Lateral diffusion along a grain boundary results in the normal migration of the boundary and a composition change in the wake of the boundary (Fig. 6). Of course, no liquid phase is present, but the grain boundary (in CIGM) and the liquid film (in LFM) each act as a mobile high-diffusivity path of limited thickness.

The process is widespread in binary solid systems. It is capable of producing substantial intermixing in situations where none would otherwise be present (i.e. where the processes of diffusion in a static boundary, and in the adjacent crystal lattices act "in series").

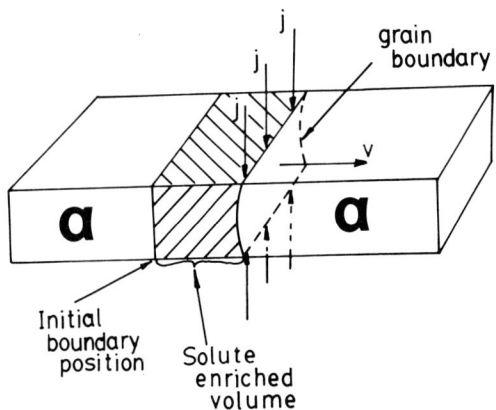

Fig. 6. Idealized solid-vapour couple, in which solute diffusion along a grain boundary leads to boundary motion, and to the solute enrichment of the swept volume.

The process was first recognized by den Broeder (1972) in solid-solid systems, and independently by Hillert and Purdy (1978) for the solid-vapour system Fe-Zn. It was originally thought that a prerequisite for chemically-induced grain boundary migration was the essential "freezing out" of lattice diffusion (Cahn and Baluffi, 1981); however, it was soon determined (Tashiro and Purdy, 1983) that the process could be induced in the system Al-Zn, under conditions where substantial lattice diffusion occurred ahead of the migrating boundary. A mechanism for coupling boundary motion to volume diffusion, through the solute-field coherency strain, analogous to that for liquid-film migration was therefore proposed for this system. Indeed, Hillert (1983) and Handwerker (1988) have opined that the driving force for CIGM lies in the coherency strain term, whatever the magnitude of the nominal volume penetration, Dv/v. (D_v is the volume diffusion coefficient, v the boundary velocity.).

The similar origins of driving forces for LFM and (high temperature) CIGM was recently confirmed in a series of elegant experiments by Rhee and Yoon (1988). For the same Mo-Ni-Co-Sn specimens, the velocities of both liquid films and grain boundaries showed a parabolic dependence on atomic misfit, consistent with equation 1, Fig.7.

Thus, even allowing for important differences between the processes, the coherency strain field undoubtedly supplies the driving force for grain boundary migration in cases where volume diffusion penetration of the parent grain is greater than atomic dimensions. There remains much to learn about high temperature CIGM, and about the storage of elastic energy in the product phase; however, the main question of current interest concerns the lower temperature process, in which the estimated diffusion penetration (Dv/v) ahead of the migrating boundary is of order 1 nm or less. It is clearly naive to assume that the transition is abrupt between the high-diffusivity grain boundary and the crystal lattice. The exchange between the boundary and the immediately adjacent lattice sites is undoubtedly easier than that within the lattice, so we might consider that the first layer of the crystal has its composition changed, and is subject to a consequent stress, in cases where volume diffusion is negligible (e.g. Fe-Zn). Unfortunately the relevant exchange frequencies are not simply accessible to experiment.

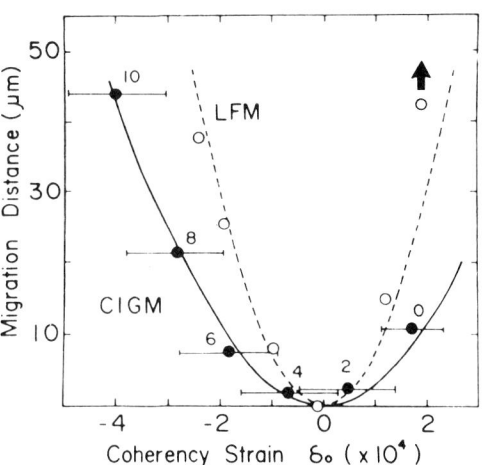

Fig. 7. Observed variation of average migration distances of grain boundaries (CIGM) and liquid films (LFM) for the same Mo-Ni-Co-Sn specimens, as a function of atomic misfit δ_o, after Rhee and Yoon (1988).

An alternative description would simply allow the vanishing of the volume diffusion field, and the development of a free energy step, or chemical driving force, at the boundary, as originally envisioned by Hillert and Purdy (1978).

The question requires further study. We will return to this and similar problems in a later section.

c) Discontinuous precipitation. The phenomena of CIGM has obvious relevance to a discussion of discontinuous precipitation, just as the study of LFM yields insight in CIGM.

It is often observed that at the initiation of discontinuous precipitation (DP), an initially static boundary bows out between precipitates, and feeds solute to them (e.g. Michael and Williams, 1986). In these cases, CIGM is clearly involved in the early stages of discontinuous precipitation. One proposed mechanism (among many) for the initiation of boundary movement in both CIGM and DP involves the differential storage of elastic energy in volume diffusion fields on either side of an initially static grain boundary (Tashiro and Purdy, 1987). Elastic anisotropy is then required for initiation in the absence of strain accommodation by crystal defects and/or precipitate traction bias (Purdy, 1988).

Once the steady discontinuous process has developed, it is again observed that the grain boundary bows out between growing precipitate lamellae, (Fig.8) indicating that a local driving force exists to overcome the capillary term. In effect, the study of the details of discontinuous precipitation provides us with a microscopic laboratory for the further investigation of CIGM; the descriptions of both phenomena must be internally and cross-consistent.

As in CIGM, the question naturally arises in the study of discontinuous precipitation concerning the effect of volume diffusion in front of the transformation interface: Is some minimum penetration of the parent phase necessary for the coupling of the lateral diffusion process with interface migration?

Figure 9 summarizes some estimated volume diffusion distances (again estimated as Dv/v) for both CIGM and DP in Al-Zn alloys. To the extent that Dv/v is a reasonable estimate of this quantity, the figure suggests that both the fully

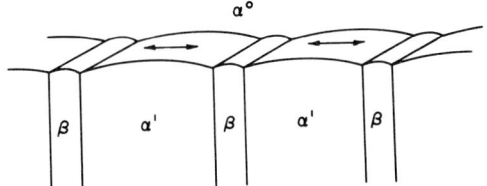

Fig. 8. Schematic discontinuous precipitation front, showing the interfacial diffusion path, and the sense of curvature of the grain boundary (the α^o/α interface).

relaxed and the unrelaxed processes can exist in the same alloy system, and at the same temperatures.

Fig. 9. Estimated volume diffusion penetration of the parent grain for Al-Zn alloys, which are subject to both discontinuous precipitation and CIGM.

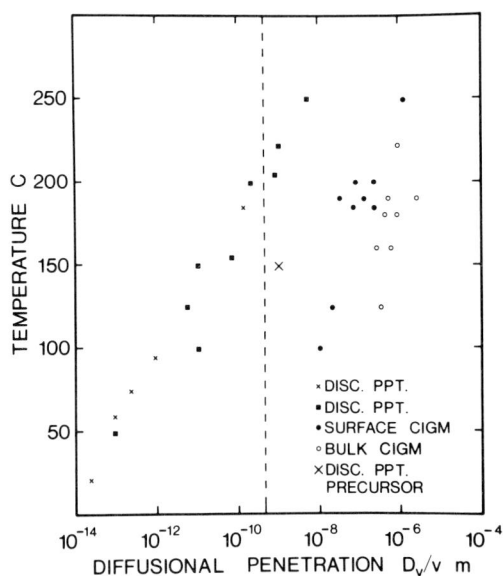

We have recently undertaken a series of in-situ studies of steady discontinuous precipitation in Al-Zn (Tashiro and Purdy, 1988). Previous investigations provided information on interfacial diffusion coefficients, and on solution thermodynamic data for this system (Solorzano, Purdy and Weatherly, 1986; Rundman and Hilliard 1967). The STEM in-situ studies permit the simultaneous acquisition of growth kinetic information, and high resolution imaging and microanalysis.

A characteristic of this system is the fast continuous decomposition of the supersaturated matrix, a process that partially replaces chemical free energy by strain and gradient free energies everywhere within the parent crystal, and effectively reduces the driving force for the discontinuous reaction.

We have assumed that the parent phase has decomposed via this competing process until it has reached a coherent equilibrium. Then the remaining driving forces for the formation of the discontinuous product lie in the reductions of strain energy and surface energy attending the reaction. The net driving force is determined from the original chemical free energy change, less the computed free energy losses due to coherent decomposition. Because the composition of the product α phase is a variable, the local chemical force free energy change across the α^o/α grain boundary is also a variable; it is this variation in chemical free energy difference which is balanced against variations in curvature to give the computed shapes of Fig.10. This treatment contains the implicit assumption that

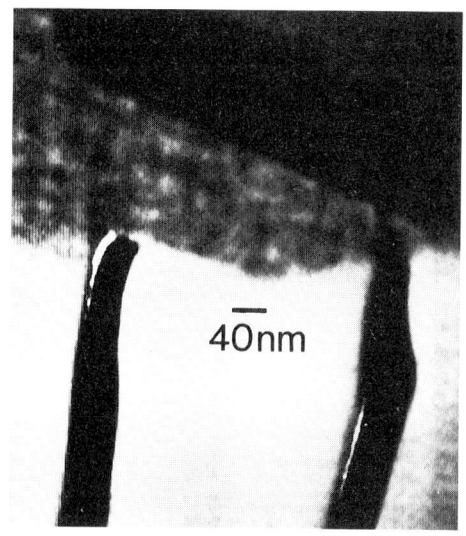

Fig. 10. Computed and observed shapes of a discontinuous precipitation front in an Al-22 At.% Zn foil. The STEM image is to be compared with the three computed shapes, each corresponding to a different amount of free energy lost in continuous decomposition of the parent phase, i.e., 70 to 80 joules/mol.

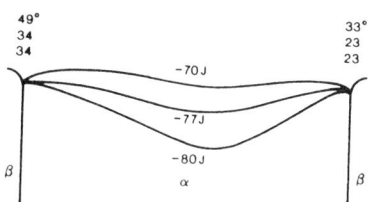

the interfacial mobility is very high; however, the results are not much different if realistic grain boundary mobilities are included. The agreement between observed and calculated morphologies is considered promising, and lends support to the idea that the remaining chemical free energy difference acts across the interface, unmoderated by volume diffusional relaxation. Since the continuously formed coherent precipitates, imaged in high resolution, are seen to intersect the transformation interface, we are further encouraged to believe that volume diffusion relaxation effects are negligible.

The analysis described above treats the parent phase as homogeneous, but with reduced supersaturation; in effect, the modulations due to the coherent transformation are considered to posses a fluctuation wavelength that is sufficiently small that is is quickly relaxed within the grain boundary. It is true that the lateral grain boundary diffusion distance and the associated relaxation time to the discontinuous precipitation reaction are much larger than the intersected coherent transformation wavelength.

Note that the discontinuous transformation is driven, in part, by strain energy in the coherently segregated parent crystal. Thus even here, where volume diffusion relaxation in the advance of the interface is thought to be negligible, solute field (coherency) strains play a major role in determining transformation kinetics and morphology.

These observations are representative of a low-temperature regime. However, until recently there was no convincing evidence for the discontinuous precipitation in velocity-temperature regimes where volume diffusional relaxation is expected at the transformation front. Lee, Baik and Yoon (1987) have now shown that discontinuous precipitation in Ni-Mo-Fe is suppressed (along with LFM and CIGM) when the atomic misfit is reduced to zero, thus providing strong circumstantial evidence for the solute-field strain term in high-temperature discontinuous precipitate growth. Their work is similar to, but more quantitative than, Bohm's earlier observation (1961) that discontinuous precipitation in binary copper-based alloys occurs mainly when the atomic misfit is large.

Many copper-based systems display discontinuous precpitation at temperatures where Dv/v is small compared with atomic dimensions. It is tempting, based on Bohm's result, to ascribe the driving force for all such transformations to the solute-field coherency term. However, such an association would neglect the probable role of coherency strain in initiating discontinuous precipitation, as discussed earlier.

What is needed is a critical test to distinguish the effects of a solute field ahead of the transformation front. One possible approach is considered in a later section.

d) <u>Strain energy density caused by dilational eigenstrains and applied stress</u>.
In many cases, we are interested in the growth of a daughter phase(s), D, which is embedded in an originally homogeneous parent phase, P, with composition C_o. From an elastic viewpoint this suggests that we make the "stress-free" datum correspond to P of composition C_o.

For an irregularly shaped precipitate with different elastic properties than the matrix phase, the elastic problem is analytically intractable. If we can assume that the parent and daughter phase are elastically isotropic and identical then the situation is more amenable to analysis.

Let S represent the stress-free dilational transformation strain (eigenstrain) which would occur in the absence of constraint. (S is assumed to be dilational strain with 3S representing the fractional volume change at a point.)

If $\tilde{e}_{ij}$ represents the total strain, then $e_{ij} = \tilde{e}_{ij} - S\delta_{ij}$ is equal to the elastic strain. The stress σ_{ij} is found by applying Hooke's law to e_{ij}. In addition to these "locked-in" components, there may be an applied stress σ^A_{ij}. The strain energy density is thus given by $(\sigma_{ij} + \sigma^A_{ij})/2 \ (e_{ij} + e^A_{ij})/2$. This local energy density can influence the course of the phase transformation. It is an intensive parameter.

The elastic density is decomposed into $W_o = 1/2\sigma_{ij}\tilde{e}_{ij}$, $W_{intA} = \sigma^A_{ij}\tilde{e}_{ij}$ and $W_A = \sigma^A_i e_{ij}/2$. (There is no factor of 1/2 in W_{int} because $\sigma^A_{ij} e_{ij} = \sigma_{ij} e^A_{ij}$.) Since W_A is spatially uniform, it is inconsequential as a biasing factor on the chemical equilibria. The term W_{int} represents the interaction between applied and locked-in stresses. The term W_0 represents the elastic energy density arising from the eigenstrains S.

The dilational eigenstrain S can be broken down into coherency and misfit components. Depending upon the amount of volume diffusion, the composition C of the parent phase may change near the growing daughter phase. If η is the linear coefficient of expansion (caused by solute atoms in P) then the so-called coherency strain, S_c, is given as

$$S_c = \eta(C-C_0) \qquad [2]$$

The sign of η will depend on the particular system.

In addition to the coherency strain, there is also likely to be a misfit strain, S_m, arising from the fact that the daughter and parent phases are likely to have different molar volumes. Thus,

$$S_m = \frac{V_m^D - V_m^P}{3V_m^P}, \qquad [3]$$

where V_m^D represents the volar volume of the daughter and V_m^P represents that of the original phase phase of composition C_0.

To obtain explicit expressions for the strain energy density we must assign some average shape to a suitably representative test region. If the microstructure is equiaxed then it is natural to choose a spherical shape. This, of course, has the added advantage of yielding simple expressions for strain energy density.

The strain energy density W_0 is in general not continuous across a transformation interface. If the interface is located at a radius r_* then we let W_0^- and W_0^+ represent the densities on either side of r_* in the daughter and parent respectively. These are given by the formulae

$$W_0^- = Y\{\hat{S}_m^2 - 2\hat{S}_m e_* + e_*^2/\alpha\}, \qquad [4]$$
$$W_0^+ = Y\{\hat{S}_c^2 - 2\hat{S}_c e_* + e_*^2/\alpha\},$$

where e_* is the tangential strain at r_* and

$$e_* = 3\alpha \, r_*^{-3} \int_0^{r_*} S_m z^2 dz,$$

$$\alpha = \frac{1}{3} \frac{1+\nu}{1-\nu},$$

$\hat{S}_m, \hat{S}_c$ = S evaluated at r_*.

For thermodynamic purposes, we require only the discontinuous portion of W_0. Thus, in [4], the term e_*^2/α does not bias chemical equilibrium because it is continuous across r_*.

An expression analogous to [4] can be also found which expresses the interaction between the locked in stresses and an externally applied stress t_{33}^A corresponding to uniaxial loading in the x_3 direction. Thus, if $\Delta W_{int} = W_{int}^+ - W_{int}^-$,

$$\Delta W_{int} = Y\{\hat{S}_c - \hat{S}_m\} t_{33}^A f(\phi) \qquad [5]$$

where $f(\phi)$ varies from approximately 1 at the x_3 axis to -1 at the equator

$\phi = 90°$) and ϕ is measured from the x_3 axis.

The expressions in [4] and [5] represent intensive quantities and are used in discussions based on local equilibrium at a transformation interface. The total strain energy arising from a "locked-in" strain S is equal to YS^2 integrated over the volume and thus in the expression for the strain energy (which is an extensive quantity) there is no interaction between S_m and S_c. The same is true for the interaction between the applied stress, S_c and S_m (ie. overall there is no interaction energy between a "locked-in" and an applied stress). In a global sense, whether or not the applied stress hinders or abets the nucleation and growth will depend on the interaction with the loading device as discussed by Eshelby (1957).

e) <u>The effects of externally applied stress on discontinuous precipitation</u>.
Sulonen, in 1964, published the results of a series of experiments which demonstrated the effects of an applied tensile stress on the rate of discontinuous precipitation in a Cu-Cd alloy. For this system, the velocities of discontinuous fronts with normals parallel to the tensile axis were decreased by the application of stress (the parallel reaction), conversely, an acceleration was found for those boundaries whose normals were perpendicular to the tensile axis (the "perpendicular reaction"). Sulonen interpreted these observations in terms of the interaction of the applied stress with a solute-field coherency stress. That is, he proposed that a volume diffusion field exists ahead of the transformation front whose strain field couples with an applied stress. He conceived of the transformed region as stress-free, using the analogy of a void to describe the stress distribution in the neighbourhood of the transforming grain.

More recently, Hillert (1972) examined Sulonen's results, and showed how they could be interpreted in terms of a treatment which presumed a local equilibrium between the coherently stressed parent grain (containing a solute field) and a solute-depleted daughter grain whose free energy is raised by a capillary term. This analysis amounted to an alternative theory for discontinuous precipitate growth in a sample with no applied stress, one which was subsequently extended to include the effects of externally applied stress. Based on the assumption that the discontinuous product is strain-free, Hillert was able to obtain quantitative agreement with Sulonen's data for the deceleration of the parallel reaction for the low-stress regime. The agreement was lost for higher stresses, but Hillert noted that this was perhaps due to the assumption that the product phases are unstressed.

For a radially symmetric annular transformed region, we distinguish two limiting cases: 1) The effect of misfit of the tranformed region is neglected, and the interaction of the applied tensile stress with a solute field in the parent phase is assessed as a function of orientation of the transformation interface. In contrast to Hillert's assumption, the transformed region is subjected to the applied stress. Then the strain energy density W+ just outside the transformed region contains an additive angle-dependent quantity, linear in the applied stress t_{33}:

$$t_{33} f(\phi) \eta (C_i^+ - C_o) . \qquad [6]$$

C_i^+ is the concentration at the interface in the parent α^o.

Upon insertion into the free energy balance, this leads to a suppression of the growth rate for parallel interfaces and to an acceleration of growth for interfaces with normals perpendicular to the stress axis for the case of copper-cadmium, (Cd atoms are larger than those of the copper solvent).

Thus, Sulonen's results are reproduced to a good approximation. However, before concluding that the coherency strain provides the driving force for discontinuous precipitation at this relatively low temperature, we should consider an alternative model.

2) In a second treatment, the solute field ahead of the transformation front is suppressed, but the annular transformed volume is assigned a misfit. An angular dependence in the strain energy density difference across the interface is found; with the form

$$t_{33} \, f(\phi) V \cdot (C_o - \bar{C_i})$$ [7]

where $\bar{C_i}$ is the interfacial composition in the daughter α, and V is a function of the molar volumes of the product and parent phases

$$V = (V_m^\beta - V_m^{\alpha o})/V_m^{\alpha o}$$

Thus a positive misfit in the transformed region will have the same effect on growth as a negative misfit in the solute field outside the transformed volume. For the case of Cu-Cd, Sulonen's results are not sufficient to differentiate between the two possible causes of the observed stress dependence. However, there may be systems for which Sulonen's methods can be used to detect the presence of a solute field, if one exists.

(f) <u>The massive transformation</u>. The solute 'spike' reappears, somewhat unexpectedly, in the composition invariant massive transformation, commonly thought to be diffusionless. Here, recent experiments has shown solute-enriched (and solute stabilized) δ-ferrite films at the intersections of massive γ grains in a quenched stainless steel (Singh, Purdy and Weatherly, 1986). It is clear that such systems have a strong tendency to equilibrate locally; a solute spike is capable of forming at the expense of the local interfacial free energy difference. Hillert (1984) treated the massive transformation as one in which the driving force can derive in part from the strain energy in the parent phase solute field, in close analogy with CIGM.

A more complete description would include both misfit and coherency strains, as outlined in section (d). For the special case where the misfit can be neglected, the solute field strain energy can be associated entirely with the parent phase, and the situation is as illustrated in Fig. 11. Note that we have required that the two interfacial chemical potential differences $\Delta\mu_A$ and $\Delta\mu_B$ be of the same sign, although this need not always be so (Langer and Sekerka, 1975, Kirkaldy, 1987). The chemical potential differences are related to the interfacial fluxes, j^{int}, by

$$\begin{bmatrix} j_A^{int} \\ j_B^{int} \end{bmatrix} = - \begin{bmatrix} L_{AA} & L_{AB} \\ L_{BA} & L_{BB} \end{bmatrix} \begin{bmatrix} \Delta\mu_A \\ \Delta\mu_B \end{bmatrix}$$ [8]

In principal, knowledge of the interfacial fluxes is equivalent to a knowledge of the interface velocity, and specification of the symmetric kinetic matrix [L] along with thermodynamic and elastic parameters is sufficient to determine the transformation interface velocity for given undercooling. Of course, $\Delta\mu_A$ and $\Delta\mu_B$ must be adjusted so as to maintain the global composition invariance of the transformation. The amounts of the driving force used in diffusional dissipation and in the interface reaction are shown on the Figure as ΔF_d and ΔF_r. Thus, it seems that both solute field and misfit strains will play significant roles in the high-temperature massive transformation.

Again the questions arises: At what point, if any, is the solute field lost, and how is a transition to a low-temperature massive transformation mode accomplished?

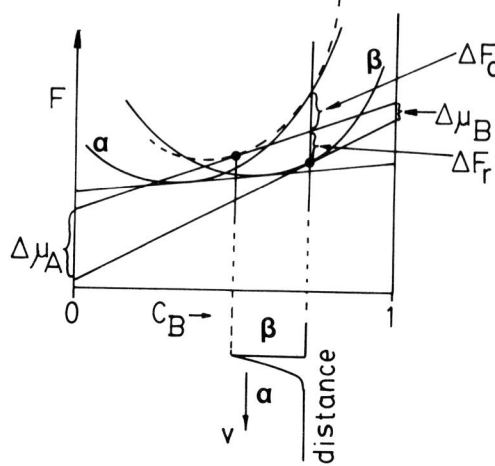

Fig. 11. Schematic free energy curves for the massive transformation of super cooled α to β. The free energy of the coherently strained α phase is shown as a broken line.

CONCLUSIONS

In this contribution, we have examined several ways in which elastic energies can influence nucleation and growth of a product phase. We have seen that there exists a range of problems in the literature of phase transitions, for which it is difficult to obtain information or clear quantitative inference, concerning near-boundary diffusion behaviour. This class of problem includes several considered here, but it is in fact considerably broader than the scope of the previous sections.

Take for example the question of alloying element diffusion during phase transformations in ternary steels, as reviewed by Hillert and Ågren (1988) in their contribution to this symposium. In that problem an alloying element "spike" in the parent phase is required for local equilibrium during unpartitioned growth. As transformation rates increase, a related decrease is expected in the volume diffusion penetration of the substitutional alloying element in the parent phase. At some stage, it is thought that local equilibrium is lost, and that a transition occurs to a 'paraequilibrium' state, in which the fast-diffusing species (carbon) is in equilibrium at the interface, but the slower diffusing alloying element is immobile during the passage of the interface. The point at which the transition from full equilibrium occurs has been a focus of debate over the past few decades.

It is interesting to note that neither solute-field strain energies nor gradient energy terms have been invoked in these treatments. As the scrutiny of the near-interface diffusion behaviour intensifies, these continuum concepts may afford fresh insights into the problem of alloying element distribution. It is possible, too, that analyses which recognise the discrete nature of the system, or perhaps analyses which represent an extension of solute-drag theory (Cahn, 1961; Hillert, 1968), will be of value.

For all of the transformations considered here (excepting liquid film migration), a transition is envisaged in which a high temperature near-local-equilibrium transformation is replaced by a lower-temperature mode, in which a major component of the driving force acts directly across the interface, and in which interfacial dissipation processes become significant. Hard evidence for such transitions is at present sparse.

The study of such near-boundary phenomena presents some interesting experimental challenges. Perhaps the effects of applied stress studied in carefully chosen systems will enable the further understanding of this important class of phenomena.

On a more general level, we are encouraged by recent progress in the coupling of the mechanics of solids with thermodynamics and with kinetic theory. The elastic energy term is generally accessible to analysis, and it is capable of influencing processes which occur near equilibrium, as well as those processes which are clearly far from equilibrium; the solute field, for example, is capable of generating energy densities of the same order as the chemical terms, and it can therefore exert a profound effect on the course of phase transitions in solids.

ACKNOWLEDGEMENTS

This research was supported by the Natural Sciences and Engineering Research Council of Canada. One of us (G.R.P.) gratefully acknowledges the friendship, counsel, and inspiration freely provided by his mentor and colleague, Jack Kirkaldy, for more than a quarter-century.

REFERENCES

Bohm, H. (1961). Z. Metallk. 52, 564.
Crum, M.M. (1940) quoted by F.R.N. Nabarro. Proc. Roy. Soc. 52, 90.
Eshelby, J.D. (1957). Proc. Roy. Soc. A241, 376.
Cahn, J.W., and R. Baluffi (1981). Acta Metall. 29, 493.
Cahn, J.W., and F.C. Larche, (1986). In "Solute-Defect Interaction", eds., Saimoto, Purdy, Kidson, Pergamon, 1.
Cahn, J.W. (1961). Acta Metall. 9, 795.
den Broeder, F.J.A. (1972). Acta Metall. 20, 319.
Handwerker, C. (1988). To be publsihed in "Diffusion Phenomena in Thin Films", ed., D. Gupta, Noyes Publications.
Hillert, M. (1969). In "The Mechanism of Phase Transformations in Crystalline Solids", The Institute of Metals, Monograph No. 33.
Hillert, M. (1972). Metall. Trans. 3, 2729.
Hillert, M. (1983). Scripta Metall. 17, 237.
Hillert, M. (1984). Metall. Trans. A, 15, 411.
Hillert, M., and G.R. Purdy (1978). Acta Metall. 26, 333.
Hosford, W.F. and S.P. Agarwal (1975). Metall. Trans A 6A, 487.
King, A.H. (1987). International Materials Reviews 32, 173.
Kirkaldy, J.S., K.K. Shrivastava, and G.R. Purdy (1983). Scripta Metall. 17, 655.
Kirkaldy, J.S., (1987). Scripta Metall. 21, 953.
Langer, J.S., and R.F. Sekerka (1975). Acta Metall. 23, 1225.
Lee, K.-R., Y.-J. Baik, and D.K. Yoon (1987). Acta Metall. 35, 2145.
Lorimer, G. (1968). In "Proc. 4th European Regional Conf. on Electron Microscopy", Tipografia, Poliglota, Vaticana, Rome, 491.
Michael, J.R., and D.B. Williams (1986). In "Interface Migration and Control of Microstructure", eds. Pande, Smith, King, Walter, ASM, 73.
Mott, N., and F.R.N. Nabarro (1940). Proc. Roy. Soc. 52, 86.
Perovic, V. (1980). Ph.D. Thesis, McMaster University.

Perovic, V., G.R. Purdy, and L.M. Brown (1979). Acta Metall. 27, 1075.
Perovic, V., G.R. Purdy, and L.M. Brown (1981). Acta Metall. 29, 889.
Perovic, V., and G.R. Purdy (1983). Scripta Metall., 17, 1305.
Purdy, G.R. and J.R. Dryden (1988). Unpublished research.
Purdy, G.R. (1988). Proceedings, "Int. Conf. Phase Transformations '87" to be published by The Institute of Metals.
Rhee, W.-H. and D.K. Yoon (1987). Submitted for publication.
Rundman, K.B., and J.E. Hilliard (1967). Acta Metall. 15, 1025.
Shewmon, P.G. (1969). "Transformations in Metals", McGraw-Hill, p. 211.
Singh, J., G.R. Purdy, and G.C. Weatherly (1985). Metall. Trans. A 16, 1363.
Solorzano, I.G., G.R. Purdy, and G.C. Weatherly (1986). Acta Metall. 32, 1709.
Stobbs, M., and G.R. Purdy (1978). Acta Metall. 26, 1069.
Sulonen, M.S. (1960). Acta Metall. 8, 669.
Sulonen, M.S. (1964). Acta Metall. 12, 749.
Tashiro, K., and G.R. Purdy, G.R. (1983). Scripta Metall. 17, 455.
Tashiro, K., and Purdy, G.R. (1988). Unpublished research.
Voorhees, P.W., and W.C. Johnson (1986). J. Chem. Phys. 84, 5108.

The Role of Creep in Diffusional Transformations

BY J. P. HIRTH

*Metallurgical Engineering Department,
The Ohio State University, Columbus, Ohio 43210*

Abstract

Many types of diffusional transformation involve creep of the matrix phase as a process in series with the diffusion of reactants. These include discontinuous precipitation, some classes of continuous precipitation, and internal reactions such as oxidation. Here, work on displacement reactions, diffusional barrier breakdown, and internal oxidation are summarized. While bulk diffusion is sufficiently slow for creep to be rate controlling in some cases, the system adjusts in all cases so that reactant diffusion is rate controlling.

Preface

It is a pleasure for me to participate in honoring Professor J. S. Kirkaldy. His innovative thinking, often leaping outside the bounds of conventional wisdom, has been a stimulus to me throughout my career. He has also been a valuable colleague in reviewing papers, serving on technical committees, and speaking at symposia. Finally, I have enjoyed his companionship on the trail and his voice in song around the campfire. May his multifarious activities continue for years to come!

1. Introduction

In a continuing research program, we have studied a number of systems in which the product phase intrudes into the matrix phase in the course of a diffusional phase transformation. This intrusion requires mass flow of the matrix, a series process, to accommodate the growth. The mass flow in principle can occur by plastic flow, by volume diffusional creep, or by other creep processes. In a comparison of diffusional resistances, or equivalently the magnitudes of free energy dissipation [1], one finds cases where volume diffusion should be slower than the reactant diffusivity and, hence, potentially rate controlling. Yet, as discussed in the following development, the systems that we have studied always exhibit reactant diffusion control to a very good approximation. In succession, we consider an aggregate displacement reaction, a periodic interwoven aggregate displacement reaction, a diffusional barrier reaction, and a case of internal oxidation.

2. Displacement Reactions

The displacement reaction between cuprous oxide and iron at 1000°C to yield iron oxide and copper [2] produces the rod-shaped aggregate product shown in Fig. 1. The model for the reaction is shown in Fig. 2. Of interest in connection with the mini-max principles of Kirkaldy, the system has an added degree of freedom, the ϕ/θ ratio in Fig. 2, that is adjusted to maximize the rate of dissipation of free energy in the process. In the present case, the diffusivity of O in Cu is slow compared to the diffusivity of iron cations and electrons, so the system selects ϕ/θ.

At the reaction front oxygen atoms, cations and electrons combine to form magnetite and extend the oxide rods into the copper phase. Thus, there must be local accommodation flow of copper to make space for the growing oxide. Yet, as indicated in Table 1, the kinetics of growth and the predictions of θ and ϕ are in agreement, with values calculated on the basis of no accommodation constraint; that is, within the scatter of data for diffusivities used for the calculations. The agreement of the magnetite/wustite growth length ratio is not as good, but this may be related to the non-ideal growth shapes of the rods. Hence, there is no indication in the results of a flow resistance associated with creep of the copper.

A second displacement reaction [3] is that between iron and nickel oxide to form austenite and spinel. The interwoven product has a Liesegang structure as shown in Fig. 3. The model for the reaction, Fig. 4, requires austenite to grow into the NiO phase with mass flow accommodation of the oxide. Yet, as shown in Fig. 5, the reaction product grows parabolically and, Table 2, the rate constants agree with those for oxidation of iron within the scatter in the data for the latter case. This again implies that the accommodation process does not significantly influence the overall reaction rate.

3. Diffusional Barrier Reaction

A study of a copper-silver-nickel triple [4] provided a prototype for the breakdown of a diffusion barrier. As illustrated in Fig. 6 for a reaction at 760°C, copper initially diffuses through the silver layer. When the copper reaches the Ni layer the Ni-Ag interface suffers an interface instability. Subsequently, dendrites of the Ni-Cu phase intrude into the silver and eventually the barrier breaks down when the dendrites reach the Cu-Ag interface. As illustrated in the model of Fig. 7, accommodation flow of Ag is required for the penetrating growth of the Ni-Cu phase. The kinetics in this case, expressed as a variation of ξ_y (see Fig. 7) as a function of time are complex as shown in Fig. 8. Yet, again the accommodation flow of silver has a negligible influence on the rate of growth of Ni-Cu protrusions, which is controlled by Cu diffusion through the silver layer.

4. Internal Oxidation

In all of the above examples, the distance over which accommodation flow must have occurred remained constant as the reaction proceeded. Hence, the creep kinetics would be expected to vary linearly with time, so that even the fact that the kinetics were parabolic with time indicated no role of creep. In the final example of internal oxidation, both the reactant diffusion and the accommodation creep processes would be expected to vary parabolically with time.

The system studied [5] was the internal oxidation of Ag-In alloys at 500-700°C. The local volume increase of 48 percent associated with indium oxide requires accommodation flow of the matrix. As internal oxidation proceeded, first silver nodules and eventually a continuous silver layer formed on the surface, indicating that accommodation was occurring by a matrix creep process with the faster diffusing silver atoms dominating the process.

A typical result is shown in Fig. 9. As indicated there, the expected accommodation creep process, Nabarro-Herring creep, was slower than the diffusion of the oxygen reactant, so the creep process was expected to be rate controlling. Yet the data, obtained with or without superposed external creep strain, agreed well with the oxygen diffusion rate control process. Calculations were then made for pipe diffusion accommodation and, as shown in Fig. 9, such a process with a dislocation density of 10^{11} cm/cc could accommodate the flow so that the rate approached that controlled by oxygen diffusion.

Subsequent microhardness measurements gave values consistent with a dislocation density of 10^{11} cm/cc. A mechanism consistent with these results is that in the initial nucleation of the oxide the large volume change is partially compensated by dislocation motion. The created dislocations provide the path for pipe diffusion and, by the mini-max principles, are stabilized at a fixed density to maximize the accommodation process.

5. Concluding Remarks

In all of the systems considered, and in some cases unexpectedly, the matrix accommodation processes are so rapid that they have no effect in the rate controlling kinetics, which in all cases correspond to diffusion of reactants. This need not always be the case. Candidate systems for accommodation rate control would be those where the diffusing reactants are rapidly diffusing interstitial atoms while accommodation creep requires substitutional diffusion. An example of such a system is the graphitization of cast iron, which appears to be controlled by volume diffusion of iron [6]. Also, internal hydriding of Ni-Ti alloys and Ag-Li alloys, while thermodynamically favorable at elevated temperatures, is completely suppressed because the accommodational flow process is so slow as to be negligible [7].

Acknowledgement

The author is grateful for the support of this work by the Office of Naval Research under Grant N00014-85-K-0196.

References

1. W.A. Tiller, J. Electrochem. Soc., 127, 625 (1980).

2. G.J. Yurek, R.A. Rapp and J.P. Hirth, Metall. Trans., 4, 1293 (1973).

3. C. Tangchitvittaye, J.P. Hirth and R.A. Rapp, Metall. Trans., 13A, 585 (1982).

4. C.S. Lin, R.A. Rapp and J.P. Hirth, Metall. Trans. 17A, 933 (1986).

5. S. Guruswamy, S.M. Park, J.P. Hirth and R.A. Rapp, Oxidation of Metals, 26, 77 (1986).

6. M. Hillert, this conference.

7. Y. Shueh, R.A. Rapp and J.P. Hirth, research in progress.

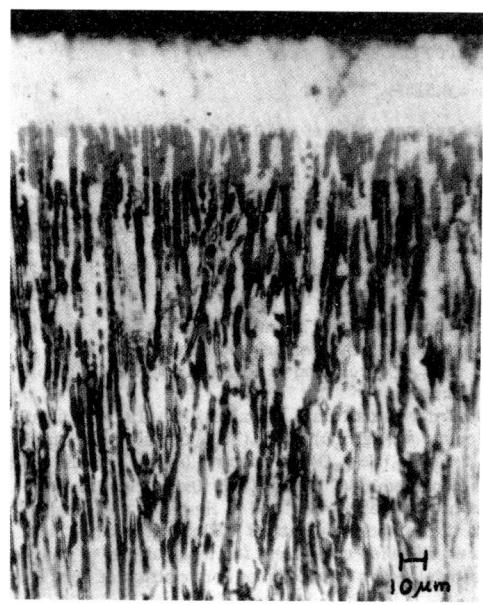

Fig. 1 Product zone of Fe/Cu$_2$O reaction after five hours at 1000°C, selective etched to remove wustite. White phase is copper, rods are wustite (black and dark grey areas) and magnetite (light grey phase adjacent to copper layer at top). Dark region at very top is unreacted cuprous oxide.

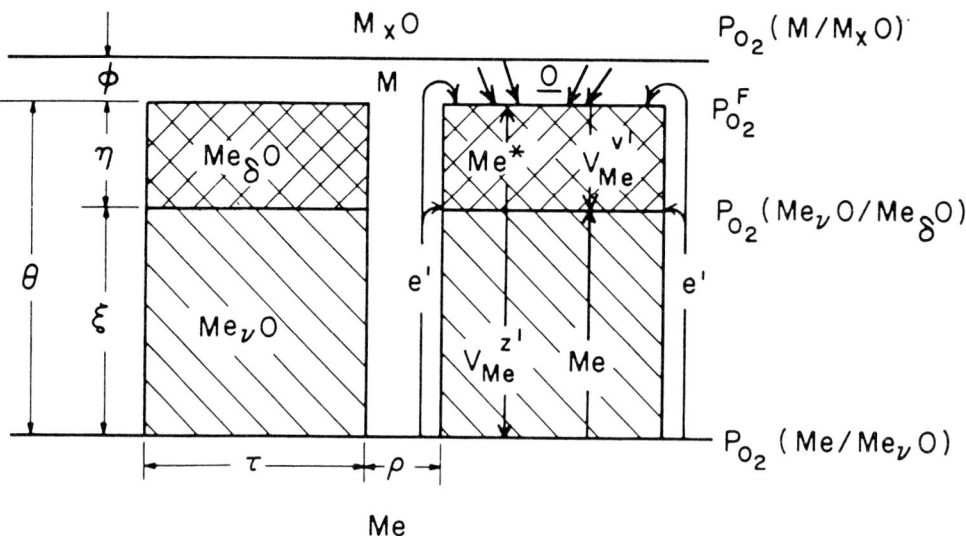

Fig. 2 A model for the growth of rod-aggregate displacement reaction products.

225

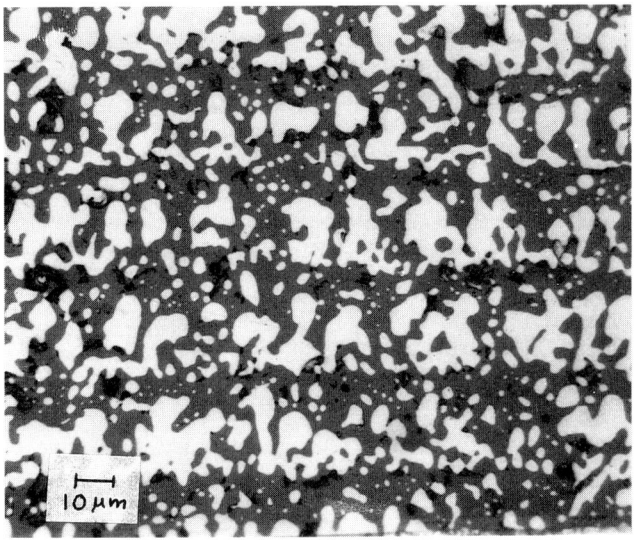

Fig. 3 Product of Fe/NiO reaction after eleven hours at 1000°C. White phase is austenite, dark phase is spinel.

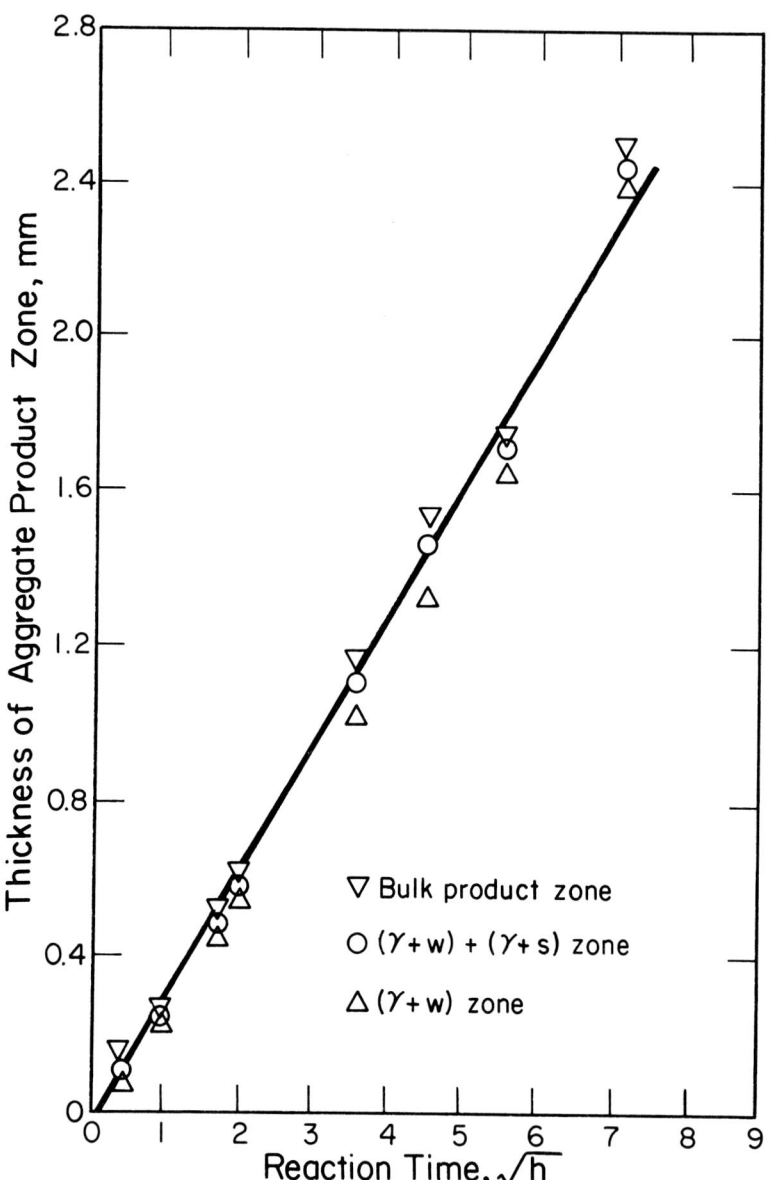

Fig. 4   Reaction product thickness as a function of square-root of time.

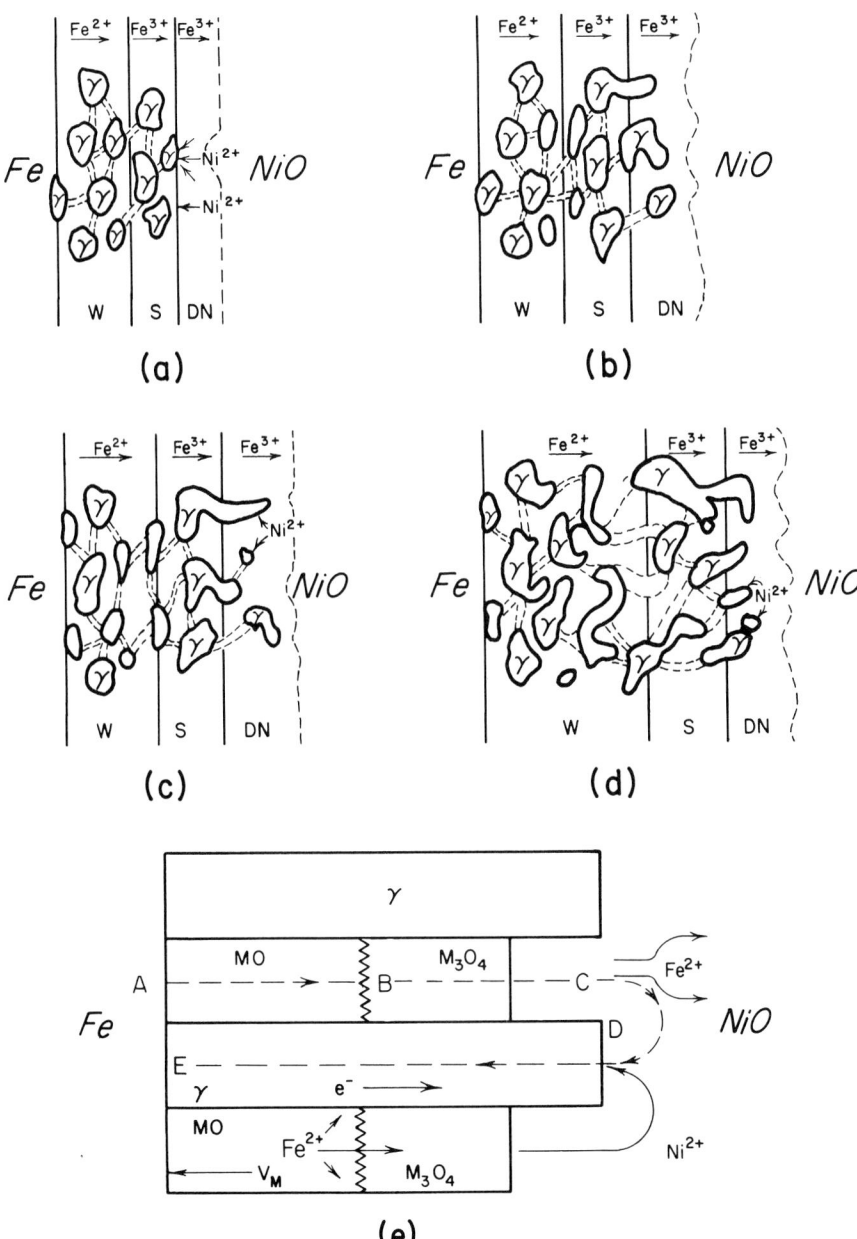

Fig. 5 Model for growth of the Fe/NiO displacement reaction product.

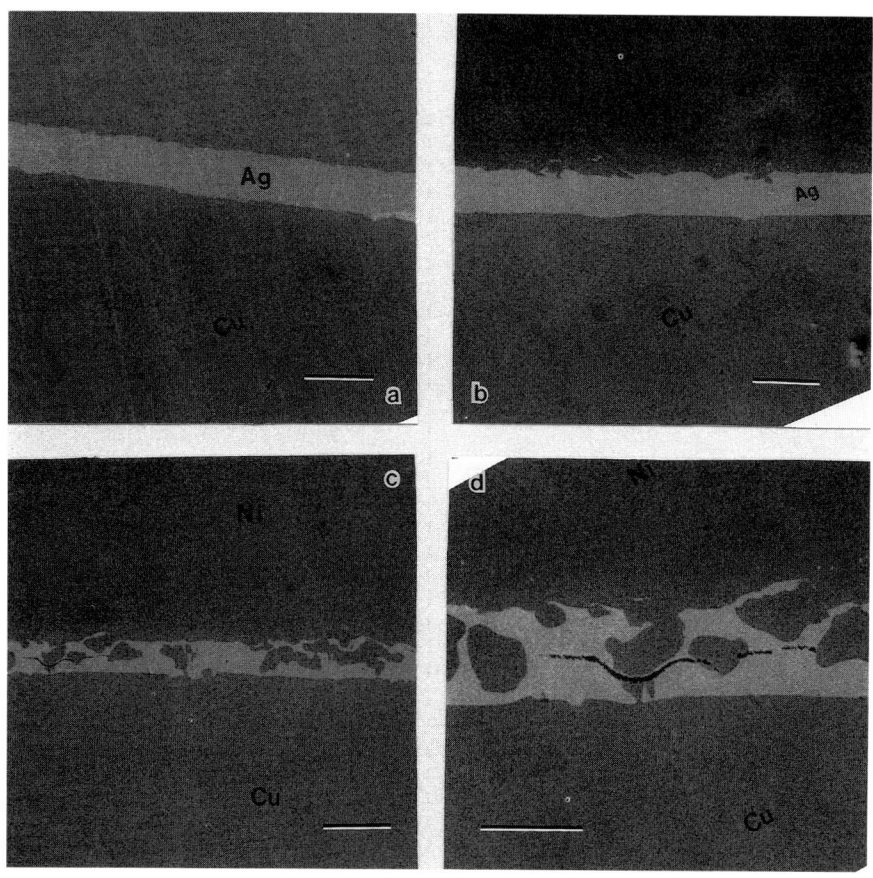

Fig. 6 Morphology of Cu-Ag-Ni triple with 25μm Ag layer, heated at 760°C for a. 12h, b. 1 day, c. 5 days and d. 5 days.

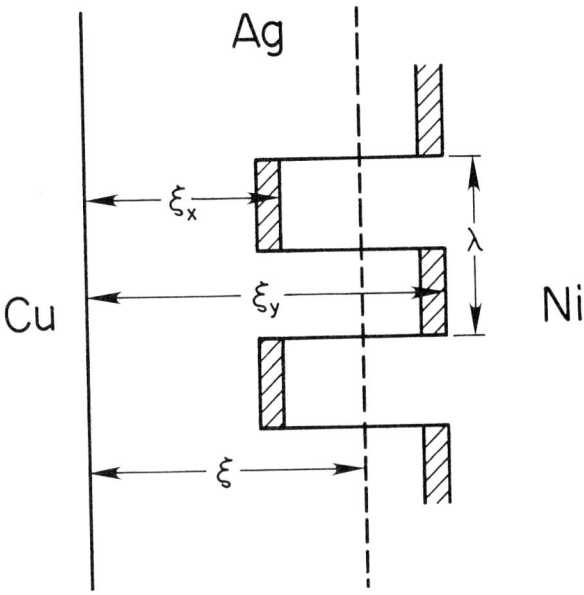

Fig. 7 Model for growth of Ni-Cu phase protrusions into Ag layer.

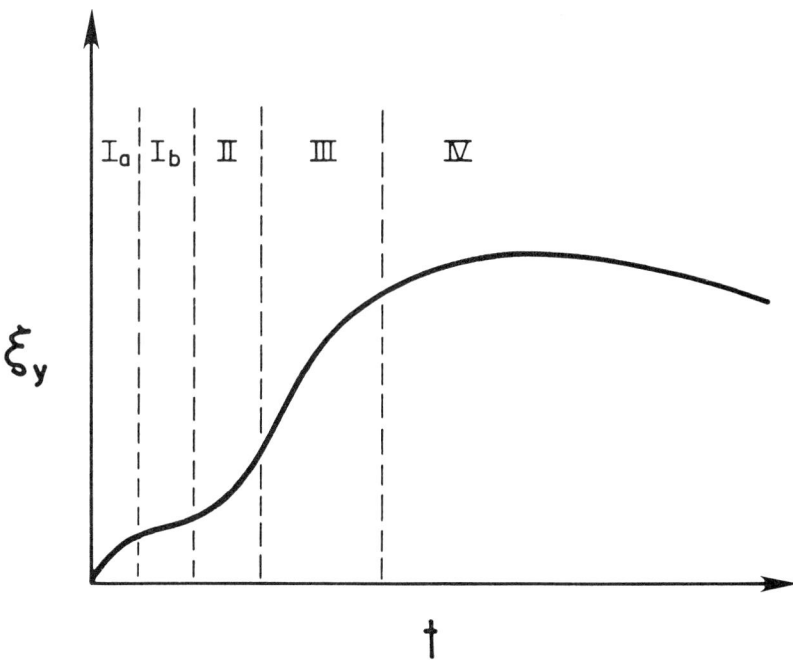

Fig. 8 Stages for growth of Ni-Cu protrusions as a function of time.

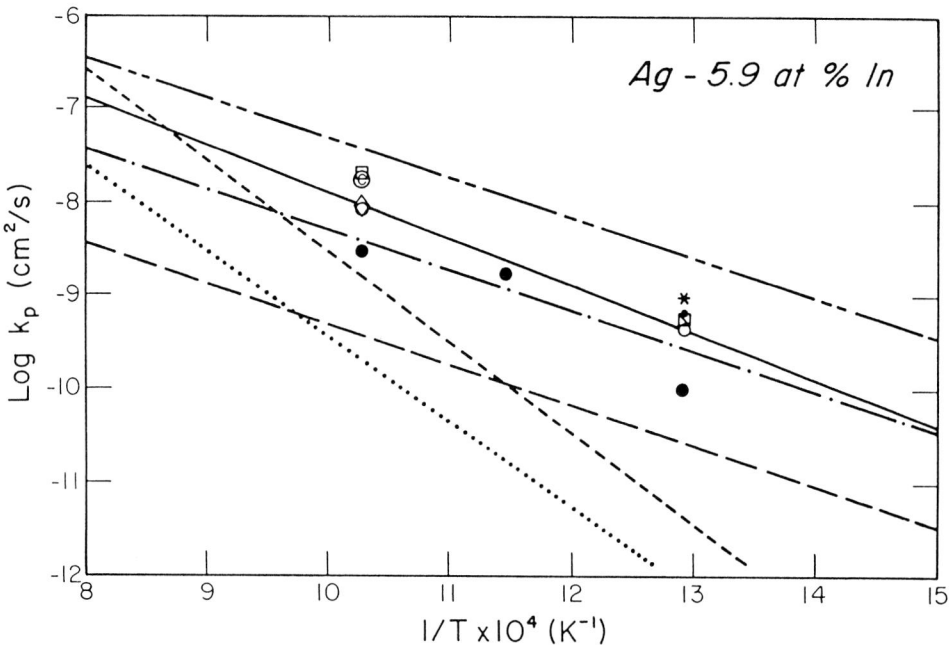

Fig. 9 Plot of oxidation rate constant k_p versus 1/T for the alloy Ag-5.9at.% In alloy. ____ Oxygen diffusion control; --- $\rho = 10^{10}$ cm/cc, _._ $\rho = 10^{11}$ cm/cc, _.._ $\rho = 10^{12}$ cm/cc. , No load (TGA); , No load; , $\dot{\epsilon} = 8.5 \times 10^{-7}$; *, $\dot{\epsilon} = 2 \times 10^{-6}$; , $\dot{\epsilon} = 2.4 \times 10^{-7}$; , $\dot{\epsilon} = 1 \times 10^{-7}$.

Dislocation Patterns and Deformation Bands

ELIAS C. AIFANTIS

*MM Program, Department of Mechanical Engineering–Engineering Mechanics,
Michigan Technological University, Houghton, MI 49931*

ABSTRACT

The methods of stability theory and nonlinear analysis are employed in order to explain the origin and evolution of dislocation patterns as well as the related phenomena of persistent slip bands, shear bands, and Portevin-Le Chatelier bands. A framework is developed for classifying and properly interpreting the hierarchy of plastic instabilities and their manifestations in the description of the macroscopic stress-strain curve. Preliminary results have been obtained which, among other things, provide the range and temperature dependence of the wavelength of the ladder structure of the persistent slip bands, in agreement with experimental observations. Moreover, they suggest the structure and thickness of shear bands, as well as the periodicity and velocity of Portevin-Le Chatelier bands.

KEYWORDS

Dislocations; patterns; plasticity; persistent slip bands; selforganization.

BACKGROUND

We outline a new interdisciplinary method for the fundamental understanding of the mechanical behavior of materials with emphasis on the tendency of deformation to organize itself in dislocation patterns and shear bands. The method is based on a recently developed framework for dislocations [1] viewing plasticity as a dynamical nonlinear dissipative process similar to other interesting and well studied nonlinear phenomena of hydrodynamics, chemistry, and biology [2]. The key issue is to identify the microscopic carriers of plastic deformation, the nonlinear interactions among themselves as well as the lattice, and then write the appropriate differential equations modelling their temporal and spatial evolution [3]. The deformation is thus obtained as a derivable quantity and its response to an externally applied agency, such as stress, is fundamentally established rather than apriori assumed as is the case in the usual constitutive equations or stress-strain relations approach [4].

The physical challenge is to capture and conveniently express the nonlinear interactions of dislocation species as they move, multiply, annihilate, and cluster in response to applied stress [1,3]. The mathematical task of the required nonlinear analysis is greatly facilitated through the newly developed tools of bifurcation and stability theory; in particular, via the method of slow mode dynamics [5] or center manifold theorem [6]. In fact, it is this feature of physical and mathematical modelling of the behavior near the instability that makes the proposed method unique and so much more promising than existing approaches to deformation. For example, it was only recently that a convincing dynamical theory was published [3,7] for the occurrence of the vein and persistent slip band structures routinely observed in fatigue experiments [8].

Specifically, it can be shown that uniform dislocation populations become unstable versus inhomogeneous ones when the stress reaches a threshold value. Indeed, the dislocations organize themselves in spatially periodic structures characterized by an intrinsic wavelength. This spatial selforganization and periodicity of the corresponding dislocation pattern arises as a dynamical instability due to terms modelling respectively the motion and production/annihilation of dislocation species. In fact, the two novel aspects embodied in the dislocation theory of [1d,1e,7c] are the introduction of gradient dependent terms in the equation of motion and nonlinear source terms in the equation of species conservation, both of which were absent from standard theories of continuously distributed dislocations [9]. This leads to the derivation of partial differential equations of the reaction-diffusion type for the dislocation populations [1,3,7].

Interestingly, the theory predicts wavelengths of the same order as those experimentally observed. Moreover, the analysis shows that due to continuous symmetry breaking a constant shift or glide of the entire pattern can easily be realized without relaxing back to its original configuration, this being in accordance with experimental observations [8]. Stationary phase fluctuations are also shown to exist leading to layer-splitting of the patterned structure and random development of super defects, also experimentally observed [8]. Moreover, the theory can provide a quantitative description of the competition between the ladder-like structure of persistent slip bands and the rod-like structure of the surrounding veins both of which are metastable beyond threshold and it shows that persistent slip bands can nucleate and travel as propagating fronts within the veins [3a]. Finally it can be shown that when secondary slip is assumed to compete with primary slip, stable labyrinth structures are generated [3c].

The fascinating problem of dislocation pattern formation triggering the initiation and growth of macroscopic shear bands is considered next. It can be shown that the inhomogeneous evolution of dislocation structures manifests itself in the form of higher order strain gradients in the macroscopic constitutive equation for the flow stress and the yield condition [1c,7c,7d]. The higher order strain gradients make it possible to dispense with the classical difficulties of uniqueness, stability, and mesh-size dependence in finite element calculations resulting from the loss of ellipticity in the govering equilibrium equations within the postlocalization regime. Indeed, this is due to the fact that classical approaches do not contain any characteristic length; an ingredient incorporated into the present approach through the introduction of second order strain gradients. The aforementioned results pertain to a single macroscopic shear band occurring in an isothermal specimen deforming under quasistatic or nearly stationary conditions. Similar difficulties exist in relation to the structure and velocity of propagating shear bands observed in specimens deforming in the Portevin-Le Chatelier (PLC) regime. This becomes clear in previous works dealing with modelling and mathematical aspects of the PLC effect [11] where an adequate qualitative treatment of the phenomenon, is provided but a quantitative analysis of the problem is absent.

Initial results obtained within the presently proposed gradient approach to the localization and patterning of deformation are quite encouraging. The structure and thickness of shear bands for hyperelastic materials can be obtained by introducing second order deformation gradients into the strain energy function [12a]. Similarly, the structure and thickness of shear bands for rigidly plastic materials is obtained by introducing second order strain gradients into the yield condition [12b]. Moreover, it is now possible to quantitatively capture the spatio-temporal periodicity of deformation as it manifests itself in the repeated passage of PLC bands through a specimen deforming under constant strain or stress rate conditions. In particular, the period, width, and velocity of such bands in viscoplastic materials, under constant stress rate conditions is predicted by introducing a second order strain gradient into the expression for the flow stress [7c,7d, 12b].

We conclude this section with a brief review of very recent works by other investigators who also begun to re-examine the problem of dislocation patterning and deformation bands by employing nonlinear methods. Among them, we distinguish recent efforts in France [13], Germany [14], Czechoslovakia [15], India [16], and United States [17]. Some of these works [13a, 13d, 15c] are similar to the present reaction-diffusion scheme and gradient approach to plasticity, while others [14a, 14b, 16] maintain the nonlinear character of the process but suppress the spatial dependence. Some of them [15a,b] are based on continuum arguments of microscopic elasticity theory and dislocation theory and others utilize molecular dynamics [17a,b,c] or cellular antomata [13b] techniques for dislocation species. A historical perspective on previous more classical approaches to dislocation patterning and the heterogeneity of plastic deformation [18-24] can be found in the lists of references quoted in the aforementioned most recent works. Similarly, there have been recent works, [25a,b,c] introducing higher order strain gradients into their mathematical structure in order to deal with the problems of ill-posedeness and shear band thickness in the postlocalization regime. However, there are important differences between these works and the present approach pertaining to the nature and precise form of the gradient dependence, as well as the exact character of nonlinearity and the definition of the shear band boundaries.

DISLOCATION PATTERNS AND PERSISTENT SLIP BANDS

As mentioned earlier the present approach to dislocation patterning is dynamical in nature. It differs from previous static procedures resting upon minimization arguments for the elastic strain energy associated with the dislocation species [e.g. 19,20,24]. In fact, self-organized dislocation structures are not usually at a low energy state and therefore minimization of a functional (strain energy or Lyapunov functional) is not always a feasible procedure. Moreover, computation of such functional must depart from standard textbook formulas on dislocation self-energies and interactions based on microscopic elasticity. This is especially true in situations of very dense dislocation structures, such as dipoles, multipoles, and small dislocation loops, where the long-range dislocation interactions can decay much more rapidly than $1/r$. In addition to screening effects, the core energies are often important for such dense situations as well as for the case of small dislocation loops. Nonlocal effects introduced, for example, through nonlocal elasticity could play a dominant role here and standard dislocation expressions such as the one for the Peach-Koehler force become unusually complex. Finally, even though dislocation reactions are central in determining the stability of any dislocation pattern, an attempt to describe such reactions through microscopic elasticity arguments seems to be a formidable task.

As a result of the aforementioned difficulties, an alternative procedure for describing the motion, interaction, and production/annihilation of dislocations has been proposed [e.g. 1c, 1e, 7c]. For example, in the case of one family of continuously distributed, infinite, parallel edge dislocations, it can be shown that the appropriate differential equations determining their dynamics are given by

$$\text{div } \underline{T}^D = \hat{\underline{f}} \quad ; \quad \frac{\partial \rho}{\partial t} + \text{div } \underline{j} = \hat{c} \quad , \tag{1}$$

where $\underline{T}^D$ is the dislocation stress, $\hat{\underline{f}}$ is the interaction force between the dislocation population and the lattice (including the Peach-Koehler force and the drag resistance), ρ is the density of dislocations, $\underline{j}$ is their flux, and the term $\hat{c}$ represents dislocation generation (positive or negative). Equation $(1)_1$ expresses conservation of momentum for the dislocated state, while $(1)_2$ expresses conservation of effective mass. It can be shown [1e, 7c] that equations (1) become identical to those that can be derived within the standard framework for continuously distributed dislocations [e.g. 9d] if the terms $\underline{T}^D$ and $\hat{c}$ are set equal to zero. These terms, however, are quite important in the present problem of pattern formation. In fact, the internal stress $\underline{T}^D$ models short range mutual dislocation interactions and core effects and the source term $\hat{c}$ represents dislocation dipole or multipole creation and annihilation.

It follows that a convenient method (even though quite different than the classical one) is now available for describing the dynamics of very dense dislocation populations. This is accomplished on the basis of (1) together with appropriate constitutive assumptions for the dislocation forces ($\underline{T}^D$, $\hat{\underline{f}}$) and dislocation reactions ($\hat{c}$). In fact, the point of view is advanced here that the terms $\underline{f}$, ρ, $\underline{j}$ maintain their classical meaning as in the traditional theory for continuously distributed dislocations [e.g 9c,d]; the term $\hat{c}$ preserves its kinetic meaning as in the usual approach of dislocation dynamics [e.g 18, 21]; but the new term $\underline{T}^D$ introduced to model short-range interaction and core effects of very dense dislocation structures should be described by a nonlocal constitutive equation. [It is noted that the term "short-range" is loosely used to describe situations for dense populations and the length scales involved here are quite smaller than the corresponding ones in dislocation arrays and pileups].

In one dimension, for example, the following nonlocal expression is expected to hold in general for the appropriate component τ^D of $\underline{T}^D$

$$\tau^D = \int \pi \, [x - x' \, ; \, \rho(x)] \, \rho(x') dx'. \tag{2}$$

This equation must be supplemented by a corresponding expression for the dislocation force $\hat{f}$ which is assumed to be of the standard form

$$\hat{f} = \hat{\alpha} + \hat{\beta} j - \hat{\gamma} \tau^L \; ; \; \tau^L = \tau - \tau^D, \tag{3}$$

with $\hat{\alpha}$, $\hat{\beta}$, $\hat{\gamma}$ denoting functions of the dislocation density ρ, τ^L being the effective stress or the difference between total (τ; $\partial\tau/\partial x = 0$) and dislocation ($\tau^D$) stress, and j being the corresponding component of dislocation flux. [It is easily recognized that the coefficient $\hat{\alpha}$ measures the lattice-dislocation interaction, $\hat{\beta}$ measures the drag associated with dislocation motion, and $\hat{\gamma}$ measures the effect of Peach-Koehler force].

On the basis of (2) and (3) and certain necessary assumptions pertaining to convergence properties of the integrand in (2), it can be shown upon substitution into the "continuum mechanics" equations (1) that the appropriate dislocation dynamics is described by the differential equation [7e]

$$\frac{\partial \rho}{\partial t} = D \frac{\partial^2 \rho}{\partial x^2} - E \frac{\partial^4 \rho}{\partial x^4} + g(\rho), \tag{4}$$

where E is positive and D may be positive or negative. Obviously, when $D > 0$ the second order term is dominant and then the fourth order term can be neglected. If, however, $D < 0$ then the fourth order term is necessary and the situation is reminiscent of that of spinodal decomposition (see also [19]) but with a nonlinear generation term $g(\rho)$. The kinetic portion of (4) can be extracted from usual dislocation dynamics arguments, while the gradient dependent portion is directly responsible for the selforganization phenomenon and the wave number selection.

Gradient dependent terms can also be generated within a "statistical mechanics" approach to the problem. In this an approach, one can postulate the existence of a chemical potential-like variable μ whose spatial gradient gives the dislocation flux. Then, in one dimension we have

$$j = -M\nabla_x \mu, \tag{5}$$

where M is the mobility, and the new notation ∇_x $(=\partial/\partial_x)$ was introduced for facilitating the subsequent presentation. A nonlocal expression for the potential μ based on microscopic elasticity [3b] can then be postulated in the form

$$\mu(x) = E_c + \int J(x-x') f(x') \rho(x') dx', \tag{6}$$

where E_c is the core energy, J is the pair interaction, and f is a distribution function taking into account the possibility for the Burgers vectors to be positive or negative. Due to the screening effect of dislocations of different Burgers vectors and the fact that we are dealing with an initially dense dislocation state, the pair interaction is effectively short ranged and (6) can effectively be approximated by the expression

$$j = -M\nabla_x [J_0 \rho + J_1 \nabla_{xx}^2 \rho], \tag{7}$$

where $J_0 \equiv \int J(x)f(x)dx$ and $J_1 \equiv \int J(x)f(x)x^2 dx$. It follows that we have again arrived at the result of (4) but through a more restrictive method.

A third, purely "kinetic approach", for generating gradient dependent terms in the equations describing the dynamics of dislocations, is to recognize the existence of point defects and write a system of three coupled equations for the density of dislocations ρ, the concentration of vacancies c_v, and the concentration of interstitials c_i as follows

$$\frac{\partial \rho}{\partial t} = g(\rho) + r_\rho (\rho, c_v, c_i) ,$$

$$\frac{\partial c_v}{\partial t} = D_v \nabla^2_{xx} c_v + r_v(\rho, c_v, c_i), \qquad (8)$$

$$\frac{\partial c_i}{\partial t} = D_i \nabla^2_{xx} c_i + r_i (\rho, c_v, c_i) .$$

It is emphasized that equation $(8)_1$, for the evolution of dislocation species does not contain gradient terms explicitly but is instead coupled with equations $(8)_{2,3}$ for the diffusive motion of point defects. The coupling between dislocations and point defects is considered through the source terms r_α ($\alpha = \rho, v, i$), while purely dislocation reactions are considered through the conventional kinetic term $g(\rho)$. Next, we note that c_v and c_i are the "fast" variables of the system and, thus, they can be "adiabatically eliminated" from (8). In fact, by assuming that r_ρ, r_v and r_i are linear forms and then setting $\partial c_v/\partial t = \partial c_i/\partial t = 0$, we can eliminate c_v and c_i from $(8)_1$, which now reduces to

$$\frac{\partial \rho}{\partial t} = g(\rho) + D \frac{\partial^2 \rho}{\partial x^2} - E \frac{\partial^4 \rho}{\partial x^4} + \ldots, \qquad (9)$$

where the order of spatial derivatives appearing in (9) depends on the power in the wave number q maintained in the Taylor's series expansion for the Fourier transforms of the concentrations $(c_v)_q$ and $(c_i)_q$.

The previously described three different approaches justifying the existence of gradient terms in the equations of dislocation dynamics did not distinguish between slow moving or immobile (trapped at obstacles) and fast moving or mobile (liberated by the action of stress) dislocations. In fact, one may not have difficulties to assign on intuitive grounds [3] a diffusivity to the immobile state resulting from the small random mobilities of trapped dislocations due to interaction with point defects, thermal effects, and localized internal stresses. It is more difficult, however, to obtain a diffusivity for the mobile state as this motion is of a drift type at the microscopic level. Nevertheless, under certain circumstances, for example in the problem of fatigue, it is possible to derive an effective diffusivity for the motion of mobile dislocations [3c, 7a].

To see this more clearly, let us consider the case of cyclic deformation in one dimension (along the slip plane) and write the equation of conservation of effective mass $(1)_2$ for both positive ρ_m^+ and negative ρ_m^- mobile dislocations

$$\frac{\partial \rho_m^+}{\partial t} + \nabla_x j_m^+ = \frac{b}{2}\rho_i - \frac{\gamma}{2}\rho_m^+ \rho_i^2,$$

(10)

$$\frac{\partial \rho_m^-}{\partial t} + \nabla_x j_m^- = \frac{b}{2}\rho_i - \frac{\gamma}{2}\rho_m^- \rho_i^2,$$

where $j_m^\pm = \rho_m^\pm v_m^\pm$ and a definite expression for the source terms $\hat{c}_m^+$ and $\hat{c}_m^-$ was adopted. The first term in the right hand side of (10) denotes production of mobile dislocations at the expense of immobile with a rate determined by the coefficient b. [The structure of b, in turn, depends on the effective stress τ^L which plays here the role of a bifurcation parameter for the problem]. The second term in the right hand side of (10) denotes the trapping of newly created mobile dislocations $\rho_m^\pm$ by immobile dipoles ρ_i^2 at a rate determined by the coefficient γ. [For simplicity, we have neglected bias in the interaction between positive and negative mobile dislocations with immobile ones]. Next, we assume that in the present case of cyclic deformation we have

$$v_m^+ = -v_m^- = \bar{v}(t) + \varepsilon V(x,t),$$

(11)

where $\bar{v} = \bar{v}(t)$ is an average space independent dislocation velocity and $V(x,t)$ is a nonhomogeneous contribution dependent on internal stress and obstacle distribution ($\varepsilon \ll 1$).

It follows that (10) can be re-written in terms of the total mobile dislocation density $\rho_m = \rho_m^+ + \rho_m^-$ and the difference $\delta = \rho_m^+ - \rho_m^-$

$$\frac{\partial \rho_m}{\partial t} = -\bar{v}\nabla_x \delta + b\rho_i - \gamma\rho_m \rho_i^2,$$

(12)

$$\frac{\partial \delta}{\partial t} = -\bar{v}\nabla_x \rho_m - \gamma\rho_i^2 \delta,$$

where contributions resulting from the inhomogeneous part of $v_m^\pm$ are left out for the moment. By further assuming that in the present case of cyclic deformation the average velocity $\bar{v}$ is periodic, that is $\bar{v} = v_o \cos \omega t$ where v_o is the amplitude and ω the frequency of the fatigue test, we can adiabatically eliminate the "fast" variable δ [3c, 7a] by solving (12)$_2$ for δ and then substituting into (12)$_1$ to obtain

$$\frac{\partial \rho_m}{\partial t} = D_m \nabla_{xx}^2 \rho_m + b\rho_i - \gamma\rho_m \rho_i^2,$$

(13)

where the effective diffusivity D_m is now explicitly given by the relation

$$D_m = \frac{v_o^2}{2\gamma\rho_{io}^2},$$

(14)

with ρ_{io} denoting a uniform solution of the immobile state. The particular form of the nonlinearity assumed in the expression for $\hat{c}_m^{\pm}$ is not important in the derivation. Moreover, the effect of the nonhomogeneous contribution $V(x,t)$ in (11) can result, under certain circumstances, into the renormalization of the effective diffusivity D_m and the bifurcation coefficent b. [This is, for example, the case when $V(x,t)$ is sinusoidal in space].

Having thus elaborated upon the issue of gradient dependent dynamics for dislocation species, we proceed with a brief account of an elementary diffusion-reaction scheme utilized in [3,7] to discuss the periodicity of persistent slip bands. The coupled system of diffusion reaction populations ρ_i and ρ_m reads

$$\frac{\partial \rho_i}{\partial t} = D_i \nabla_{xx}^2 \rho_i + g(\rho_i) - b\rho_i + \gamma \rho_m \rho_i^2 ,$$

(15)

$$\frac{\partial \rho_m}{\partial t} = b\rho_i - \gamma \rho_m \rho_i^2 + D_m \nabla_{xx}^2 \rho_m .$$

By defining new coefficients $a \equiv -g'(\rho_i^o)$ and $c \equiv \gamma \rho_i^{o\,2}$ where (ρ_i^o, ρ_m^o) is a uniform steady state determined by the conditions $g(\rho_i^o) = 0$ and $\rho_m^o = b/\gamma \rho_i^o$, we can perform a linear stability analysis of (15) to obtain two critical values of the coefficient b where instabilities occur. The first instability occurs for $b = b_H = a + c$ and is a Hopf - bifurcation leading to homogeneous oscillations. The second instability, which is of interest here, occurs for $b = b_p = (\sqrt{a} + \sqrt{cD_i/D_m})^2$ [note that $b_p << b_H$ as $D_i << D_m$] and is a patterning or Turing instability leading to spatially periodic solutions with wave number q_c and wavelength λ_c given by

$$q_c = \frac{2\pi}{\lambda c} = \left[\frac{ac}{D_i D_m}\right]^{1/4} . \quad (16)$$

On utilizing the result (14) for the mobile diffusivity D_m, estimating the immobile diffusivity D_i from the relation $\sqrt{D_i}/a \approx \ell$ where ℓ denotes the annihilation length for immobile dipoles, and making use of the Orowan's relation $\dot{\gamma} = b\rho_m v_o$ where $\dot{\gamma}$ denotes the plastic strain rate, we can establish the familiar inverse square root dependence of the wavelength of the ladder structure on the disloction density

$$d \equiv \lambda_c \sim \frac{1}{\sqrt{\rho_i}} , \quad (17)$$

In deriving (17) we have also assumed that the coefficient b is proportional to the plastic strain rate $\dot{\gamma}$.

It is now important to examine whether the pattern defined by (16) is stable. This is accomplished by considering the nonlinear regime where the method of slow mode dynamics is employed. Specifically, near the bifurcation point there are two time scales: one associated with the slow modes S (the eigenmodes corresponding to the eigenvalue with vanishing real part $\omega_S \approx 0$) and the fast modes R (the eigenmodes corresponding to the eigenvalue with negative real part $\omega_R < 0$). The dynamics is governed by the slow modes (slaving principle) as the fast modes are rapidly relaxing to their steady values. Then one can perform an "adiabatic elimination" of the fast modes in the Fourier space ($\partial R_q/\partial t \simeq 0$) to obtain the slow mode dynamics or amplitude equation [3,7]. By considering then small perturbations from the patterned solution given by (16) one finds that the patterned solution is stable and therefore it persists over homogeneous or random states [3a].

More information from the slow mode dynamics equation can be extracted by considering a two-dimensional problem, that is by allowing a small mobility of the trapped dislocations in the y direction on the x-y slip plane. It then turns out [3,7] that the slow mode dynamics equation reads

$$\frac{\partial \sigma}{\partial t} = [\varepsilon - d_x (q_c + \nabla_x^2)^2 + d_y \nabla_y^2] \sigma - v\sigma^2 - u\sigma^3, \qquad (18)$$

where $\varepsilon = (b-b_p)/b_p$ denotes the departure from the bifurcation point, the order parameter-like variable σ is a linear combination of ρ_i and ρ_m and the remaining parameters in (18) relate directly to the original coefficients of the problem. On searching in (18) for patterned solutions of the form $\sigma \sim R \exp[i(q_c + \phi)x]$ with the amplitude $R = R(x,y,t)$ and the phase $\phi = \phi(x,y,t)$ being slowly varying functions in space and time, we can show that (18) admits preferred steady-state modulations of the form

$$\sigma_o = 2R_o \cos(q_c x + \phi_o); \qquad R_o = \sqrt{\frac{\varepsilon}{3u}}, \quad \phi_o = \text{const.} \qquad (19)$$

The stability of this solution is then tested by examining the behavior of small amplitude and phase fluctuations of the form $R = R_o + \tilde{R}$ and $\phi = \phi_o + \tilde{\phi}$. The result is [3] that the amplitude R obeys a relaxation dynamics, relaxing to its steady state value R, while the phase ϕ obeys a diffusive dynamics of the form

$$\frac{\partial \phi}{\partial t} = D_x \nabla_{xx}^2 \phi + D_y \nabla_{yy}^2 \phi, \qquad (20)$$

where the new diffusivities D_x and D_y are related to the previous parameters of the problem. It is noted that the steady state solutions of this equation admit singularities which have been shown to relate to layer splitting and other phase symmetry breaking processes [3,7], also observed experimentally [8e]. More details on dislocation pattern formation in fatigue and a description of the situation in three dimensions where the competition between the vein and the ladder structures becomes a central feature of the deformation process can be found in [3,7].

STATIONARY SHEAR BANDS AND TRAVELLING PLC BANDS

In this final section we examine the problem of initiation, growth, and periodicity of macroscopic deformation bands. The appropriate variable here is not the dislocation density but the plastic strain itself. However, the inhomogeneous evolution of the underlying microstructure (e.g. point defects or dislocations) enter implicitly into the problem in the form of higher order strain gradients which modify the standard constitutive equations of (homogeneous or nearly homogeneous) continuum mechanics and alter the qualitative character of the governing differential equations.

To make this claim more clear we consider the generation of second order strain gradients into the one-dimensional constitutive equation describing a viscoplastic material $\sigma = h\varepsilon + f(\dot{\varepsilon})$ [where σ is the stress, ε is the strain, and f is a nonconvex function of the strain rate $\dot{\varepsilon}$]. Under PLC conditions, the dislocations (which are accounted here for through Orowan's relation for the strain rate $\dot{\varepsilon} \sim b\rho v$) interact strongly with the solute population c which, as usual, obeys a diffusive dynamics. We can thus expect that, in general, the system is described by the equations

$$\sigma = h\varepsilon + f(\dot{\varepsilon}) + g(\varepsilon,c),$$

$$\dot{c} = D\nabla_{xx}^2 c + r(c,\varepsilon), \qquad (21)$$

where the term $g(\varepsilon,c)$ represents, in a phenomenological way, the extra stress generated by the solute-dislocation interaction, and $r(c,\varepsilon)$ is a measure of the trapping/freeing mechanism of solute atmospheres in the dislocation cores. However, the solute concentration c is the fast variable of the system and can thus be "adiabatically" eliminated. Specifically, by assuming that g and r are linear forms, taking the Fourier transform of $(21)_2$ and then solving for c_q ($\dot{c}_q \approx 0$) we can show that for "sufficiently large space scales" $(21)_1$ is reduced to

$$\sigma = h\varepsilon + f(\dot{\varepsilon}) + c\nabla_{xx}^2 \varepsilon, \qquad (22)$$

where c is a positive constant and the strain hardening modulus h is renormalized.

A nonlinear gradient dependent expression for the flow stress $\tau = \kappa(\gamma)$ in one dimension can formally be obtained by considering the effect of an internal variable α (say a measure of the dislocation density and their average flight distance). On assuming that the internal variable α obeys a diffusive-like evolution, then the steady state of the system is described by the following set of equations [1c]

$$\tau = \kappa(\gamma) - \mu(\gamma)\alpha; \qquad \nabla_x \tau = 0,$$

$$\nabla_{xx}^2 \alpha = \lambda(\gamma)\alpha, \qquad (23)$$

where a linear dependence on α was assumed in $(23)_{1,3}$ and the equilibrium equation $(23)_2$ for the stress τ in one dimension is also utilized (quasistatic conditions). It is then possible under certain conditions to eliminate α from (23) to obtain the following nonlinear differential equation

$$a(\gamma) \nabla_{xx}^2 \gamma + b(\gamma) [\nabla_x \gamma]^2 = \kappa(\gamma) - \kappa_o, \qquad (24)$$

where $\tau = \kappa(\gamma)$ can be a nonconvex (single loop) or softening type curve, κ_o is a constant (equal to the applied stress at the boundary) and the coefficients $a(\gamma)$ and $b(\gamma)$ relate directly to the original material functions of the problem. It is shown below that a special form of equation (24) and equation (22) can conveniently be utilized to capture respectively the structure and thickness of shear bands and the periodicity, width, and velocity of PLC bands. First, however we give a brief discussion on the critical conditions and initial stages of localization, also from the point of view of linear stability analysis.

Traditionally, critical conditions for the onset of localization were obtained as a result of a bifurcation analysis leading to the loss of ellipticity in the governing equilibrium equations. This, in fact, has led to serious mathematical difficulties associated with ill-posed problems occurring in equations of changing type. As a result of introducing gradient dependent terms into the constitutive structure, the governing differential equations remain elliptic and the aforementioned difficulties associated with uniqueness, stability, convergence and mesh-size dependence in finite element analyses, are overcome. Moreover, classical bifurcation analyses are not suited for time dependent problems. In particular, viscoplastic and rate-sensitive materials cannot be easily analyzed. An alternative approach is to identify the onset of localization with the onset of instability in the linearized equations. In this approach we assume exponential fluctuations from a homogeneous state of the form, for example,

$$\underset{\sim}{v} = \underset{\sim}{L}_o \underset{\sim}{x} + \underset{\sim}{v}_o e^{iqz + \omega t}, \qquad (25)$$

where $\underset{\sim}{v}$ is the velocity, $\underset{\sim}{L}_o$ and $\underset{\sim}{v}_o$ are constants (with $|v_o| \ll 1$), $z = \underset{\sim}{n} \cdot \underset{\sim}{x}$ with $\underset{\sim}{n}$ denoting the unit normal to the assumed planar localized zone, q being the wave number, and ω denoting the rate of growth of the fluctuation (the eigenvalue of the problem).

Expansions similar to (25) are also written for the back stress α, the temperature, and all other internal variables of the system. On substituting these expressions into the governing equations we arrive at an eigenvalue equation of the form

$$\psi(\omega) = \det[\underset{\sim}{A}] = 0, \qquad (26)$$

where $[\underset{\sim}{A}]$ depends on the material functions and the stress state. Instability occurs when $\text{Re}[\omega] > 0$ and this, in turn, determines the critical conditions, as well as the corresponding wavelength.

As an example, we consider incompressible rigidly plastic materials of the form

$$\sigma = -p\mathbf{1} + 2\mu D \quad ; \quad \mu = \tau/\dot{\gamma},$$
$$\tau = \kappa(\gamma) - c\nabla^2 \gamma, \qquad (27)$$

where c is a constant and p denotes the pressure. Equation (27) is the standard constitutive form for rigidly plastic materials, while equations $(27)_2$ for the flow stress is motivated by the analysis leading to (24). For plane strain conditions $D_{11} = -D_{22}$ it turns out that $\underset{\sim}{v}_o = v_o \underset{\sim}{\nu}$ where $\underset{\sim}{\nu}$ is the unit vector in the direction of the shear band and $\nu = (\cos\theta, \sin\theta)$ with θ being the angle between the direction of the band and the principal direction of D (say, the one with the largest eigenvalue). In this case (27) becomes

$$\omega = -\frac{h + cq^2}{\mu \cot^2 2\theta} \quad , \quad h = \kappa'(\gamma); \tag{28}$$

suggesting that the critical conditions are

$$\theta_{cr} = \frac{\pi}{4} \quad , \quad h \leq h_{cr} = 0 . \tag{29}$$

This analysis predicts a vanishing wave number, a fact physically undesirable. However, if a fourth order gradient term $\bar{c} \nabla^4 \gamma$ is added in the yield condition, a definite wave number $q_c = \sqrt{c/4\bar{c}}$ is obtained, while $(29)_1$ remains the same and $(29)_2$ is replaced by $h \leq h_{cr} = c^2/4\bar{c}$.

As mentioned earlier, the important problem of shear band thickness cannot be addressed within the classical analyses for isothermal shear bands under quasistatic conditions. This is due to the fact that these analyses do not contain any characteristic length, a feature which is implicitly incorporated in the gradient approach. In fact, it can be shown [7c,7d,12a,12b] that shear band widths can be calculated within the gradient approach from the solution of a nonlinear equation for the strain distribution. To see this for the particular case of rigidly plastic materials described by (27) we search for solutions of the form

$$\underset{\sim}{v} = \underset{\sim}{L}_o \underset{\sim}{x} + f(z)\underset{\sim}{v} \quad ; \quad z = \underset{\sim}{n} \cdot \underset{\sim}{x} \quad , \tag{30}$$

where $f(z)$ is a finite amplitude specifying the velocity distribution in the direction perpendicular to the band. Equation (30) is assumed to hold in the postlocalization regime with $\underset{\sim}{v}$ determined from $(29)_1$. Equation (30) suggests that $\gamma = \gamma(z)$ and then the only nontrivial equilibrium equation $\partial\sigma_{12}/\partial z = \partial\tau/\partial z = 0$ gives

$$c\gamma_{zz} = \kappa(\gamma) - \kappa_o \quad ; \quad \dot{\gamma} \geq 0, \tag{31}$$

where the flow stress $\kappa(\gamma)$ is a softening type curve. Following methods advanced by Aifantis and Serrin (see references quoted in [12a] we find the following explicit solution to (31)

$$z = \bar{z} + \int_{\gamma(\bar{z})}^{\gamma(z)} \left\{ \frac{2}{c} \left[\int_{\gamma_1}^{\gamma} [\kappa(\gamma) - \kappa_o] d\gamma \right] \right\}^{-1/2} d\gamma \quad , \tag{32}$$

valid for $-\infty < z < z^*$ and being symmetric with respect to z^* defined by $\gamma_z(z^*) = 0$ and $\gamma_2 = \gamma(z^*)$. The condition of existence of such solutions is given by the relations

$$\int_{\gamma_1}^{\gamma_2} [\kappa(\gamma) - \kappa_o] d\gamma = 0 \quad , \quad h = \kappa'(\gamma) \leq 0 . \tag{33}$$

The profile of $\gamma(z)$ according to (32) is a smooth symmetric curve having a maximum at the center of the shear band, while $(33)_1$ simply expresses the fact that within the present structure, patterned solutions can develop only in the softening regime. Finally, condition $(31)_2$ is used to define the region of continuing plastic flow and therefore the plastic flow boundaries. This, in effect, defines the shear band boundaries and the corresponding construction is outlined in [12b].

We conclude this section with a brief presentation of a fully nonlinear analysis pertaining to the periodicity and spacing of the PLC bands. As discussed in detail in [11] the appropriate constitutive equation in one dimension, for materials exhibiting the PLC effect, is of the form

$$\sigma = h\varepsilon + f(\dot{\varepsilon}) , \qquad (34)$$

where σ and ε are the axial stress and strain and h is the strain hardening modulus. The function f is the viscous part of the flow stress and is assumed to be a single loop (nonconvex). When the applied stress rate $\dot{\sigma}_o$ is such that the corresponding homogeneous steady state solution $\dot{\varepsilon}_s$ $(= \dot{\sigma}_o/h)$ lies in the negative slope region $\dot{\varepsilon}_1 < \dot{\varepsilon}_s < \dot{\varepsilon}_2$, the homogeneous solution becomes unstable and a periodic succession of shear bands cross the specimen with constant velocity. To quantify the above discussion we assume a gradient dependent flow stress of the form given by (22), that is

$$\sigma = h\varepsilon + f(\dot{\varepsilon}) + c\varepsilon_{xx} . \qquad (35)$$

With this simple modification to (34) and on assuming a travelling wave solution of the form $\dot{\varepsilon} = Z(x - Vt)$, (35) can be written as

$$Z_{\eta\eta} + \mu f'(Z)Z_\eta + (Z - Z_s) = 0 , \qquad (36)$$

where $\eta = -\sqrt{h/c}\,(x-Vt)$, $\mu = V/\sqrt{ch}$. Equation (36) is the well-known Lienard's equation, a classical example of relaxation oscillations. According to LaSalle's theorem (see references quoted in 12b) a stable periodic solution exists for $\dot{\varepsilon}_1 < \dot{\varepsilon}_s < \dot{\varepsilon}_2$. Moreover, the natural speed of the travelling wave is given by the relation $V^* = 2\sqrt{ch}/|f'(Z_s)|$. More details on this problem together with appropriate numerical results illustrating the periodicity of the PLC bands, their velocity, as well as their direct influence in obtaining serrated and staircase stress-strain curves are given in [7c,d, 12b].

REFERENCES

[1] a) Bammann, D.J. and E.C. Aifantis. Acta Mechanica, **45**, 91.
   b) Aifantis, E.C., J. Mat. Engng. Tech., **106**, 326 (1984).
   c) Aifantis, E.C., in: Mechanics of Dislocations, edited by E.C. Aifantis and J.P. Hirth (ASM, Metals Park, OH, 1985), pp. 127-146.
   d) Aifantis, E.C., in: **Physical Basis and Modelling of Finite Deformations of Aggregates**, edited by J. Gittus, S. Nemat-Nasser, and J. Zarka (Elsevier Appl. Sci. Publ. London-New York, 1986), pp. 283-323.

- e) Aifantis, E.C., in **Mechanical Properties and Behavior of Solids: Plastic Instabilities**, edited by V. Balakrishnan and C.E. Bottani (World Scientific, Singapore, 1986), pp. 314-353.

[2] a) Nicolis, G. and I. Prigogine. **Self-Organization in Nonequilibrium Systems**, Wiley, New York, 1977.
- b) Swinney, H.L. and J.P. Gollub: **Hydrodynamic Instabilities and the Transition to Turbulence**, Springer, Berlin, 1981.
- c) Nicolis, G. and F. Baras: **Chemical Instabilities: Applications to Chemistry, Engineering, Geology and Materials Science**, Reidel, Dordrecht, 1983.
- d) Hlavacek, V., **Dynamics of Nonlinear Systems**, Gordon and Breach, New York, 1985.
- e) See the series of volumes entitled "Synergetics", by H. Haken, Springer, Berlin. Also: H. Haken, **Synergetics**, 2nd Ed., Springer, Berlin, 1978.
- f) D. Walgraef (editor), **Patterns, Defects, and Microstructures in Nonequilibrium Systems (Applications to Materials Science)**, Proceedings of a NATO ARW Workshop, Austin TX, March 1986, Martinus-Nijhoff, Dordecht, 1987.
- g) Newell, A.C. and J.A. Whitehead, J. Fluid Mech. **38**, 279 (1969). L. Segel, J. Fluid Mech. **38**, 203 (1969).

[3] a) Walgraef, D. and E.C. Aifantis, Int. J. Engng. Sci. **23**, 1351, 1359, 1365, (1985).
- b) Walgraef, D. and E.C. Aifantis, J. Appl. Phys. **58**, 688 (1985).
- c) Walgraef, D. and E.C. Aifantis, "Dislocation Patterning in Fatigued Metals: Labyrinth Structures and Rotational Effects", Int. J. Engng. Sci. **24**, 1789 (1986).

[4] C. Truesdell and W. Noll, **The Non-Linear Field Theories of Mechanics**, in: Handbuch der Physik, edited by S. Flugge, Vol. III/3, Springer-Verlag, Berlin, 1965.

[5] a) Haken, H., **Synergetics**, 2nd Edn., Springer, Berlin, 1978.
- b) Walgraef, D., G. Dewel, and P. Borkmans, Adv. Chem. Phys. **49**, 311 (1982). D. Walgraef, P. Borkmans, and G. Dewel, in: **Phase Transformations**, edited by E.C. Aifantis and J. Gittus (Elsevier Appl. Sci. Publ. London-New York, 1985) pp. 29-54.
- c) Newell, A.C. and J.A. Whitehead, J. Fluid Mech. **38**, 279 (1969).
- d) Swift, J. and P. Hohenberg, Phys. Rev. **A15**, 315 (1977).
- e) Di Prima, R.C. and H.L. Swinney, **Hydrodynamic Instabilities and the Transition to Turbulence**, Springer, Berlin, 1979.
- f) Mannvile, P., J. Physique **44**, 563 (1983).

[6] Guckenheimer, J., J. Moser, and S. Newhouse, **Dynamical Systems**, Birkhauser, Boston, 1980.

[7] a) Aifantis, E.C., Mater. Sci. Eng. **81**, 563 (1986). Also in: **Low Energy Dislocation Structures**, edited by M.N. Bassim, W.A. Jesser, D. Kuhlman-Wilsdorf, and H.F.G. Wilsdorf, Elsevier, Sequoia, Lausanne, 1986.
- b) Walgraef, D., C. Schiller, and E.C. Aifantis, in: Patterns, Defects, and Microstructures in Nonequilibrium Systems, edited by D. Walgraef (Martinus-Nijhoff, Dordrecht, 1987), pp. 257-269.
- c) Aifantis, E.C., Int. J. Plasticity **3**, 211 (1987).
- d) Aifantis, E.C., in: Const. Relations and their Physical Basis, edited by S.I. Anderson et al (8th Riso Int. Symp. Roskilde 1987), pp. 205-219.
- e) Aifantis, E.C., in: Dislocations in Solids, edited by H. Suzuki et al (Tokyo University Press, 1985) pp 41-47.

[8] a) Laird, C. and D.J. Duquette, in: **Corrosion Fatigue**, edited by O.F. Devereux, A.J. McEvily, and R.W. Staehle. (NACE, Houston, 1982) pp. 88-117.

- b) Winter, A.S., Philos. Mag. **30**, 719 (1974).
- c) Essman, U. and H. Mughrabi, Philos. Mag. **A40**, 731 (1979).
- d) Mughrabi, H., in: **Continuum Models for Discrete Systems 4**, edited by O. Brulin and R.K.T. Hsieh (North-Holland, Amsterdam, 1981), pp. 241-257.
- e) Tabata, T., H. Fugita, M. Hiraoka, and K. Onishi, Philos. Mag. **A47**, 841 (1983).
- f) Ackerman, F., L.P. Kubin, J. Lepinoux, and H. Mughrabi, Acta Metall. **32**, 715 (1984).
  N.Y. Jin and A.T. Winter, Acta Metall. **32**, 1173 (1984).

[9]
- a) Kroner, E., and G. Rider, Zs. Phys. **145**, 424 (1956).
  E. Kroner, Kontinuumstheorie der Versetzungen und Eigenspannungen, Springer, Berlin, 1958.
- b) Hollander, E.F., Czech. J. Phys. **B10**, 469, 474, 551 (1960).
- c) Mura, T., Philos. Mag. **8**, 843 (1963).
- d) Kosevich, A.M., JETP **42**, 152 (1962), Soviet Phys. JETP **15**, 108 (1962). A.M. Kosevich, in: **Dislocations in Solids**, edited by F.R.N. Nabarro (North-Holland, Amsterdam, 1979) pp. 33-141.

[10]
- a) Hill, R. and J.W. Hutchinson, J. Mech. Phys. Solids **23**, 239 (1975).
- b) Rice, J., in: **Theoretical and Applied Mechanics**, edited by W.T. Koiter (North-Holland, Amsterdam, 1976) pp. 207-220.
- c) Bazant, Z.P. and T. Belytschko, in: **Constitutive Laws for Engineering Materials: Theory and Applications**, edited by C.S. Desai et al (Elsevier Appl. Sci. Publ. 1987), pp. 11-13.
- d) Willam, K., E. Pramono, and S. Sture, in: **Constitutive Laws for Engineering Materials: Theory and Applications**, edited by C.S. Desai et al (Elsevier Appl. Sci. Publ., 1987), pp. 249-269.

[11]
- a) Kubin, L.P. and Y. Estrin, J. Physique, **47**, 497 (1986).
- b) Estrin, Y. and L.P. Kubin, in: **Phase Transformations**, edited by E.C. Aifantis and J. Gittus (Elsevier Appl. Sci. Publ., London-New York, 1986), pp. 185-202.
- c) Kubin, L.P., K. Chihab, and Y. Estrin, "Nonuniform Plastic Deformation and the Portevin-Le Chatellier Effect" in: **Patterns, Defects, and Microstructures in Nonequilibrium Systems**, edited by D. Walgraef, (Martinus-Nijhoff, Dordecht, 1977) pp. 220-236.

[12]
- a) Triantafyllidis, N. and E.C. Aifantis, J. Elasticity **16**, 225 (1986).
- b) Zbib, H. and E.C. Aifantis, On the Localization and Postlocalization Behavior of Plastic Deformation, Parts I-III, Res Mechanica (in press).

[13]
- a) Brechet, Y., Doctorate Thesis, University of Grenoble [Also: Lecture at **CNRS Symposium on Nonlinear Problems in Metallurgy**, Aussois, France, September 1987].
- b) Lepinoux, J. and L.P. Kubin, "The Dynamic Organization of Dislocation Structures: A Simulation", Scripta Met. **21**, 833 (1987). [Also: J. Lepinoux, Doctorate Thesis, University of Poitiers and Lecture at **CNRS Symposium on Nonlinear Problems in Metallurgy**, Aussois, France, September 1987].
- c) Kubin, L.P. and Y. Estrin, "Strain Nonuniformities and Plastic Instabilities" in: **Proc. Europhysics Study Conf. on Mechanisms and Mechanics of Plasticity**, Editions de Physique (in press).
- d) Aifantis, E.C., Y. Estrin, and L.P. Kubin, "The Phenomenon of Repeated Yielding: Temporal and Spatial Organization", preprint.

[14]
- a) Estrin, Y. and L.P. Kubin, Acta Met. **34**, 2455 (1986).
- b) Bocek, M., "The Dynamics Stability of Plastic Flow", preprint. [Also: Lecture at **CNRS Symposium on Nonlinear Problems in Metallurgy**, Aussois, France, September 1987].

[15]
- a) Kratochvil, J. and S. Libovisky, Scripta Met. **20**, 1625 (1986).

b) Kratochvil, J., "Dislocation Pattern Formation in Metals", in: **Proc. Europhysics Study Conf. on Mechanisms and Mechanics of Plasticity**, Editions de Physique (in press).

c) Kratochvil, J., "Dislocation Structure Instability and Fatigue," in: **Proc. Int. Colloq. on Basic Mechanisms in Fatigue of Metals**, Bruo, Czechoslovakia (in press).

[16] a) Ananthakrishna, G.,and D. Sahoo, J. Phys. D: Appl. Phys. **14**, 2081 (1981).

b) Ananthakrishna, G. and M.C. Vasakumar, J. Phys. D. Appl. Phys **15**, 2171 (1982).

c) Ananthakrishna, G., Bull. Mater. Sci. **6**, 665 (1984).

[17] a) Amodeo, R.J. and N.M. Ghoniem, "A Review of Experimental Observations and Theoretical Models of Dislocation Cells and Subgrains, Res Mechanica (in press) [Special Issue on "Material Instabilities" edited by E.C. Aifantis, D. Walgraef, and J. Gittus].

b) Amodeo, R.J. and N.M. Ghoniem, "Dynamic Computer Simulation of the Evolution of a One-Dimensional Dislocation Pile-up", Int. J. Eng. Sci (in press).

c) Ghoniem, N.M., Lecture at **CNRS Symposim on Nonlinear Problems in Metallurgy**, Aussois, France, September 1987.

[18] Li, J.C.M., J. Appl. Phys. **33**, 2958 (1962).

[19] Holt, D.L., J. Appl. Phys. **41**, 3197 (1970).

[20] a) Gittus, J.H., Acta Met. **22**, 789 (1974).

b) Gittus, J.H., Phil. Mag. **34**, 401 (1976); **35**, 293 (1977); **39**, 829 (1979).

[21] Sandstrom, R., Acta Met. **25**, 897, 905 (1977).

[22] Gibeling, R.C. and W.P. Nix, Acta Met **28**, 1743 (1980).

[23] a) Argon, A.S. and W.C. Moffat, Acta Met. **29**, 293 (1981).

b) Argon, A.S. and S. Takeuchi, Acta Met. **29**, 1877 (1981).

c) Argon, A.S. in: **The Inhomogeneity of Plastic Deformation** (ASM, Metals Park, Ohio, 1973) pp. 161-189.

[24] Kuhlmann-Wilsdorf, D., Mater. Sci. Eng. **55**, 79 (1982). See also the collection of papers in the bound volume: Low Energy Dislocation Structures, edited by M.N. Bassim, W.A. Jesser, D. Kuhlmann-Wilsdorf, and H.F.G. Wilsdorf, Elsevier, Sequoia, Lausanne, 1986.

[25] a) Bazant, Z.P., ASCE J. Eng. Mech. **110**, 1693 (1984).

b) Coleman, B.D. and M.L. Hodgdon, Arch. Rat. Mech. Anal. **90**, 217 (1985).

c) Schreyer, H.L. and Z. Chen, J. Appl. Mech. **53**, 791 (1986).

The Influence of Spatial Distributions on Metallurgical Processes

BY G. BURGER*, E. KOKEN†, D. S. WILKINSON† AND J. D. EMBURY†

*Kingston Research Labs, Alcan International, P.O. Box, Kingston, Ontario.
†Department of Materials Science and Engineering,
McMaster University, Hamilton, Ontario.

INTRODUCTION

In modelling a variety of processes in physical metallurgy, whether they have been associated with phase transformations or mechanical response, it is usual to describe the material microstructure by means of global parameters such as the average volume fraction of second phase particles or the mean particle size. There are, however, many processes which are spatially inhomogeneous in character, such as the nucleation of recrystallization or the accumulation of microstructural damage. These demand descriptions of the microstructure capable of including distribution effects in a quantitative manner, for example, the quantification of the degree of clustering, or of percolation processes. It is germane to note that there exists a broad literature associated with both the description of space filling polyhedra and the clustering of points in distributions, the body of which has provided the formal basis on which to quantify both microstructure and the geometric relationship of events occurring in these structures.

Before developing a quantitative description of microstructure it is important to emphasize that the present work considers the distribution of <u>discrete</u> microstructural features such as particles, pores, voids, etc. Interconnected features are not described. The first section outlines this development, while the second and third sections deal with applications of these descriptions to the problems of recrystallization and damage accumulation.

ANALYSIS OF SPATIAL DISTRIBUTION USING THE DIRICHLET TESSELLATION

We begin by considering particles as points in a structure. The simplest description of such a point pattern distribution is in terms of nearest neighbour distances. The distribution of nearest-neighbour distances provides information only on a very local scale however, i.e. between two particles. Its use as an indicator of particle clustering is thus restricted from providing any information on aspects of the distribution involving clustering of three or more particles, i.e. a description of the larger scale character of the particle distribution cannot be obtained.

The distribution of near-neighbour distances would provide information on a larger scale. However, the problem here is one of defining the neighbours of any given particle. A radial distribution function could be obtained, but this would be dependent upon the origin chosen. The shape of the distribution and the conclusions drawn would vary between different particles chosen as a reference location and many analyses would be required. Clearly the effort required by this type of analysis would be considerable, and the comparisons made between different distributions employing this approach would be complex.

An alternative to these distance approaches is to subdivide the material into cells constructed about each particle or discrete feature, and to subsequently characterize the geometry of these cells. The method used in subdividing the material into cells could employ a physically-based criterion, or it may be purely geometrical. In this description we consider only the latter. For any three dimensional particle or point pattern it is possible to divide the region into cells, the so-called "Voronoi polyhedra", such that each cell contains that part of the region closer to its point than to any other. The Voronoi polyhedra may be constructed from the planes which bisect perpendicularly the lines joining any given particle to those around it (Rogers, 1964). The inner envelope of such planes about a given particle defines the Voronoi polyhedron constructed about that particle. The geometry of each of these cells is thus intimately related to the local distribution about the particle which it surrounds. This tessellation procedure results in a space-filling subdivision of volume. Note that it also results in a unique criterion for defining the neighbours of any given particle.

Microstructural characterization is usually concerned with plane sections taken through a material, resulting in a two-dimensional spatial distribution of exposed particles which is related to the underlying 3-dimensional distribution. This relationship however, is not unique, in that the 3-d distribution is not necessarily obtainable (in other than an approximate level) from a limited number of sections through the material. However, the approach to characterizing the distribution of discrete microstructural features in a section plane does provide useful information in predicting material properties. It is the characterization of microstructure in two-dimensional sections which we will discuss here.

The subdivision of a plane into regions about each particle is similar to that used to construct Voronoi polyhedra in three dimensions. It results in a tessellation of space known as the "Dirichlet tessellation" (Wray et al., 1983; Spitzig et al., 1985; Burger, 1986). The construction of a polygon, or Dirichlet cell about a given **point** is illustrated in Fig. 1. Each cell consists of that region in the plane closer to its point than to any other.

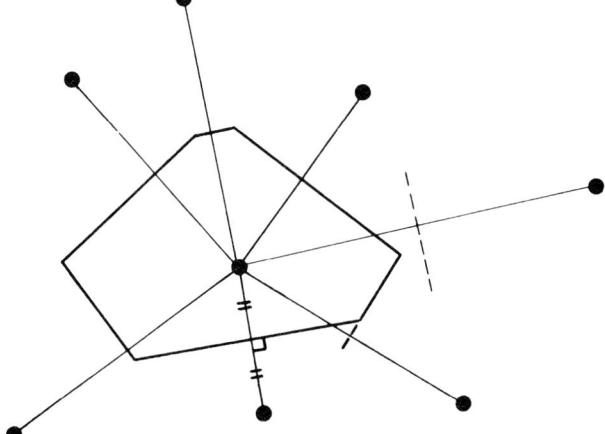

Fig. 1: Construction of a Dirichlet cell about a given point. It is given by the inner envelope of lines which perpendicularly bisect the lines joining the given point to others in the dispersion.

The influence of the particle distribution on the nature of the tessellation can be marked, as illustrated in Figs. 2a-c. Here a tessellation constructed about a pseudo-random distribution of points (Fig. 2a) is compared with a more periodic (Fig. 2b) and a more clustered (Fig. 2c) distribution. A more complete range of point distributions with varying degrees of periodicity or clustering has been analyzed elsewhere (Burger, 1986; Burger et al.,1988).

For any given tesselation, a large number of parameters can be calculated and used to characterize the associated point distribution. Table I summarizes qualitatively some of the results of an analysis of a range of point distributions (Burger, 1986). It lists the relative sensitivity of the mean and standard deviation of various parameters to deviations in the point distribution from randomness, both towards periodicity and towards clustering. Note that indicators which are sensitive to clustering may be only

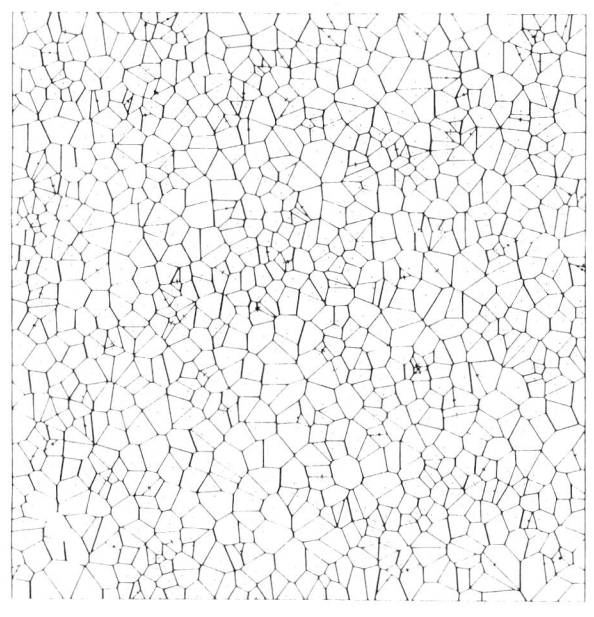

(a)

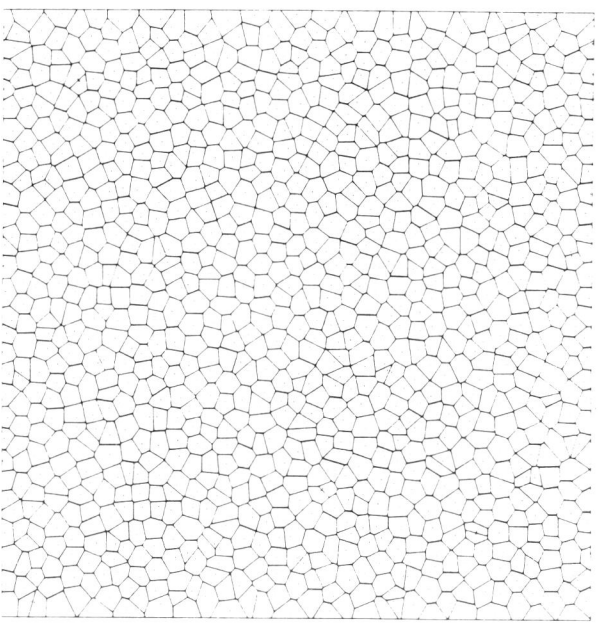

(b)

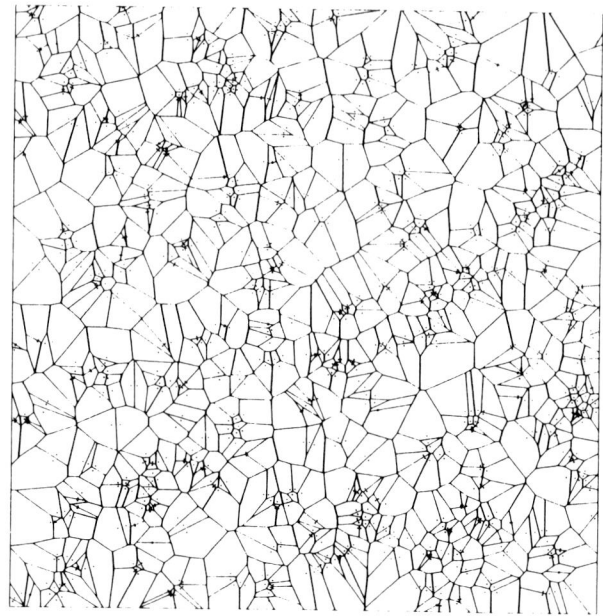

Fig.2: Dirichlet tessellations constructed about three different computer-generated point distributions. Fig. 2a is pseudo-random, and 2b and 2c are characteristically more periodic and clustered respectively, relative to the pseudo-random distribution.

weakly sensitive to periodicity in a distribution, and vice versa. For example, the standard deviation of the near neighbour spacing is sensitive to periodicity, but not to clustering. There are, however, several parameters related to cell area and cell aspect ratio which are sensitive to clustering.

Although the nearest-neighbour distance distribution is found to be a relatively weak indicator of clustering or periodicity in a distribution (see Table I), it can be used successfully to distinguish between dispersions. A series of point dispersions (such as those given in Fig. 2), were generated ranging in character from strongly periodic to strongly clustered (Burger, 1986). If we then look at the fraction of

TABLE I SENSITIVITY OF TESSELLATION PARAMETERS TO THE DEGREE OF PERIODICITY OR CLUSTERING IN A DISTRIBUTION

		RELATIVE SENSITIVITY TO	
		PERIODICITY	CLUSTERING
1.	NEAREST-NEIGHBOUR DISTANCE		
	i) average value	WEAK	WEAK
	ii) standard deviation	WEAK	WEAK
2.	NEAR-NEIGHBOUR DISTANCE		
	i) average value	VERY WEAK	VERY WEAK
	ii) standard deviation	STRONG	WEAK
3.	NUMBER OF SIDES PER CELL		
	i) average value	NO SIGNIFICANT DEPENDENCE	
	ii) standard deviation	WEAK	MILDLY STRONG
4.	CELL AREA		
	i) average value	NO DEPENDENCE	NO DEPENDENCE
	ii) standard deviation	WEAK	STRONG
5.	CELL ASPECT RATIO		
	i) average value	WEAK	STRONG
	ii) standard deviation	WEAK	STRONG

points in the distribution which are contained within clusters, employing a critical separation distance to discriminate whether or not any two points are clustered, distributions may be characterized on a **relative** basis. This is done in Fig. 3, in which the critical separation distance is varied, and the corresponding fraction of points in the distribution associated with the clusters determined.

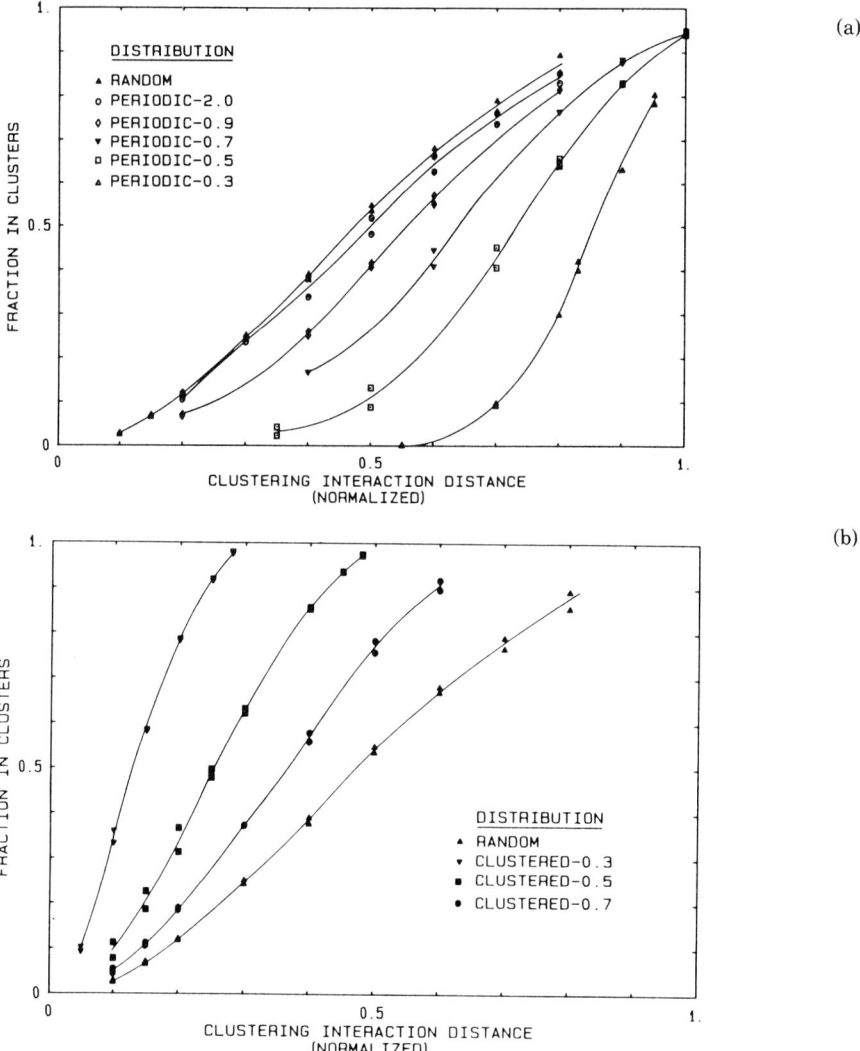

Fig. 3: Analysis of clustering in the generated point distributions. Results are given as the fraction of points in the distribution which have at least one point within the given clustering interaction distance. Note that the distances have been normalized by d, and that for each value of α and β, two distributions were generated.
(a) Analysis of random and characteristically periodic distributions. Values of α are indicated
(b) Analysis of random and clustered distributions. Values of β are indicated.

This approach could be used to define a standard set of point distributions for comparisons with others. However, one mst be aware that a nearest-neighbour distance criterion does not provide information on the large-scale character of the distribution and is thus of limited value as an index of clustering. A point distribution may have an associated clustering curve similar to that obtained from a distribution used in generating Fig. 3. However, the size and distribution of clusters in the two cases could be quite different.

As examples of the use of the Dirichlet tessellation, the microstructures from two common types of phase transition have been analyzed. Figure 4 shows a directionally solidified rod eutectic and its tessellation, which is close to regular in form. Figure 5 shows the microstructure of a spheroidized 1045 carbon steel,

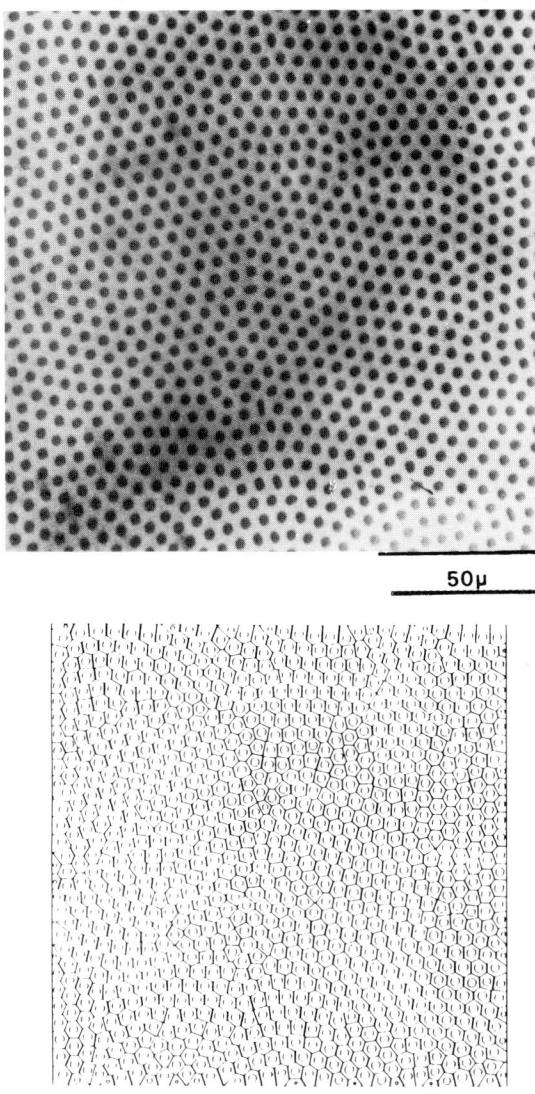

Fig. 4: Micrograph and associated tessellation for a Mn-Sb rod eutectic microstructure, viewing the rods end-on.

in which clustered regions of carbides are observed. Here, the Dirichlet cells are constructed from the inner envelope of lines which are perpendicular to the lines joining the particle centroids, bisecting them at a point midway between the particle surfaces.

A clustering analysis employing a distance-related clustering parameter could be used to characterize these particle distributions (as was done for the point distributions in Fig. 3). However, if both the particle size and spacing are considered in the clustering parameter, then the criterion for clustering could be based on a critical local particle area fraction. The microstructures of Figs. 4 and 5 are compared in this manner in Fig. 6, where the clustering parameter used is defined by the particle spacing-to-radius

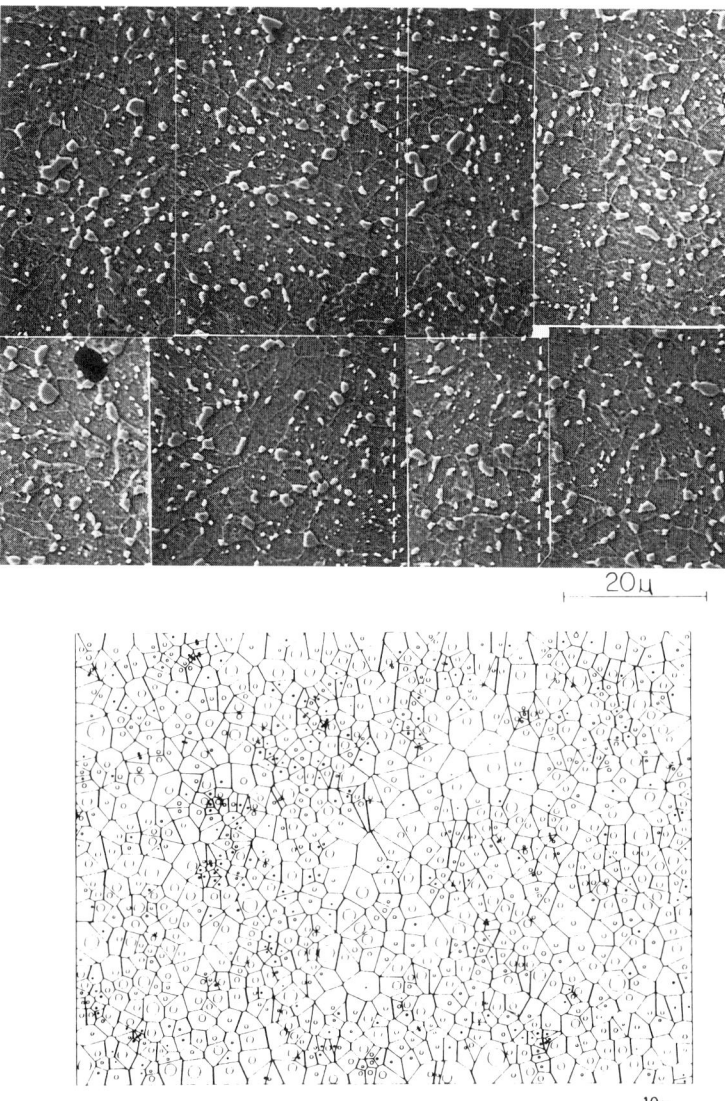

Fig. 5: SEM micrograph of the carbide distribution in a spheroidized 1045 steel, and the associated tessellation constructed about carbides from a portion of the field.

ratio. This approach is of value for comparing microstructures with different average second-phase volume fractions.

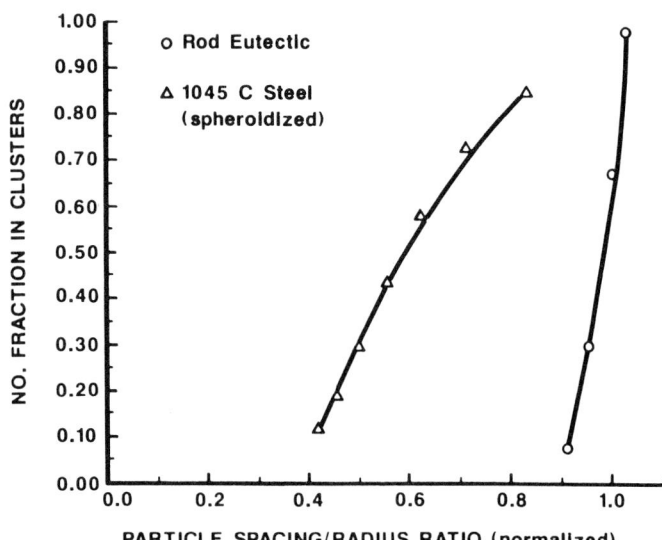

Fig. 6: Clustering analyses for microstructures given in Figs. 4 and 5. Results are given as the number fraction of particles which are clustered with at least one other particle, at a given particle spacing/radius ratio. This parameter has been normalized by the average value for each microstructure so that comparison can be made.

Conclusions Regarding the Tessellation Approach

1. The tessellation construction can be implemented on a computer, so that useful information regarding the spatial distribution of a large body of microstructural data can be analyzed.
2. A single clustering index or parameter which uniquely describes a spatial dispersion of particles exhibiting clustering is not amenable to development, because of clustering which can arise on several scales.
3. Further work is necessary in developing the tessellation approach towards application to the characterization of larger scale parameters of distributions, for example, to the distribution of clusters.

The Dirichlet tessellation could also find useful application in the following two areas:

i) Particle coarsening in microstructures containing a finite volume fraction of second phase dispersion. Observations suggest that local clustering can influence particle coalescence, altering the particle size dependence of coarsening (Burger, 1986; Brown, 1985).
ii) Describing the distribution of variations in local free volume in amorphous materials, notably, in metallic glasses. Argon (1979) has considered the influence of local dilatation on the onset of shear localization in metallic glasses.

ROLE OF PARTICLE DISTRIBUTIONS IN RECRYSTALLIZATION IN TWO-PHASE ALLOYS

Introduction

It is well established that inhomogeneous plastic flow occurring in the vicinity of hard second phase particles can lead to the creation of severe lattice rotations and subsequent nucleation of recrystallization at the particles. This is often termed particle stimulated nucleation. A number of detailed studies have resulted in a comprehensive description of the role of particle size, and the level of imposed strain on the nature of the structure and local lattice rotations in the vicinity of individual particles (Humphreys, 1978, 1979a,b, 1980, 1983; Porter and Humphreys, 1979). In addition, more macroscopic viewpoints have indicated the role of particles stimulated nucleation on texture development (Jensen et al. 1985; Jensen and Humphreys, 1986).

It is of importance to consider that in most two-phase systems the density of recrystallization nuclei is much less than the density of second phase particles. Thus not only the particle size, but the local spatial distribution of particles may be of importance in determining which particles act as nuclei for recrystallization. The influence of particle distribution poses a difficult problem both in terms of quantitative assessment of the appropriate microstructural parameters, and modelling of the mechanism of particle stimulated nucleation. It is of interest to note that similar problems arise in relation to the process of void nucleation in ductile fracture where the interaction between closely spaced particles plays an important role in attaining the critical condition for decohesion of the interface.

In the current experiment, studies were conducted on single crystals containing well-characterized second-phase particles in order to avoid nucleation at grain boundaries. A two-dimensional analysis was used in which the particles were regarded as points, and the structure characterized using the Dirichlet tessellation.

Characterization of Particle Distributions

For a given array of points which represent the second phase particles observed in a plane section two methods were used to characterize the distribution.

The first is a window technique in which the area of interest is divided into N^2 sub-areas, and a window of area $1/N^2$ is scanned over the array, and the number of particles in each window is recorded. The density of particles in each window can be visualized as a prism located at the site of the window, and of height proportional to the local density of particles. The distribution of prism heights can be compared to various multiples of the average particle density and the location of the prisms to the location of new grains so that the probabilities of grain nucleation can be computed for various local values of particle density.

The spatial distribution of the particles was also studied by Dirichlet tessellation technique. Tessellations were constructed by analysing a series of point dispersions obtained from micrographs taken at four randomly selected locations.

Results

Optical metallography and SEM observations were used to determine both the spatial distributions of the second phase particles, and the location of the new grains formed by particle stimulated nucleation on annealing after 40% deformation. The results indicated that particle stimulated nucleation was associated with groups of second phase particles as shown in Fig. 7.

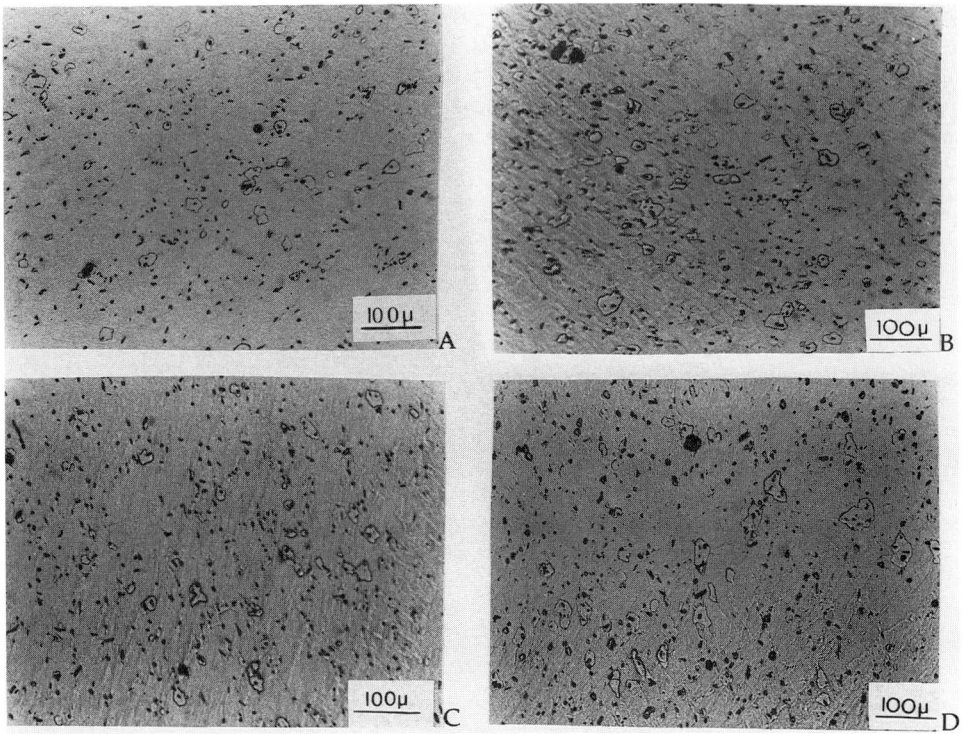

Fig. 7: Optical micrographs illustrating the nucleation of recrystallization associated with a group of particles.

Various fields were examined in order to reveal a) the underlying distribution of particles, and b) the resulting location of new grains. The Dirichlet tessellations associated with the particles were determined as described in an earlier section, and are illustrated in Fig. 8. The associated field of new grains was superimposed as shown in Fig. 9 showing in a more pictorial sense the spatial correlation of the nucleation events with clusters of particles.

Analysis of a number of areas using the window method to determine the local density of particles and its correlation with grain nucleation on annealing can be expressed as a cumulative probability diagram, as shown in Fig. 10. The results suggest that in order to produce particle stimulated nucleation the local density of particles needs to be of order 2-3 times the average particle density.

Discussion

The objective of the present work was to provide a view of particle stimulated nucleation extending from microscopic observations of the detailed dislocation structures produced around the second phase particles to macroscopic observations which correlate the spatial relationship between the sites of recrystallization and the local particle density.

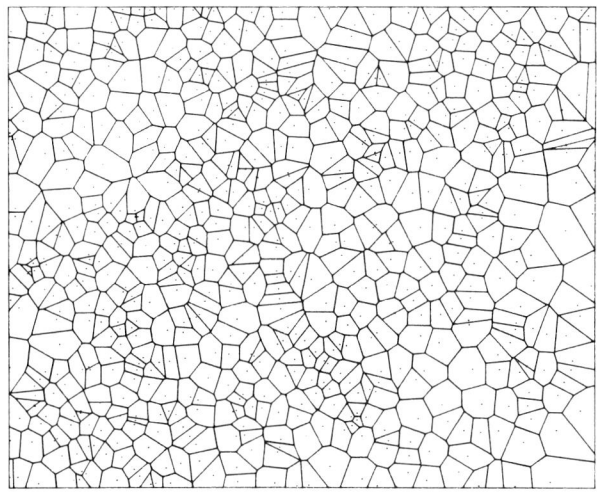

Fig. 8: A Dirichlet tessellation illustrating the cell pattern associated with particles.

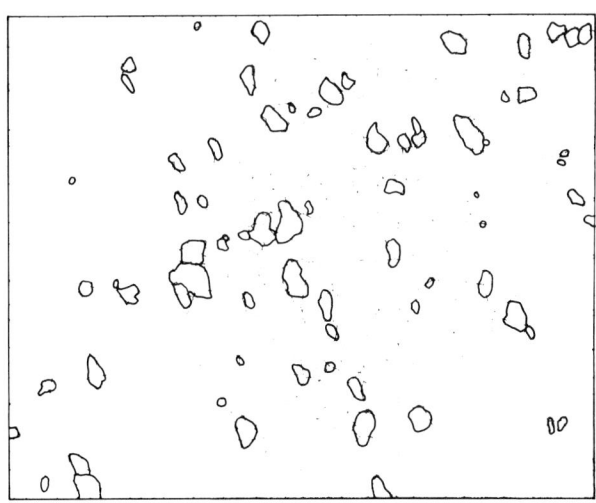

Fig. 9: A Dirichlet tessellation showing the superimposed locations of new grains on the corresponding cell pattern.

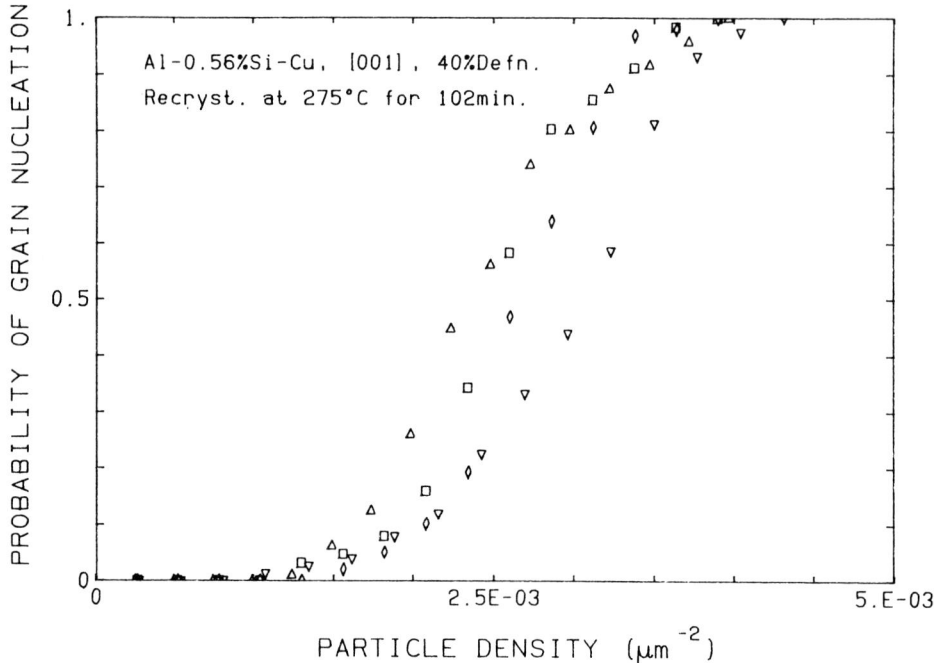

Fig. 10: A cumulative probability diagram showing the local particle density and its correlation with grain nucleation.

Clearly it must be emphasized that the study has some intrinsic limitations in that it deals with single crystals in which the deformation modes are likely to be simpler than in those pertaining to the polycrystals and also the assessment of particle distributions by 2 dimensional sections clearly does not consider the role of particles located in the perpendicular direction. However, despite these limitations, some broad conclusions can be drawn:

a) The description of the heterogenous structure around second phase particles demands that cognizance be taken of the local lattice rotations produced in groups of cells close to the particle rather than considering a single misorientation which can be related to the rotation of the particle relative to the matrix.
b) The heterogeneity of the structures in two phase systems appear to depend strongly on interaction between particles. This may reflect the overlap of the local plastic zones around the particles in the manner originally outlined by Argon et al. (1975).
c) Particle stimulated nucleation appears to involve only a small fraction of the particles in the distribution. The results of this study indicate that the nucleation is dependent on particle clustering rather than simply particle size or overall average strain level.

The results described here provide some new insight into the problem of particle stimulated nucleation. The Dirichlet cell method of describing the distribution of particles may provide a basis for developing simulations of recrystallization in two phase systems by considering a range of nucleation times which are based on the measured distribution of second phase particles and simulations of this type are currently being developed. These may provide descriptions of the evolution of grain size distributions which can be related to the recent models of Saetre et al. (1986).

APPLICATION OF THE DIRICHLET TESSELLATION TO MODELS FOR CREEP FRACTURE

In previous sections we have described how the Dirichlet tessellation can be used to analyze the degree of inhomogeneity in microstructure, and how it can be used to analyze experimental data. We now turn our attention to the possibility of using these concepts to **model** the effect of inhomogeneity on processes such as grain boundary creep fracture.

It has long been known that creep cavitation is inhomogeneous. In particular, it is apparent that the extent of cavitation varies among different grain boundary facets. This has led to the development of models for "constrained" cavity growth (Dyson, 1976, 1979; Rice, 1981; Cocks and Ashby, 1982), in which it is assumed that only a fraction of the available grain boundary facets participate in the cavitation process. Because of the strain produced by cavitation, load shedding must occur from regions which cavitate to those which do not. The net effect is a decrease in the rate of global damage accumulation.

A similar effect must occur on a single grain boundary facet if the distribution of cavities is non-uniform. This is illustrated in Fig. 11. If the grains act as rigid slabs then they must separate at a uniform rate ($d\delta/dt$) as cavitation occurs. This constrains and couples the diffusive fluxes around each cavity. Load is transferred from those cavities which grow most easily (i.e. those which are clustered together) to those which cannot (i.e. those which are isolated from their neighbours).

In modelling this effect we need to identify, for any given arrangement of voids on a facet, the diffusion zone around each cavity. This is the area of grain boundary which must be supplied with atoms by diffusion from that particular cavity. The Dirichlet tessellation provides an obvious means of doing this. Figure 12 shows the distribution of Dirichlet cell areas for two different void arrays, one random and the other clustered with a cluster parameter of 0.5. A wide distribution of cell areas is found, especially for the clustered arrays. Thus a broad distribution of cavity growth rates may be expected. The cell area A, enters naturally as a scaling parameter into models for cavity growth. Each void grows at a rate given by

$$\frac{df}{dt} = h(A, f)(\sigma - \sigma_s)^p$$

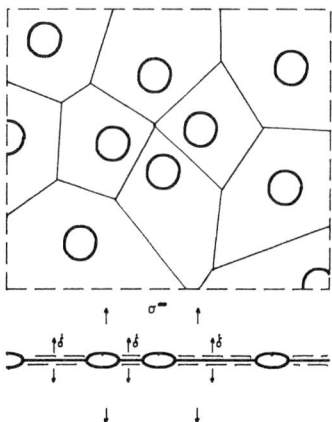

Fig. 11: A schematic diagram illustrating the effect of a non-uniform void distribution on void growth, when a uniform rate of grain boundary deposition is imposed.

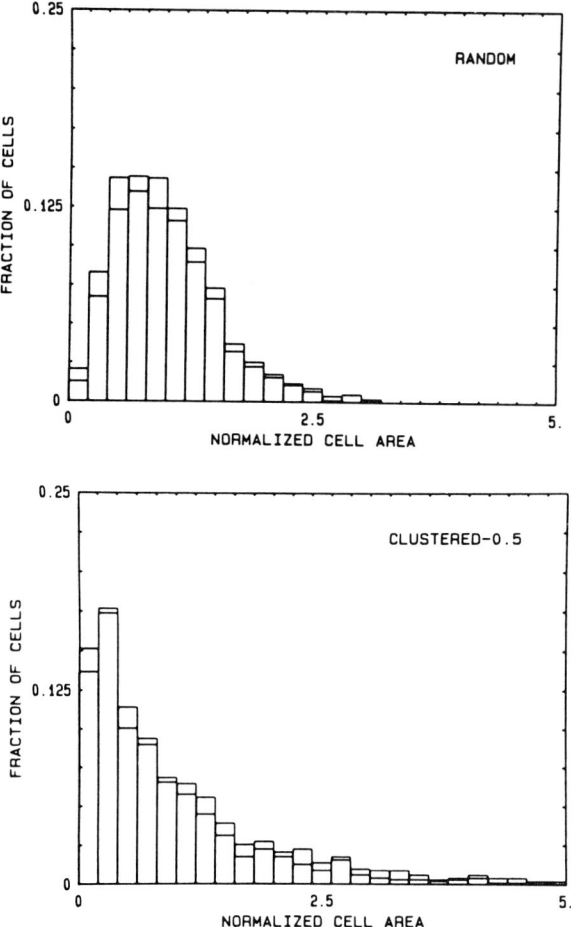

Fig. 12: The distribution function for Dirichlet cell areas is shown for two arrays, one random the other clustered.

where

$$f = \frac{\pi r^2}{A}$$

is the local void area fraction, and r is the void radius. The function h and the parameter p depend on the mechanism of void growth, while σ_s is the sintering stress. For an inhomogeneously distributed array of voids, (df/dt) is different for each void. However, the set of equations for (df/dt) are coupled by the need to keep (dδ/dt) uniform, and by the equation of mechanical equilibrium. The first condition, for rounded voids, gives

$$\frac{d\delta}{dt} = \frac{2r\theta_s}{1-f}\frac{df}{dt}$$

where θ_s is a function of the equilibrium dihedral angle at the void-grain boundary interface. The second condition is given by

$$\sum_{i=1}^{N} \sigma_i A_i = \sigma^{\infty} A_b$$

for an array of N cavities on a facet of area A_b. Here σ^∞ is the far-field applied stress.
Solutions to this set of coupled equations can be developed for individual void growth mechanisms. Since this development is published elsewhere (Wilkinson, 1988) only the results are presented here.

We are particularly concerned to know the extent to which an inhomogeneous array of voids constrains the damage accumulation rate. We define the global damage f_G as the fraction of the grain boundary facet which is cavitated, i.e.

$$f_G = \frac{\sum_{i=1}^{N} \pi r_i^2}{A_b}$$

Now we can compare, for $f_G \ll 1$ at least, the damage accumulation rate (df_G/dt) at the same value of f_G for uniform and non-uniform void arrays. The result depends on the void growth mechanism. For example, if we consider the case in which void growth is controlled by grain boundary diffusion, then

$$\frac{\dot{f}_G}{\dot{f}_{un}} = \frac{\ln\left(\frac{1}{f_G}\right) \sum_{i=1}^{N} a_i}{\sum_{i=1}^{N} \ln\left(\frac{a_i}{f_G}\right) a_i^2}$$

where a_i is the normalized cell area ($a_i = A_i/A_b$). Similar results are obtained for different mechanisms (Wilkinson, 1988).

The development of artificially random and clustered distributions as described in a previous section, can be used to analyze these results. Table II lists values of ($\dot{f}_G/\dot{f}_{un}$) for three different void growth mechanisms. Two results are given for each condition, each being the result of a different simulation. It is clear that there is some scatter in the results due to the finite number of particles (1024) in each simulation. However, there is also a clear and significant decrease in the damage rate for all three mechanisms as the degree of clustering increases. Further analysis shows that the radial growth rate of individual voids scales as the area the cells in which they sit. Thus the distribution in Fig. 12 also represents the distribution of growth rates. It is therefore clear that a large variation in local void growth rate can be expected. This has important implications for the interpretation of some experimental results. For example the void number density is usually found to increase with time during creep. This is often interpreted as evidence of continuous void nucleation. However, it may also be due to the growth of voids to an observable size at different rates due to distribution effects.

TABLE II. Effect of Void Distribution on the Global Void Growth Rate

Cluster parameter	G.b. Diffusion		Creep	Surface Diffusion	
	$f_G = 10^{-4}$	$f_G = 10^{-3}$	$f_G = 10^{-3}$	$f_G = 10^{-4}$	$f_G = 10^{-3}$
random	0.756	0.747	0.857	.828	.828
	0.719	0.709	0.834	.801	.800
0.7	0.602	0.589	0.755	.709	.709
	0.601	0.589	0.746	.703	.702
0.5	0.436	0.422	0.606	.514	.514
	0.399	0.385	0.570	.551	.551
0.3	0.208	0.197	0.337	.321	.321
	0.220	0.209	0.372	.294	.294

A non-uniform void distribution has other implications. In particular, it is clear that void coalescence must occur gradually. This has generally been assumed to be unimportant. However, this turns out not to be the case (Wilkinson, 1987a,b). As voids coalesce and the void density decreases, the distance between the remaining voids can become quite large. This slows down diffusion processes and the final coalescence may involve a transition to a new mechanism. Models for this process (Wilkinson, 1987a,b) lead to predictions for the stress and temperature-dependence of the failure time, which differ from those predicted by models based on a uniform array of voids. Moreoever, there is experimental evidence to support these models.

In summary, the concept of the Dirichlet tessellation has been used to adapt models for creep cavitation to the case of a non-uniform arrangement of voids. The effects are significant, both in terms of the constraint on global damage rate and the distribution of individual void growth rates, and in terms of the effect on predicted failure times.

ACKNOWLEDGEMENTS

The authors are grateful to NSERC and CANMET for research support, and to their colleagues Professor R. Sowerby, Dr. N. Chandrasekaran and Mr. G. Robertson at McMaster, and Dr. Owen Richmond at Alcoa and Dr. E. Arzt of Stuttgart for valuable discussions in the course of this work.

REFERENCES

Argon, A.S. (1979) Acta Met. **27**, 47.
Argon, A.S. Im,J. and Safoglu, R. (1975) Metall. Trans. **6A**, 825.
Brown, L.C. (1985) Acta Met. **33**, 1391.
Burger, G. (1986) Ph.D. Thesis, McMaster Univ., Canada.
Burger, G., Wilkinson, D.S., and Embury, J.D. (1988) to be published.
Cocks, A.C.F. and Ashby, M.F. (1982), Prog. Mater. Sci., **27**, 189.
Dyson, B.F. (1976), Metal. Sci. J., **10**, 349.
Dyson, B.F. (1979), Can Met. Quart., **18**, 31.
Humphreys,F.J. (1977) Acta Met., **25**, 1323.
Humphreys, F.J. (1978) Metals Forum, 1, 123.
Humphreys,F.J. (1979) Acta Met., **27**, 1801.
Humphreys, F.J.,(1979) Metal Sci., **13**, 136.
Humphreys, F.J., (1980) Proc. 1st Riso Int. Symp. Recrystallization and Grain Growth of Multi-Phase and Particle Containing Materials, p. 35.
Humphreys, F.J. (1983) Proc. 4th Riso Int. Symp. on Deformation of Multi-Phase and Particle Containing Materials, p. 41.
Humphreys, F.J. and Juul Jensen, D. (1986) Proc. 7th Riso Int. Symp. on Annealing Processes – Recovery, Recrystallization and Grain Growth, p. 93.
Juul Jensen,D. , Hansen,N. and Humphreys, F.J. (1985) Acta Met., **33**, 2155.
Porter,J.R. and Humphreys,F.J. (1979) Metal Sci., **13**, 83.
Rice, J.R., (1981), Acta Met., **29**, 675.
Rogers, C.A. (1964)Packing and Covering, Cambridge Mathematical Tracts **54**, Cambridge Univ. Press.
Saetre, T.O., Hunderi, O. and Nes, E. (1986) Acta Met., **34**, 981.
Spitzig, W.A. Kelly, J.F. and Richmond, O. (1985)Metallography, **18**, 235.
Stewart,A.T. and Martin, J.W. (1970) J. Inst. Metals, **98**, 62.
Stewart,A.T. and Martin, J.W. (1975) Acta Met., **23**, 1.
Wilkinson, D.S. (1987a), Acta Met. **35**, 1251.
Wilkinson, D.S. (1987b), Acta Met., in press.
Wilkinson, D.S. (1988) Acta Met., in press.
Wray, P.J., Richmond, O. and Morrison, H.L. (1983) Metallography, **16**, 39.

Thermologistics: A Science of Pattern Formation

J. S. KIRKALDY

Institute for Materials Research
McMaster University

ABSTRACT

The problem of wave number selection in irreversible pattern formation is addressed via the principle of maximum path probability and its corollary minimax in the dissipation for steady state systems. A compendium of applications in the animate and inanimate universe which ranges through chemical transformations, hydrodynamics, thermalhydraulics, biology, evolution and cosmology is presented. It is demonstrated that this class of patterns maps into a Boolean inductive logic possessing a conventional syntax and rules of inference. The patterns of nature, mathematics and language are thus seen to be isomorphic in a much stronger sense than previously considered. Thermologistics is offered as a viable route to epistemology and a theory of artificial intelligence.

HISTORICAL PERSPECTIVE

Several decades ago John von Neumann remarked in the context of automata as spontaneous pattern formers: "I have been trying to justify the suspicion that a theory is needed, and that very little of what is needed exists yet ... if found, it is likely to be similar to two of our existing theories: formal logics and thermodynamics" (1966). If von Neumann had survived to complete his program it is likely that we would be much more advanced than we are in the science of pattern formation and the related technology of pattern recognition. Nothwithstanding his untimely death he planted a number of seeds whose fruits bear on current developments. In particular, his theory of games and economic behaviour (1944), when considered in the form of dynamic play, and which postulates for the process of economic (i.e. human) reasoning a minimax balance between acquisitiveness and moderation, is a theory of logical inference or intellectual pattern formation. Recognizing that economic utilities and free energies are equivalent I gained considerable inspiration from this structure in developing the general minimax theory of patterns for animate and inanimate systems alike.

The materials scientist may give pause at this point and ask: what can intellectual pattern formation and pattern formation in phase transformations possibly have to do with each other? The partial answer, which I believe most can accept, is that the mutual reference is to systems which are self-organizing, which is to say sharing an underlying logic. It is one of the aims of this paper to demonstrate that the shared underlying logic is actually a classical formal logistic in the sense of possessing a syntax or grammar, valid rules of inference, and upon interpretation, semantics or meaning. In the process we will come to understand precisely what is meant by the term pattern. I must state from the outset that while related, it does not in general mean negentropy.

The science of pattern formation as we know it goes back to the late 19th century where pattern-forming oddities associated with surface tension and various kinds of chemical oscillators, including the Liesegang phenomenon (1896), were well-known. Researchers saw these as analogies, even explanations of morphogenesis in the biological world where pattern formation is ubiquitous. D'Arcy Thompson (1917) summarized these insights rather conservatively in his classic volume "On Growth and Form". Some key ideas of irreversible thermodynamics and of free boundaries (one might say, free will) were lacking, so this approach languished for many years.

In the early part of the century the actuary and biologist Alfred Lotka began interpreting population dynamics in terms of differential equations which described non-linear chemical kinetics (Lotka, 1956) and was able, for example, to explain the oscillating ecologies of the sub-arctic (Fig. 1a). These are isomorphic to the well-known dendritic phase tansformation (Fig. 1b). Important

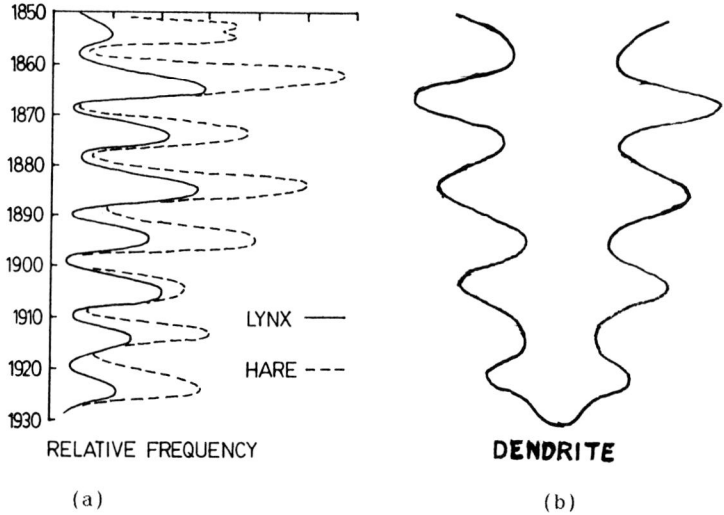

Fig. 1. Oscillations (a) in the vegetation-hare-lynx sub-arctic ecology after Hudsons Bay Company records, (b) in dendritic solidification.

mathematical problems were raised to which Volterra (1936), of screw dislocation fame, offered solutions. Turing (1952), who is noted in computer science, generalized these rate equations to diffusion-reaction processes and demonstrated in support of Lotka the role of autocatalysis in generating simultaneous temporal and spatial oscillations in a single phase system. The Belousov-Zhabotinskii reaction illustrated in Fig. 2 is the best known of these phenomena (Zhabotinskii, 1974). It should be reemphasized that these patterns refer to continuous systems, such as in the early stages of spinodal decomposition. Thom's theory of chaos (1972) and Nicolis' and Prigogine's dissipative structures (1977) involving bifurcation theory are offshoots of this approach to pattern formation. Very little of this bears directly on our thesis.

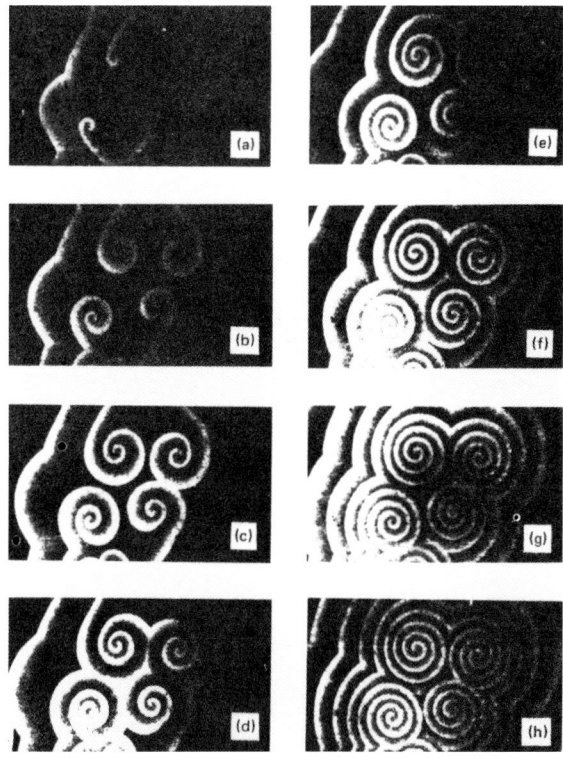

Fig. 2. Belousov-Zhabotinskii reaction. After Zhabotinskii (1974).

Lotka (1922, 1945), as a mathematical biologist, is less well known for his synthetic perspective on life processes and morphogenesis. He realized that life and ecological processes are degenerate in the sense that the kinetic equations, together with all the constraints, whether genetic or environmental, allow an infinite number of solutions. This is an implicit recognition of a free boundary situation. He therefore sought a corollary or generalization of the Second Law which would remove the internal degress of freedom. He postulated a maximum power principle together with maximal efficiency as the criteria for ecological and evolutionary development. These are still widely applied by ecologists. One can recognize in these a thinly veiled expression of a minimax in the dissipation. In 1945 he became quite explicit in recognizing the symbiosis of the plant and animal kingdoms where in his view the former tends to be minimal in the dissipation and the latter maximal. Bertrand Russell (1927) had already expressed the principle for life processes as "chemical imperialism" which implies the maximal imperative.

During completion of his thesis Prigogine began the exploration of the application of his principle of minimum entropy production. This was a generalization of principles of minimum dissipation due to Lord Thompson and Lord Rayleigh in the 19th century and Onsager in 1931. He saw the minimal principle, which he characterized as "dynamic efficiency", emerging in homeostasis of the biological individual and Lamarckism in the process of evolution (Prigogine and Wiame, 1946; Prigogine, 1955). During the 1950's and 1960's he and his associates expended much effort in generalizing the minimal principle to non-linear systems (Glansdorff and Prigogine, 1971). Finally, in 1972, in two articles in Physics Today on the Thermodynamics of Evolution (Prigogine and co-workers), he was forced to admit that maximal tendencies cannot be ignored. In 1977, the year of award of his Nobel prize, he together with Nicolis makes the statement: "Assuming that a dissipative structure has been formed we would like to find a set of thermodynamic properties characterizing the structure as uniquely as possible ... A general solution of this important problem is not yet available". In other words, the methodologies of he and his extended school had failed to answer either their own or Lotka's question, and in our view the most important question of all. As far as I can see, the Prigogine school and the derivative "Synergetics" school of Haken (1983a, 1983b), have to this date made no real progress on this central problem of wave number selection.

My own introduction to the problem area began during the period 1954-1960 with a spare introduction to the thermodynamics of irreversible processes and multicomponent diffusion via Darken and Gurry (1953) and to the beautifully patterned series of cellular solidification observations carried out by Bruce Chalmers and co-workers (1963) at the University of Toronto (Fig. 3). I immediately recognized in the latter a free boundary problem and a potential non-trivial application of the minimal dissipation principle (Kirkaldy, 1959). Clearly, an unmixed solid product stemming from a steady state cellular morphology designates a lesser dissipation than a completely mixed product stemming from a planar morphology (Fig. 4). Other materials scientists including

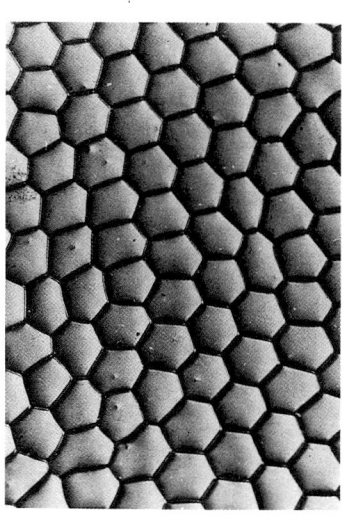
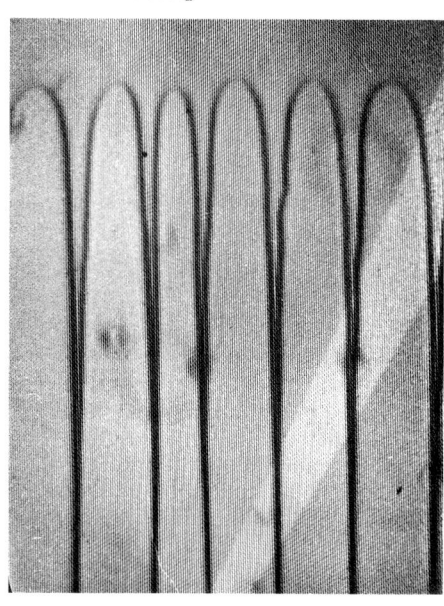

Fig. 3. Decanted binary cell interface. After Cole (1963).

Fig. 4. Cellular solidification cells in a thin film experiment. Courtesy of D. Venugopalan.

Tiller, Li, Cahn, Weart, Oriani and Kuczinski contemporaneously explored possible applications of irreversible thermodynamics, but the results were equivocal. I believe we were all trapped by the minimal syndrome. My escape came in about 1960 through reading Lotka's contributions and a non-committal conjecture of John Cahn (1959) vis-a-vis the stable spacing of discontinuous precipitates. Finally, on rereading Onsager's papers of 1931 I found that his linear formalism actually accomodated a kind of minimax in the dissipation, perturbations to thermodynamic state functions or forces regressing to a minimum dissipation and perturbations which interrupt or deflect fluxes relaxing to a maximum. Fig. 5 shows the

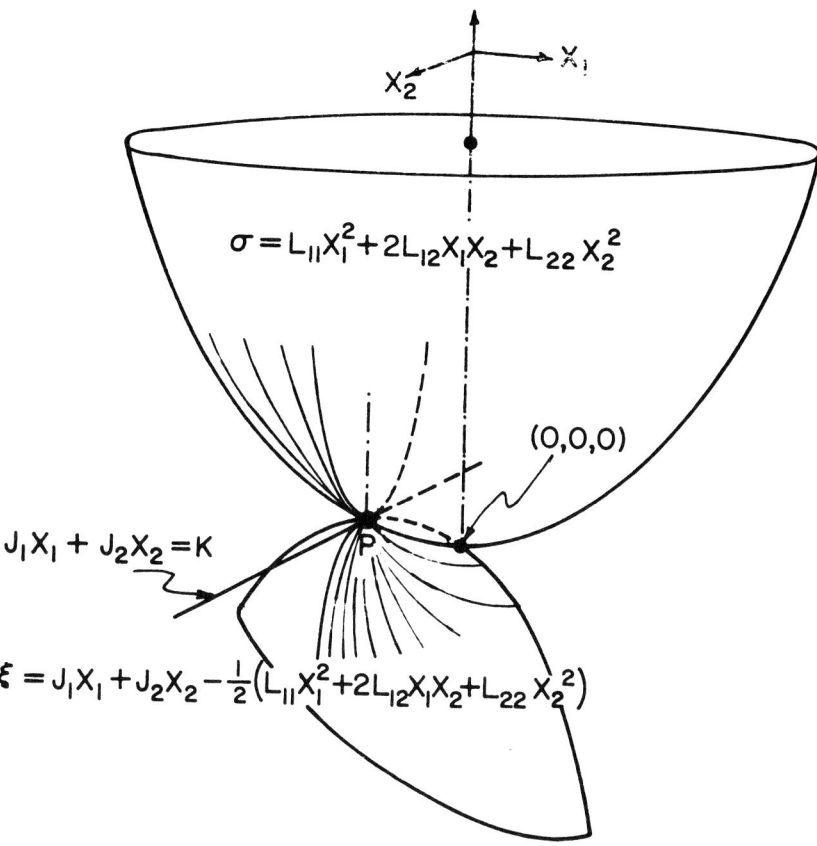

Fig. 5. Surface in the dissipation defined by the two-dimensional minimax in Onsager's linear dissipation theory. J's and X's are fluxes and forces. σ and ξ are dissipation functions. Arrowed lines define regression towards stability (P) on the surface.

dissipation surface for a two force system of his type (Kirkaldy, 1978). I realized at that time that a theory based on microscopic fluctuations alone was inadequate to the free boundary problem and therefore postulated that the macroscopic phase space of fluctuating pattern has similar properties to the microscopic one (Kirkaldy, 1965a, 1965b). Had I known at that time about renormalization group theories of critical phenomena, which accomodate and require compatible fluctuations of all scales of length (Wilson, 1975, 1979), I could have put my postulate on a much firmer basis.

In 1965 I published two papers in the Biophysical Journal, Thermodynamics of Terrestrial Evolution (Kirkaldy, 1965a) and Thermodynamics of the Human Brain (Kirkaldy, 1965b). It was a pleasure to discover some seven years later that I had anticipated Prigogine's article on the Thermodynamics of Evolution in many essential points, and in particular on the importance of maximality. A subsequent lengthy dialogue with Doug Chambers led to his 1973 Ph.D. thesis

"Thermodynamics of Self-Organizing Systems" (Chambers, 1973). This touched for the first time on the game theoretical and logistical connections of the minimax postulate. By this time (and as early as 1969) I had notarized documents pertaining to potential applications in the engineering of automata and artificial intelligence (Kirkaldy, 1978). In about 1970 I realized that Zener-Turnbull-Hillert type formulas (Zener, 1946; Turnbull, 1955; Hillert, 1957), which invoke the maximal propensities of pattern formation in pearlitic steels, had great potential for property prediction in low alloy steels (Kirkaldy, 1973) and during the next decade I focussed on the development of an extensive software package, CASIS, and its commercial dissemination (Feldman, 1978; Kirkaldy, 1986). This is now in use in some 30 international corporations. Towards the end of the 1970's I began to detect my earlier ideas on pattern formation, which had been widely disseminated but rejected with some abuse, emerging in the current literature. Thus in 1978, as a defensive action, I privately published a summarizing booklet entitled "Life, Logic and Bootstrap Physics". During the 1980's I have devoted considerable theoretical and experimental energies to the advancement of my old thesis. The Physical Review paper "Pattern Formation, Logistics and Maximum Path Probability" (Kirkaldy, 1984) provides a reasonably up-to-date summary of my work and the substantial supportive literature. I have only recently come to realize that Boole's monumental 1854 work "The Laws of Thought" offers powerful support for my thesis.

In the following I propose to primarily review our researches of the past decade. This includes the definitive theory of the minimax, a compendium of applications, the logistical intepretation with assists from Boole, some epistemological and cosmological implications, and for my fellow engineers, some technological implications which are on the verge of being realized.

THE MINIMAX PRINCIPLE AND ITS PARADIGMS

Thermodynamic arguments always pertain to ideals. Here it is a discontinuous system in isolation as illustrated in Fig. 6. The small subsystem is the

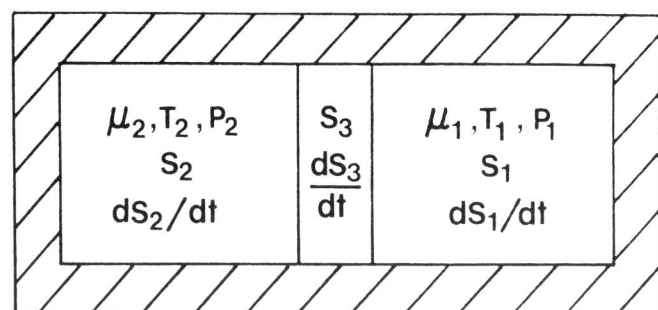

Fig. 6. Idealized "discontinuous" system to which the general minimax principle applies.

fluctuating pattern former in contact with one or more very large <u>uniform</u> reservoirs, and these define the potential for achieving a quasi-steady state. Fluctuations in pattern induced by a continuous spectrum including microscopic fluctuations produce instantaneous <u>macroscopic</u> changes in the entropy production of the small subsystem designated by $\Delta\dot{S}(0)$, and this leads to a change in the total system entropy, primarily through the non-dissipative transfer of entropy to or from the reservoirs (e.g. in the form $\dot{q}/T$ where q is the dissipative heat transferred) without change in their intensive functions. The system entropy change (from a base $S(t)$) in subsequent time τ will accordingly be on the average

$$\Delta S = \int_0^\tau \Delta\dot{S}\,dt \tag{1}$$

Note that this structure is unique to the entropy since it is the only non-conserved extensive function in isolation.

Now we examine the correlation product pertaining to the pattern fluctuation

$$C = \Delta S\, \Delta\dot{S} \geq 0 \tag{2}$$

and assume on the grounds that the probability of large pattern-destroying fluctuations is vanishingly small that its time average $\bar{C} \geq 0$ exists as a stationary parameter. Since every fluctuation has a precedent and a serial consequent it represents a truth-functional proposition or set in logic. Thus Eq. 2 can be interpreted as a logical equation and read as "the correlation set is the intersection of the sub-system set $\Delta\dot{S}$ and the system set ΔS". We will recall this intelligence in a later section. We take the existence of $\bar{C}$ as a condition "a posteriori" to the fact that a patterned quasi-steady state exists. This is in the spirit of classical statistical mechanics which assumes that an equilibrium state and a unique mean correlation between extensive and intensive correlations exists. In these terms the time rate of change of each C following the fluctuation $\Delta\dot{S}(0)$ is

$$\frac{dC}{dt} = (\Delta\dot{S})^2 + \Delta S\, d(\Delta\dot{S})/dt = (\Delta\dot{S})^2 + \Delta S\Delta(d\dot{S}/dt) \tag{3}$$

Now despite possible macroscopic changes in pattern and corresponding entropy production we insist that the stationary state be microscopically defined in accord with $d\dot{S}/dt = 0$, i.e.,

$$\frac{dC}{dt} = (\Delta\dot{S})^2 \geq 0 \tag{4}$$

This is in accord with the precepts of renormalization group theory which requires that fluctuations of all scales of length and time exist in an analytic manifold (Wilson, 1975, 1979), (e.g., self-similarity or fluctuation scale invariance). Now we can form the average over a large set n of C.

$$\frac{1}{n}\Sigma\frac{dC}{dt} = \frac{d}{dt}(\frac{1}{n}\Sigma C) = \frac{d\bar{C}}{dt} = \frac{1}{n}\Sigma(\Delta\dot{S}_i)^2 \tag{5}$$

But $\bar{C}$ like $\dot{S}$ is a time invariant of the steady state so for each and every fluctuation

$$\Delta\dot{S} = 0 \tag{6}$$

and therefore the stability point is a maximum or minimum or both in the dissipation with the corollaries via Eqs. 1 and 2, $\Delta S = C = 0$. Since correlations decrease the statistical accessibility of quantum states we can associate $\bar{C}$ in a negative monotone relation with a path probability, P. This connects with Boole's generalization of deductive to inductive or probabilistic logic. If P is taken to be a maximum as expected in an isolated system then the mean time correlation $\bar{C}$ based on Eqs. 1 and 2 must vanish and Eq.(6) again follows. Evidently, maximum path probability, the existence of a well-behaved statistical phase space and a logistical foundation are equivalent pattern selection constraints (more on this later.).

Tykodi (1967) has independently presented arguments for the minimal part of the
minimax theorem as have Paltridge (1981) and Sawada (1981) for the maximal part.

To make Eq.(6) quantifiable as an optimal principle $\dot{S}$ must be expressible as a
function of n independent order parameters, p_i. Such expressions are usually
obtained trivially as a consequence of the degeneracy of the steady state dynamical equations and therein the signature of the optimality is implicitly
determined by the initial and boundary conditions. They can be strictly applied,
however, only if the steady state is such that the initial conditions have been
statistically "forgotten". This is of course the necessary condition for the
existence of a macroscopic statistical manifold and a unique steady state. The
alternative involves hysteresis, which is very common in irreversible pattern
formation, and here unique steady states are not attainable.

In the usual convention of the variational calculus the path probabilities
defined in the foregoing arguments pertain to the same instant of time.
Consequently, the optimization may be regarded as "orthogonal" to the time axis.
In a later application of Eq.3 we shall be seeking a solution for which time
itself is an order parameter.

It will be noted that the patterned steady state with its macroscopically fluctuating dissipation function is somewhat analogous to an equilibrium critical
point with its macroscopically fluctuating density functions. Because of this
and the assumed continuity of the spectrum of microscopic and macroscopic fluctuations it is apparent that there exists a complementary approach to pattern
formation through renormalization group theory, emphasizing the interaction
between fluctuations of broadly disperse scales of length (Wilson, 1975, 1979)
This intelligence has already been noted in current approaches to "Deterministic
Chaos".

There are two irreversible systems "par excellence" for illustrating the minimax;
one in the inanimate world, the other in the biosphere. Take first a tree. It
expresses its chemical imperialism over atmospheric and earth resources with a
concomitant maximization of the dissipation by extending its leaf and root areas
so as to fill the available space. However, it does this in the most efficient
way possible through minimal length, streamlined vascular pathways and simultaneous storage of available energy in its leaves, branches and trunk. The well-
known patterns which express the unity of form and function are noteworthy.

A second paradigm for macroscopic spontaneous irreversible pattern formation in
isolation is a wood fire in still (infinite) surroundings (Fig. 7) (Kirkaldy,
1978). This element of man's technology as an artificial sun and aesthetic
accoutrement (and perfected in lamps, stoves, fireplaces and furnaces) is the
natural companion of the Carnot engine in analyzing for a comprehensive general
thermodynamic science. The quasi-steady state patterns of interest lie in the
reticulated wood surfaces, the radiant flame shapes and the billowing smoke
clouds, underlain by spasmodic von Karman vortices generated by turbulent gas
flow through the network of wood. The kinematic elements of stabilization are a
surface area limited chemical reaction, a draft or upward convection of hot gases
according to Archimedes Principle and a convergent balancing inflow of air
(oxygen). The fluctuation of pattern as seen in the "licking" of flames is
essential to the stabilization process as described above in mathematical terms.
Indeed, conditions of wood packing which abet strong hydrodynamic fluctuations
are essential to the search of the kinetic phase space for stability and therefore for successful ignition (see also later section on the maple key).

The empirics of the minimax stability point for the wood fire are particularly
transparent, for the entropy production rate near the stable quasi-steady conf-

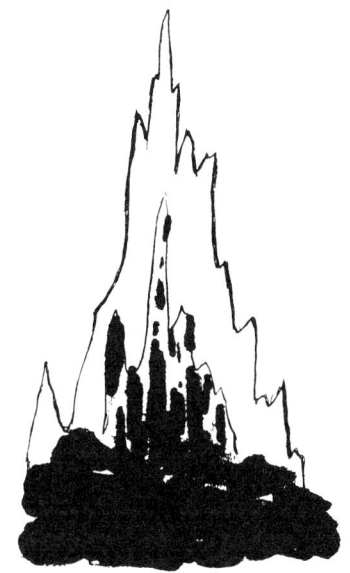

Fig. 7. Steady state configuration of a campfire.

figuration must be monotone with, and approximately proportional to the oxygen delivery rate (the draft) and to the integral brightness of the flame. if we weakly perturb the draft via a temporary restriction or an augmentation, the fire intensity returns asymptotically in both directions to its unperturbed state in a time appreciably less than natural drift times associated with non-ideal changes in surface area or other artifacts. The essential involvement of both minimal and maximal trends in the dissipation as principles of stability is thus illustrated.

A COMPENDIUM OF APPLICATIONS

We now undertake the examination of our own selection of pattern-forming systems which submit more or less quantitatively to the minimax principle or its corollaries as the criteria for pattern selection. They extend through phase transformations, hydrodynamics and life processes.

Liesegang Spirals

Both crystals and fungal roots growing dendritically from a central seed or spore in a ternary solution exhibit Liesegang rings, and apparently by accident, these occasionally degenerate into the spiral mode of the same wavelength and frequency (Figs. 8 and 9). The morphology of Fig. 3 is attractive from the theoretical point of view since one can identify an approximate quasi-steady state pattern in a rotating and radially expanding frame of reference at the growth front. Our

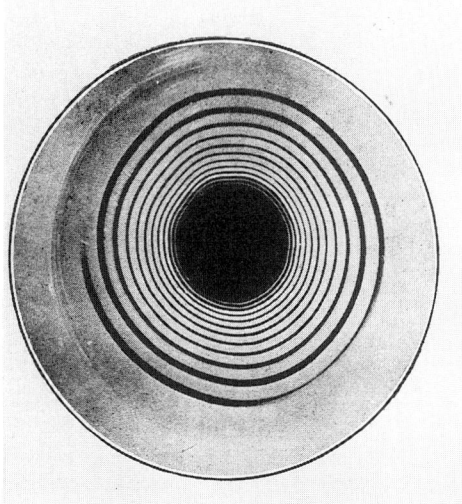
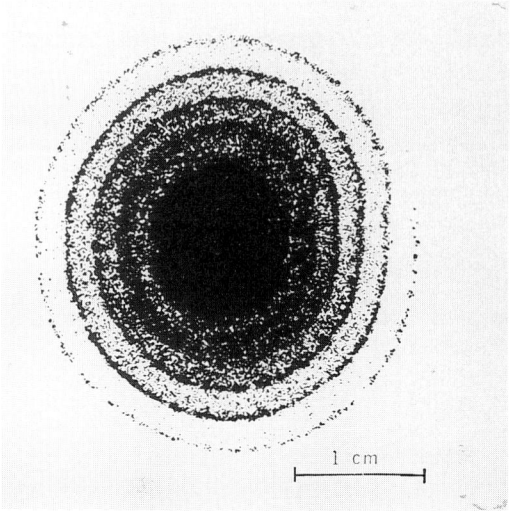

Fig. 8. Precipitation of silver chromate as a simple spiral. After Veil (1934).

Fig. 9. Archimedean spiral produced by culture zonation of Penicillium diversum. After Bourret et al (1969).

Analogously to the spinodal instability, the separated quasi stationary azimuthal (Φ) and radial (R) differential equations are

$$\frac{D_e}{\rho_{av}^2 \omega R} \left[\frac{\partial^2 R}{\partial (r/\rho_{av})^2} + \left\{ \frac{1}{\rho_{av}} + \frac{D}{16 D_e} \right\} \rho_{av} \frac{\partial R}{\partial (r/\rho_{av})} \right]$$
$$= \frac{1}{\Phi} \left[-\frac{\partial \Phi}{\partial \phi} - \frac{D_e}{\omega \rho_{av}^2} \frac{\partial^2 \Phi}{\partial \phi^2} \right] = m^2 \qquad (7)$$

with the characteristic separation constant corresponding to the solution of Eq.7

$$m^2 = -\frac{D^2}{4(16)^2 \omega D_e} - \frac{(16)^2 D_e \omega}{D^2} > 0 \qquad (8)$$

Neither m^2 nor the angular frequency ω are "a priori" determined by other conditions on the problem so we have here identified a free boundary problem analogous to the well-known pearlite spacing problem. The framing of the dissipation principle appears out of the question in this complicated case; yet nature in its usual good graces provides an alternative. It will be noted in Eq. 7 that m^2 is a monotone measure of the strength of correlation between the radial and azimuthal diffusion processes. Thus from the probabilistic version of our minimax theorem m^2 should go to a minimum to assure a maximum in the path probability. The application of this principle identifies $m^2 = 1$ and yields a unique frequency

$$\omega = \frac{D^2}{2(16)^2 |D_e|} \qquad (9)$$

which for quite reasonable values of D_e and D is in accord with experiment.

Furthermore, heuristic, temporal-spatial correlation arguments pertaining to the scaling law for the more common ring morphology lead to essentially the same values for radial wavelength and frequency. Empirically this must be the case because for the usual low pitch spirals dendritic nature cannot know locally which mode within which it resides. As the reader may have comprehended, the subtle collective action within the variants of this phenomenon cannot help but remind one of a powerful and complex survival logic. In a separate contribution we have demonstrated that it is a natural paradigm for Boole's deductive and inductive logic (see also below).

Electron-hole Drops

When an intrinsic semiconductor such as germanium or silicon is pumped with radiation frequencies in excess of the band gap (~ 0.8 eV) pairs of electrons and holes are created. In the recombination process a small fraction of the pairs adopt Bohr orbits forming what are known as excitons (they are in fact a form of positronium). In germanium, which we shall consider here, these have a radius of 1770 nm and a lifetime of about 1μ sec. The exciton gas has something like a Van der Waals characteristic so it possesses an apparently quite normal coexistence curve and critical point. For germanium this is at 6.5 K. The dense, Fermi liquid phase has a lifetime of about 40μ sec. Infra-red diagnostics show that the liquid appears as a hydrodynamically expanding cloud of typically $5\mu m$ drops in the exciton gas, which quickly recombines on passing outside the pumped region (Fig. 10). In materials terms, the process is a spinodal decomposition which goes to a steady state phase separation. Together with Patterson and Hubert (1987) I have framed the hydrodynamic model and have demonstrated that this is a

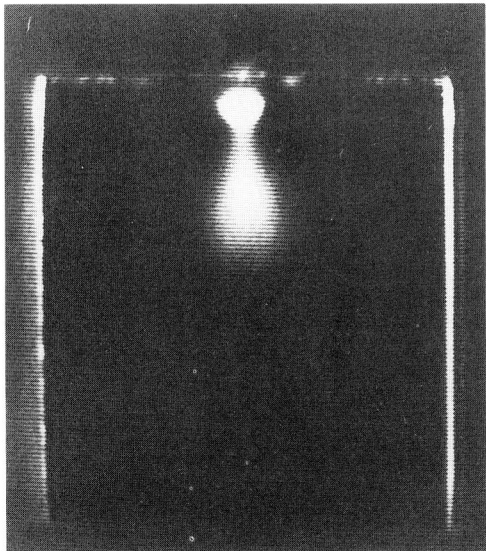
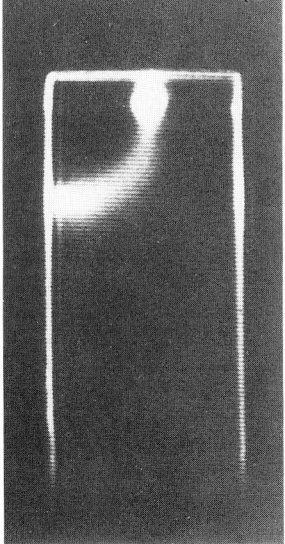

Fig. 10. Electron hole drop cloud in Ge crystal produced by laser beam focussed from above. The cloud is attracted into a highly stressed region on the edge of the crystal. After Wolfe (1983).

bivariate maximal free boundary problem. The degeneracy is in the order parameters, radius and cloud density, or the phase densities. Figures 11 and 12 give the calculated entropy production surfaces for pumping inside and just outside the gas-side spinode. In the latter case the liquid must nucleate by fluctuation across a negative entropy production barrier. The predictions con-

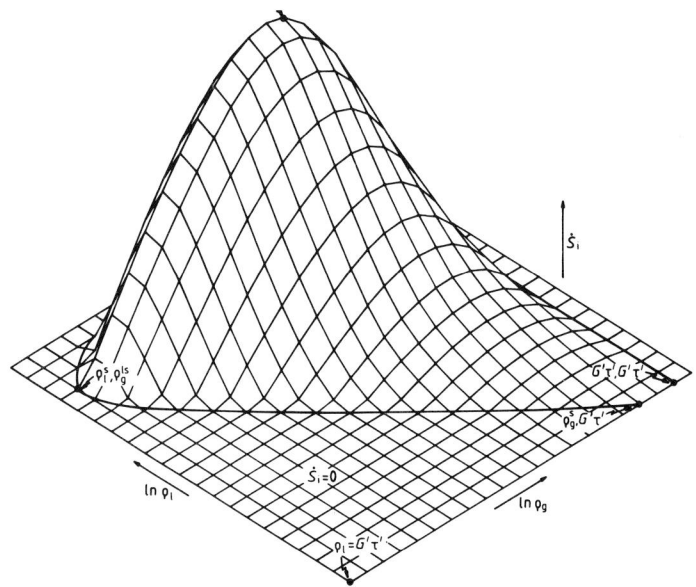

Fig. 11. A positive definite surface of entropy production corresponding to uniform pumping density index at $G'\tau'$ just inside the spinode. ρ_ℓ and ρ_g are liquid and gas densities, respectively.

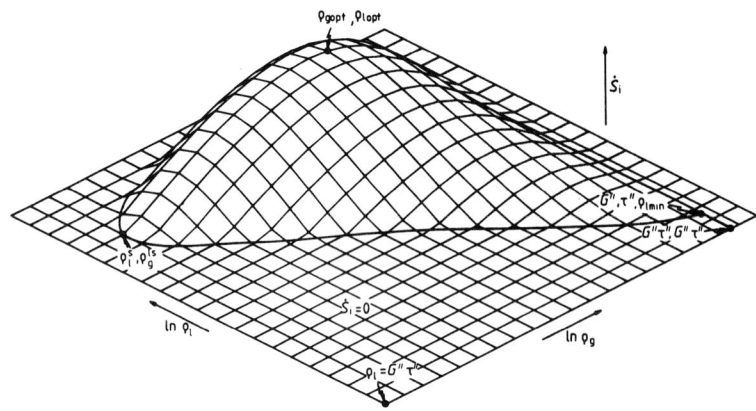

Fig. 12. A positive definite surface of entropy production corresponding to pumping at $G''\tau''$ just outside the spinode.

tain no adjustable parameters and lead to the comparisons with experiment in Figs. 13 and 14. Note that this problem is not completely free of hysteresis since the cloud drop density is a function of the laser pump rise time (Fig. 14).

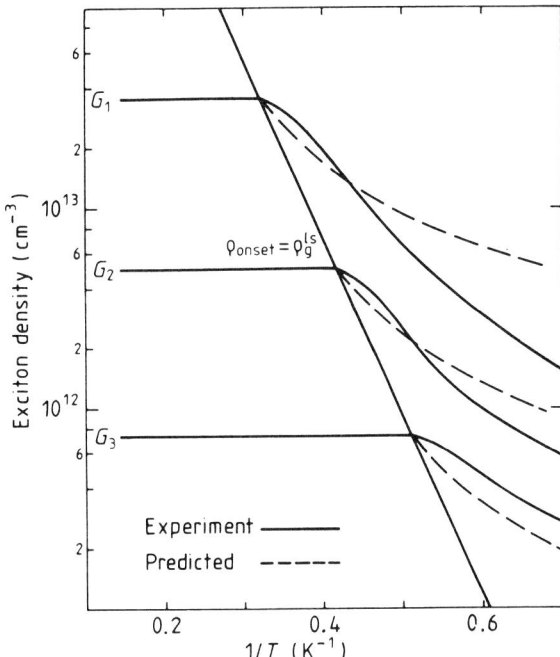

Fig. 13. Predicted and observed exciton densities within a cloud as functions of the temperature and pump power G.

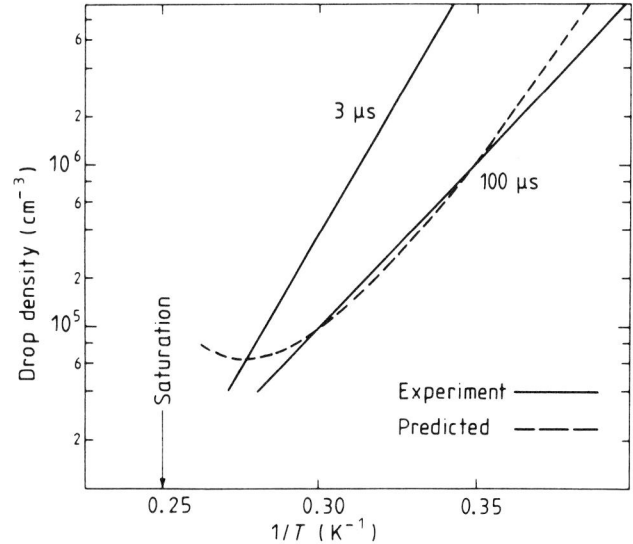

Fig. 14. Predicted and observed drop densities as functions of $1/T$. Hysteresis effects associated with rise time are apparent.

Pumped Superconductors

There is a strong sense in which a superconductor can be regarded as a binary alloy of Cooper pairs and quasi-particles (excitations). When a superconductor is externally excited by laser or tunnelling of quasi-particles from another superconductor the relative density of quasi-particles can be increased, in the extreme destroying the superconductivity. It was early speculated that there might exist a spinodal decomposition into quasi-particle rich and poor regions. This phase separation has been confirmed by infra-red diagnostics and indeed, Iguchi and Suzuki (1983) have mapped the pattern of phase separation in Sn injected from Indium (Figs. 15 and 16). Unfortunately, the early hand-waving

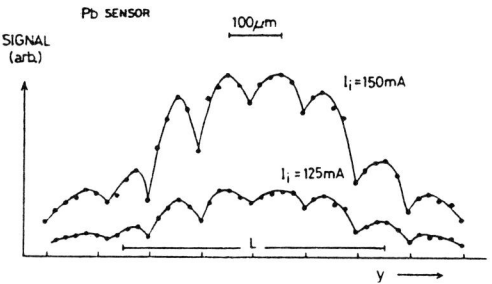

Fig. 15. Intensity of detected phonon signal as a function of position for lead sensor and different injection currents. After Iguchi and Suzuki (1983).

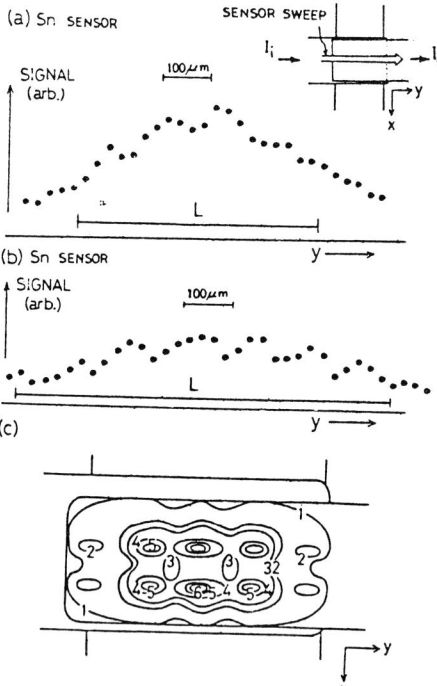

Fig. 16. Intensity of detected phonon signal as a function of position for tin sensor and different injection currents. After Iguchi and Suzuki (1983).

theories which suggested the phenomena have not stood up to detailed scrutiny. My student, Roger Patterson (1987), framed the rigorous BCS kinetic theory for this process and demonstrated that there does not exist an instability to small density perturbations within the entire range of feasible parameters.

Other possibilities have occurred to us. It may be that the spatial symmetries of the two phases are completely different so a classical nucleation and growth problem is defined. Somehow, the pattern of Fig. 16 does not seem to accord with this viewpoint. There is also the possibility that the effect is chemical-thermal in nature and somewhat analogous to cellular solidification. The observed scale of $100\mu m$ might be significant in this context. If the lifetime of quasi-particles is a non-linear function of composition then a density fluctuation will change the entropy production rate and therefore the phonon density or temperature profile. Since the latter must be directed towards the helium bath it is not hard to see how the gradient energy might interact within a dissipation optimum to product a regular pattern. A flux dependent phonon coupling to the helium bath could have similar consequences. Such eventualities if validated would not be very interesting from the physics point-of-view but would provide added support for the present approach to pattern formation.

Reflux Condensation

Very complex free boundary problems are ubiquitous to hydrodynamics and thermalhydraulics. A maximal dissipation principle has been deemed to be applicable to boiling heat transfer and the circulation of the earth's atmosphere. I, together with Venugopalan and Girard have developed an approximate theory for a thermalhydraulics free boundary problem involving a special case of reflux condensation. This is quite similar to the boiling heat transfer problem. Figure 17 defines the configuration and parameters. Figure 17(a) may be taken

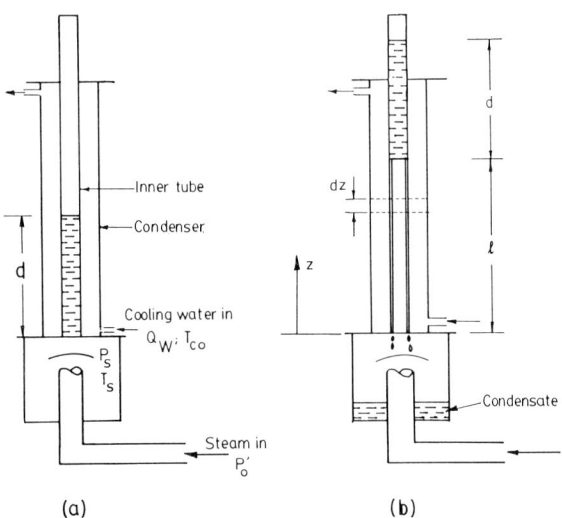

Fig. 17. Schematic of a single cooling jacket reflux condensation column. (a) unstable state with an entropy production rate of zero, (b) stationary state indicating wet gas and single phase zones.

to represent the initial condition as an unstable, flat interface equilibrium between the steam pressure in the lower plenum and a quiescent column of water. This state has essentially zero entropy production. Clearly, a slight perturbation will cause water to drop into the plenum, and steam entering the space will lift the lightened plug with condensation beginning on the thinning reflux wall of water. One, however, cannot predict that a stable but fluctuating plugged reflux configuration as in Fig. 17(b) necessarily accrues under appropriate conditions (actually in the experiments the gas column contains over 15% moisture). It is quite easy to fathom that the average cooling manifold temperature is a free variable and that therefore stability must be established by a maximum in the entropy production as a function of this parameter or of the equivalent height of the wet gas column. Figs. 18 and 19 compare theory and

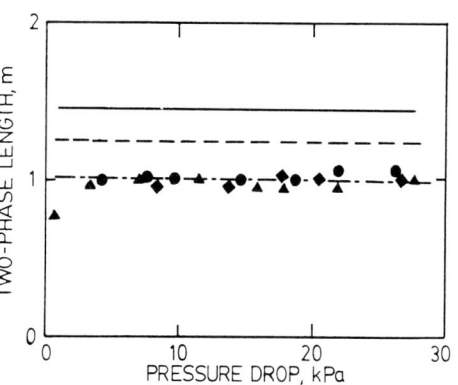

Fig. 18. Variation of the observed phase region length with pressure drop in a single cooling jacket column. Lines are calculated.

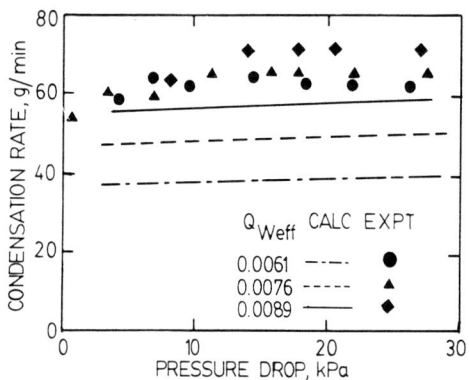

Fig. 19. Variation of the observed condensation rates with pressure drop in a single cooling jacket column. Lines are calculated.

experiment for the single manifold experiment of Fig. 17 (Banerjee and coworkers, 1983). In view of the approximations and the recognized oversights the agreement is salutary. The reader should appreciate that in normal engineering practice the configuration can be quantitatively defined via empirical correlations and dimensional analysis (Girard, 1985). The principle of analyticity which resides within the latter discipline may be seen to establish the link with the minimax theorem and logistics.

Simulations of Evolution in a Sub-Arctic Ecology

In the introduction we briefly discussed the Lotka-Volterra chemical rate equations pertaining to sub-arctic ecologies. We suspect that these "dendritic" type

instabilities have little survival merit and have therefore arisen as a result of
fairly abrupt environmental changes acting upon earlier stable populations which
have not had time to evolve. Here we are concerned with the slow evolutionary
drift of the order parameters such as hair length and leg length (phenotypes)
pertaining to the very fecund snowshoe hare population. We will accordingly
adjust the parameters to suppress oscillations.

In the present treatment (Kirkaldy and Lonergan, 1987) we express the ecological
rate equations in terms of an approximate dynamic entropy balance instead of the
conserved biomass balances of Lotka and the ecologists (Fig. 20). Following the

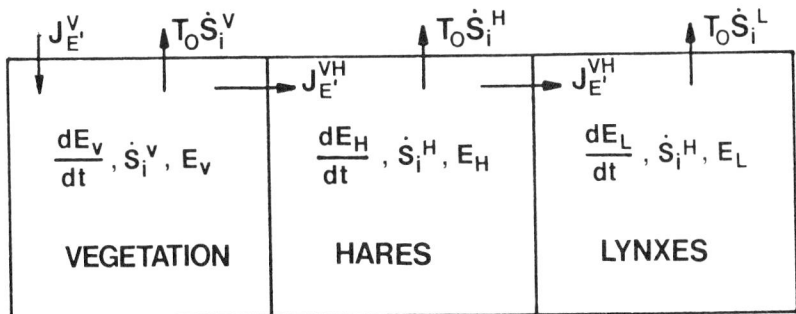

Fig. 20. Approximate free energy flow diagram for a classical preditor-prey
ecology. From this, coupled rate equations of the Lotka-Volterra type are
constructed. J represents free energy flux, E represents bioenergy and $T_0 \dot{S}$
represents an entropy production rate as a heat flow to the arctic environment.

practice of evolutionists, the slow genetic component is to be built into the
coefficients of the ecological rate equations and we do this in such a way that
phenotype drift is related to minimal and maximal propensities of the ecosystem.
For example, increasing hair length (h) of the hares tends to decrease the
entropy production through the insulating effect, while increasing leg length (l)
tends to increase it through the provision of a greater foraging range and
greater consumption. These morphological variabilities also reflect on predation
rates and forage recovery rates so must be entered appropriately into the corresponding coefficients.

Our final expression for the hare ecosystem dissipation rates was calibrated to
empirically reasonable bioenergy ratios of 100:10:1 and to current phenotypes h =
3 cm and l = 15 cm). Fig. 21 shows the saddle point surface generated with the
bivariate stability point at h = 2.2 cm and l = 33 cm and corresponding to the
biomass ratios 100:8:0.8. The simulation is accordingly self-consistent. This
structure is of course vastly over-simplified and subject to attack from many
directions by the experts.

The evolutionists with whom we have dealt seem to be quite prepared to accept a
controlling principle of maximum path probability for it is somewhat generic with
the classical Darwinian statistical analyses of Fisher (1930) and Wright (1949).
However, they find it very difficult to believe that the gene bank with its drift
under survival or environmental imperatives can be rigourously related to a
macroscopically fluctuating statistical phase space with a very long relaxation
time ($\sim 10^6$ generations) and that the optimal principle can be made much more
concrete and transparent in the entropy formalism. They have not come to grips

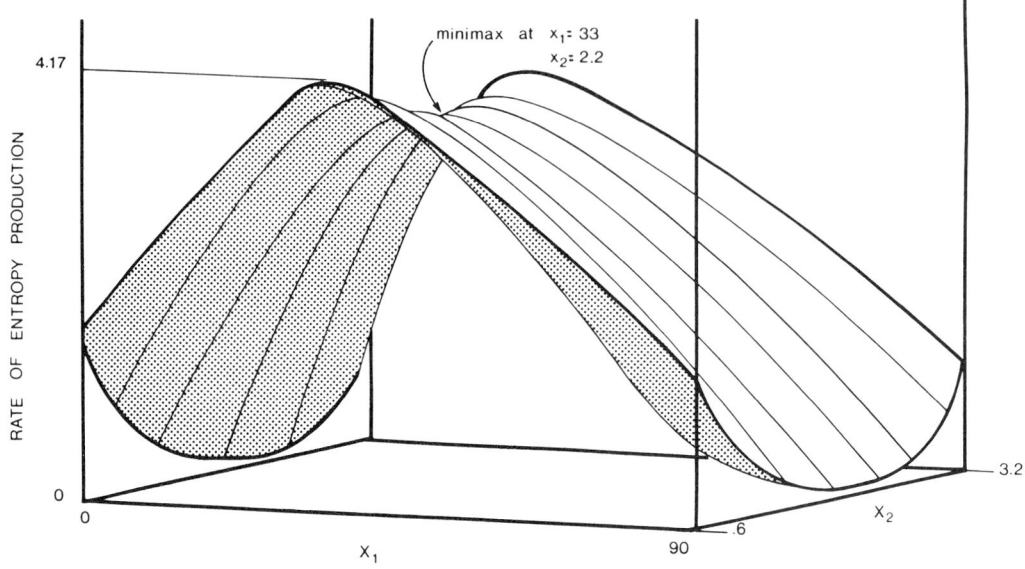

Fig. 21. Dissipation surface for the hare ecodeme demonstrating a minimax stability point. X_1 = leg length and X_2 = hair length.

with the fact that there are no known limits on time and space scales or upon the entropy concept expressing collective actions in the thermodynamics of the universe. Fluctuation scale invariance or self-similarity are sufficient principles of physics to effect the long range relationship.

Hydrodynamics of the Maple Key

Evolutionists might have been better equipped to understand the thermodynamics of their focus of research had they contemplated the general minimax nature of tree dynamics as described earlier, and related this to the very refined aerodynamic features of seed dispersal of a species such as the maple. Viewed strictly as an aerodynamic object, maple keys of the type shown in Fig. 22 fall first in a bomb-like trajectory, but through turbulent perturbation the mode transforms to the very slow propeller-like trajectory that as children we found so intriguing. In about 1960, as a grown-up child, I examined the motion of keys by high speed cinephotography and reported the results in the Canadian Journal of Physics (Kirkaldy, 1964). My observations on velocity and rotation frequency are also exhibited in Fig. 22. The integral rate of entropy production is exactly equal to the rate of loss of gravitational energy which is proportional to the velocity of fall of the center of gravity. As can be seen, the seed falls in a bomb-like trajectory until it approaches a steady terminal velocity, whence through perturbations it searches a transient trajectory ending in a highly patterned steady state of minimum entropy production. The dependent order parameters are azimuthal and dihedral angles and frequency. Note that nature has taken the trouble to provide crossed surface flutings so that a slightly turbulent milieu

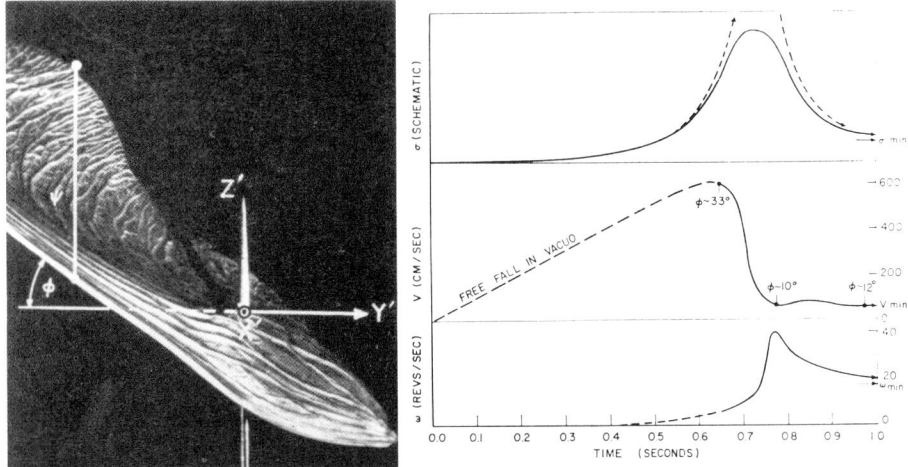

Fig. 22. Velocity and angular frequency versus time relations for the seed on the left. ϕ is the azimuthal angle. σ is the entropy production rate as estimated from the conjectured velocity curve. After Kirkaldy (1964).

or fluctuating phase space has been assured in all configurations. Now this is remarkable in itself, but consider that the function of the minimal state in a fluctuating milieu on the millisecond, micrometer scale is to increase the species domain and therefore its entropy production in a fluctuating milieu on the megameter, gigasecond scale. In terms of renormalization group theory the full phase space of causal evolutionary process must evidently contain at least these limits and all scales of length and time in between in an analytic and invariant way (Wilson, 1975, 1979).

Cellular Solidification

The pattern of Figs. 3 and 4 has been the subject of active investigation for at least forty years and still offers a vehicle for theoretical controversy (Venugopalan, 1988). This is an oriented dilute binary crystal grown from the melt in a fixed temperature gradient G at a controlled velocity v. The instability of a normal planar interface is brought on at a critical point reached via an increasing v, an increasing concentration X_o and (or) a decreasing G. A number of authors, including the writer, have seen this phenomenon as the chemical analogue of the well-known Rayleigh-Benard hydrodynamic instability in a gravitational field whose key control parameter is also the temperature gradient. Attempts have accordingly been made to employ similar mathematical methods to the problems of critical instability and marginal wave number selection, but these have not proven to be quantitatively successful. By contrast the hydrodynamic predictions of marginal wave numbers are very accurate indeed and provide good estimates for Rayleigh numbers up to ten times the critical values.

Together with Venugopalan we have discovered that the analogy is badly conceived. Firstly, the problems are topologically distinct for in the chemical case the extra degree of freedom allows dissipative non-planar states to extend continuously into the equilibrium state (v, G → 0 in an appropriate fixed ratio v/G). In the hydrodynamic case, on the other hand, there is only one control parameter G which is monotone with the dissipation and which for values $G > G_c$ signals states of instability of increasing dissipation. The unstable states are accordingly topologically isolated from the equilibrium state. In Prigogine's terms the "bifurcation" lies outside the thermodynamic branch.

One might hope to rescue the analogy by fixing one of the parameters (say G) at a strongly non-equilibrium value and consider instability in terms of the single control variable v. Unhappily, fixing G causes conjectured marginal stationary states analogous to those in the hydrodynamic problem to become overdetermined. For a trial stationary, periodic perturbation and the usual assumption of local equilibrium all liquid boundary concentration values pertaining to a single wave are determined and so by the Dirichlet principle the concentration distribution is uniquely determined. When one applies the independent and necessary local mass balance, which is also capable of fixing the boundary concentrations, the stationary problem is seen to be strongly overdetermined.

Some time ago I realized from a different point of view that one had to give up the local equilibrium constraint if cells of the observed, rather singular stationary shape are to be understood. Our most advanced current model suggests the existence of a turbulent regime at and near marginal instability (as observed) and generates in the strongly supersaturated (non-local equilibrium) stable regime degeneracies in the cell spacing and cell length which can be removed by imposing a bivariate minimum in the dissipation. Predictions are in good agreement with the observed boundary between the marginal turbulent and stable regimes (Venugopalan and Kirkaldy, 1984). This system represents one of few dissipative transformations where this minimal propensity is exhibited (cf. the campfire and the maple key). Such minima in combination with maxima are essential to the development of complete natural logics.

Pearlite-type Reactions

Two of the companion papers will deal with analogues of these well-known reactions so I don't propose to review the matter in any detail (Purdy, 1988; Young, 1988). It is sufficient for me to say that a resolution of the well-known free-boundary spacing problem is quantitatively achieved by optimization at a maximum in the dissipation (Kirkaldy, 1984). Firstly, I wish to use this problem to frame a fundamental question of principle: Can a perturbation or Le Chatelier procedure be mounted which uniquely determines the stability point and is thus equivalent to the optimal principle. This is the case for equilibrium thermodynamics, linear and non-linear irreversible thermodynamics and might be expected as well for a macroscopic phase space. Secondly, the high perfection of the patterns in some cases (Fig.23) and the conciseness of rigorous formulas provide a ready bridge to a logistical formulation.

Generally speaking, as a consequence of great complexity, procedures must be carried out for pattern formation problems in rather gross approximation as to structure and parameters so satisfactory closure between the perturbation and optimal methods is difficult to obtain. The case of isothermal eutectics or eutectoids is, however, an exception since if the terminal solutions are of discrete composition, and segregation is by volume diffusion, all aspects of the problem can be treated with good accuracy (Fig. 23). In this case the perturbation methods identify a stability point at the inflection point of the velocity versus spacing relation

$$v(s) \sim \frac{s_c}{s}(1 - \frac{s_c}{s}) \qquad (10)$$

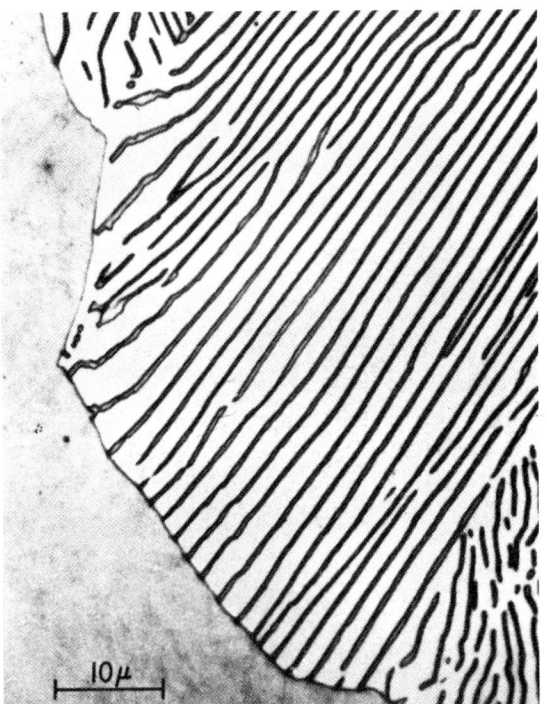

Fig. 23. Micrograph of an isothermal pearlite in steel. After Vilella (1962).

where s_c is a temperature dependent capillarity limit, which upon evaluation yields the stability point

$$s = 3s_c \qquad (11)$$

The entropy production is

$$\dot{S} \sim \frac{s_c}{s}(1 - \frac{s_c}{s})^2 \qquad (12)$$

whose maximum identifies exactly the same stability point. The existence of an equivalence in a special case has thus been illustrated. However, we have not yet been able to identify this generic correspondence in any other system.

Concerning logistics, which we will develop more fully, we recognize that formulas (10) and (12) have forms equivalent to Boole's Law of Duality (1958) within the deductive and inductive algebras.

The Cellular Eutectics

Fig. 24 presents a pattern which has not yet been studied quantitatively (Jackson and Hunt, 1966). This is a conventional forced velocity eutectic reaction which has been destabilized by a liquid impurity possessing very low solubility in both phases. The trend in this ternary system, therefore, is to deposit this impurity at the roots of the liquid intrusions just as in the regular cell structure of Fig. 4. In view of the dissipative propensities of the previous two systems

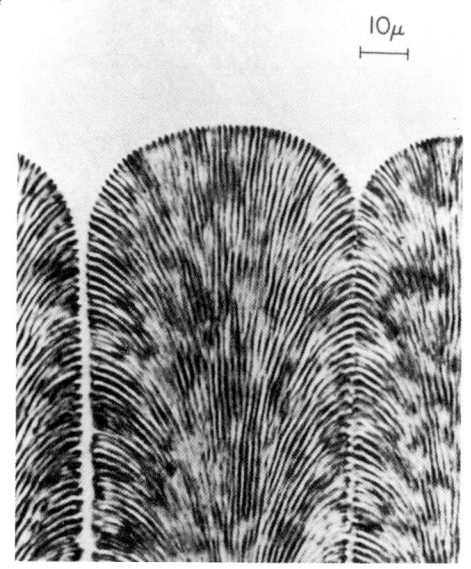

Fig. 24. Solidification of a ternary solution in a combined lamellar eutectic-cellular reaction. After Jackson and Hunt (1966).

(Figs. 4 and 23) this composite transparently exhibits a minimax in the dissipation. Because of its richness in pattern we shall use this as our most comprehensive model for mapping a pattern-forming system into a logistic.

THE LOGISTICAL INTERPRETATION OF PATTERN FORMATION

Upon examination of the previous examples one will recognize that the kinetic evolution up to a patterned steady state involves the spontaneous sorting of a more or less uniform configuration of similar microscopic thermo-kinetic objects into sets of separated, self-similar macroscopic objects in cartesian or momentum space, which is to say, it is a classification process. This is particularly evident in the composite system of Fig. 24 where a homogenous C, D, E solution is classified into C-rich, D-rich and E-rich sets. The logistical character immediately appears since classification has a Boolean syntax (1958), and this can be conveniently represented within an irreversible electrical switching system because this also possesses a Boolean syntax (Fig. 25) (Shannon, 1949; Shannon and Weaver, 1949). Our aim here is to demonstrate that these syntactical equivalences are not trivial for we can show that the minimax dynamics of pattern evolution is exactly equivalent to the rules of inference in a completely formulated logistic, and thence one can conclude that the pattern formation $\underline{is}$ a logistic and a particular pattern is a theorem or formula within that $\underline{logistic}$. Such a viewpoint has been strongly emphasized by Suzanne Langer, the philosopher and logician (1967).

Consider first the eutectoid subsystem of Fig. 23 and its degenerate formulas (10) and (12) in relation to Boolean Algebra. Boole's deductive Law of Duality (1958) has the functional form

$$x^2 = x \text{ or } 0 = x(1-x) \tag{13}$$

where x represents categorical or truth functional propositions or events, both of which can be represented as sets or classes of objects. These equations have two numerical solutions: 0, which can be interpreted as a falsehood or the empty

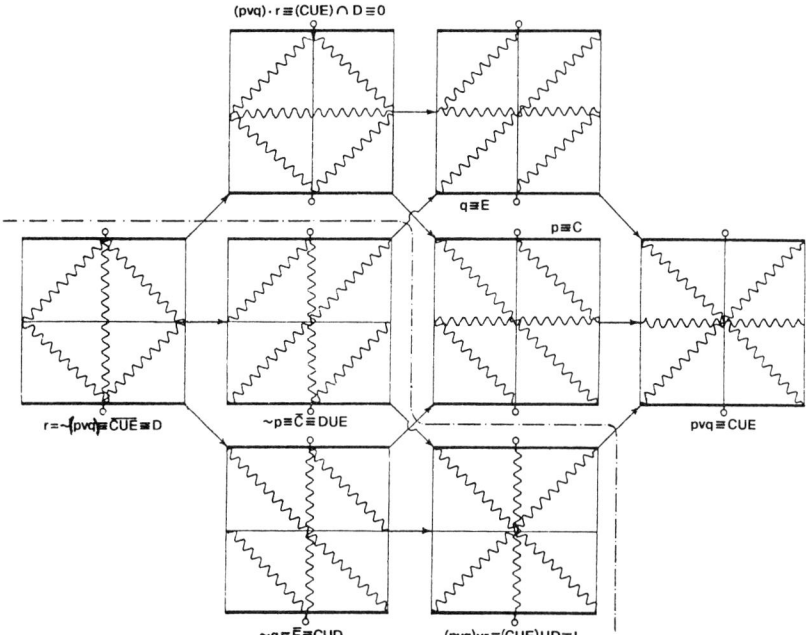

Fig. 25. A stochastic, adaptive switching system based on the minimax dissipation principle. Note that many of the nonconducting filters and intermediate configurations are omitted. The nondiagonal filters shown represent two in parallel. After Kirkaldy (1984).

set, and 1, which can be interpreted as the truth or the universal set. This, of course, is the binary algebra of the digital computer. Boole generalizes his deductive logic to a probabilistic or inductive logic by the rule of substituting probabilities, or relative truth values p of events for the corresponding events x, where the p's can take any values between 0 and 1. Eq.13 now is fully numerical and must include a non-zero left hand side, viz.,

$$P = p(1-p) \tag{14}$$

Now we see that if probability $p(x) = 0$ (a falsehood) or $p(x) = 1$ (a truth) Boole's deductive form (13) is recovered.

Next observe that the Law of Duality has, according to $x^m = x$ ($m \geq 2$), a hierarchy of equivalents

$$0 = x^n(1-x)^m \quad (n, m \geq 1) \tag{15}$$

and in particular a corresponding probabilistic formula

$$P' = p(1-p)^2 \tag{16}$$

Since in Eqs.10 and 12 $v, \dot{S}$ and s_c/s are all recognized as fluctuation averages they can be associated with P, P' and p in Eqs.14 and 16, and the logistical credentials of the eutectoid structure are thereby established. Note that the alternate boundary diffusion model of the eutectoid corresponds to n = 2 in the above. These associations are by no means trivial for in both of the above cases the probability $p = s_c/s = s_c/\infty \to 0$ defining an unpatterned or chaotic (falsehood) state with zero composite probabilities P,P', and $p = s_c/s = 1$ defining an inaccessible but maximally patterned or metastable equilibrium (true), again with zero composite probabilities P,P'.

As in ordinary discourse truths are relative to the universe chosen so we are free to transpose the maximum value of Eq.12 to

$$\dot{S} \sim P' = \frac{27}{4}\frac{s_c}{s}(1-\frac{s_c}{s})^2 = \frac{27}{4}p(1-p)^2 \qquad (17)$$

which assures that the stable maximum at $s_c/s = 1/3$ is the true (P'=1) state. Thus if we idealize the evolutionary path as proceeding from $s_c/s = s_c/\infty = 0$ up to $s_c/s = 1/3$ then we establish the equivalence of the optimization to a key rule of inference. Informally, in logic every falsehood implies every proposition including each individual truth. Thus the physical form of implication is

$$P' = 0 \to P' = 1 \qquad (18)$$

Formally, this is written as the disjunctive syllogism involving the compound propositions p and q where p is false and q is true, viz.,

> Premiss: p or q is true
> Premiss: ~p is true   (19)
> Conclusion: q is true

If according to the optimization path of $\dot{S}$ we define each member of a sequence of converging quasi-stationary spacings as true, then this syllogism applies to each step in the sequence. Furthermore, the powerful hypothetical syllogism emerges, viz., "if p implies q and q implies r then p imples r". As we shall demonstrate in the following other syllogisms require a full minimax formulation.

The richer natural system of Fig. 24 has at least a triple degeneracy or free spatial classification capability indexed by the lamellar spacing (associated with the C,E composition classification, say), the cell spacing, and the groove length (associated with the D classification). The optimum is attained in the ideal via fluctuations over the order parameters, some seeking a maximum dissipation, which therefore achieve successful pattern changes if the dissipation increases or registers no change, while others seek a minimum and achieve successful pattern changes if the dissipation decreases or registers no change (Kirkaldy, 1984). The stochastic electrical network of Fig. 25, although through oversimplification failing to represent completely the disaggregation of the spatial distributions, models the physical system in several essential respects. Here we arrange that the available energy is delivered to the network as three electrical or sound frequencies (say C,D, and E natural musical notes to emphasize the human information content) of predetermined relative intensities as the mapping of the alloy composition classifications. The stochastic elements are partly resistive band-pass filters which "diffuse" in the network according to certain "crystallographic" constraints. We choose to simulate the stochastic diffusion process by random switching in or out of multiple elements since this would be most feasible in practice, and furthermore, the Boolean syntax or discreteness of pattern transformation in Fig. 24 is brought into sharp focus. By further analogy with the natural process a successful switch opening or closure is predicated upon the global dissipation change thereby attained, some trends tending towards a maximum, others towards a minimum. In a working model we would provide a feedback mechanism from the terminals to moderate the switching process thus simulating those long-range collective interactions between the system and reservoirs which moderate the global optimum and thus assure entropy differentiability.

In reference to Fig. 25 consider a dissipative array consisting of four cells each containing several interswitchable band-pass filters. Four D-pass filters lie on the central horizontals and verticals, and these can interswitch in pairs within the northeast and southwest blocks, respectively. With two on the vertical they form a short for D and otherwise a shunt which implies complete exclusion (or equivalently, a deflection to a "D-dump"). We associate these pattern changes with minimal paths so that an acceptable transformation is one which decreases the dissipation or makes no change. This path is thus to be associated with the cellular development of the pattern in Fig. 24 and the segregation or classification of component D to the impurity "dump". Also, each of the four blocks contain a C-pass filter which lies diagonal down to the right and an E-pass filter which is diagonal down to the left. These can be interswitched within each block subject to the maximal imperative, thus corresponding to a stochastic injection or rejection of C or E lamellae in Fig. 24 by nucleation, overgrowth, or equivalent processes. The interswitching is assumed to occur at random, a return to the original state being required if the caveat for success is not satisfied. Simultaneous interswitching is not ruled out. One can now verify the partial sequence of transformation as designated by the arrows in Fig. 25. The C and E selectors tend to the short positions to increase the dissipation while the D selector tends to the shunt positions to decrease it.

The final preparatory step is to recognize that statements about class inclusion and exclusion can be mapped into propositions of the sentential calculus (Boole, 1958; Copi, 1967). For example, if classes are represented as C,D,E subject to union ($\cup$), intersection ($\cap$), and complementation ($\bar{C},\bar{D}$, and $\bar{E}$), and propositions are represented as p,q,r then we can self-consistently map according to C → p, D → r, etc., $\bar{C}$ → ∼p, $\bar{D}$ → ∼r, $\cap$ → ·, and $\cup$ → v, where ∼ is negation, · is conjunction, and v is disjunction. Now, in respect to Fig. 25 we interpret "class" as "the class which is passed by the filter" and to each class we assign a proposition according to the above specifications. Next, referring to the sequence of propositions it is recognized that a process of logical deduction has been nontrivially abstracted from the inanimate irreversible process of Fig. 24. If the first frame is taken as the premiss, then the subsequent frames following the solid arrows are the deductions therefrom or theorems. Note how the rules of inference of formal logic explicitly emerge within the sequence determined by the minimax imperative. One can easily check the following argument forms for the uninterpreted calculus.

(1) Modus ponens: If p implies q and p is true, then q is true. This holds for the precedent and antecedent of every arrow in Fig. 25.

(2) Modus tollens: If p implies q and q is true, then p is true. This appears in Fig. 25 with p and p interchanged.

(3) Hypothetical syllogism: If p implies q and q implies r, then p implies r. Apply this to ∼p,q, and pvq in Fig. 25, noting that a direct transition from ∼p to pvq is achievable via a simultaneous interswitch operation.

(4) Disjunctive syllogism: If either p or q (pvq) is true and ∼p is true, then q is true. Here compare the sequence ∼p → q → pvq in Fig. 25.

(5) Addition: If p is true, then p or q or both (pvq) is true.

Note that argument forms involving conjunction do not appear since the classes identified in this simple model are mutually exclusive (there are no intersections). The physical system, on the other hand, includes intersection or conjunction since the phases are generally impure. It also involves a much more complicated Venn diagram corresponding to potential parameter (spacing) changes and therefore many more predicate statements.

It will be noted further that in this propositional interpretation of the class structure there is a natural division between the interim and final theorems (pvq·r,p,q, and pvq) and their negations. The final theorem pvq or union of classes C E on the earlier interpretation is the C natural chord. The theorems may therefore be assigned the truth values T (true) and their negations F (false) recognizing that the "law of the excluded middle" holds for an elementary deductive system. Furthermore, this assignment assures in accord with valid arguments, that falsehoods can imply truths (indeed, every other statement), but not vice versa (Note the direction of arrows admitted by the minimax prescription). The assignment is accordingly unique.

In the foregoing I have emphasized the deductive propensities of the model by choosing discrete transitions directed by the thermodynamics. Following Boole's prescription the predicate states or events can be replaced by probabilities of such events and in this way the structure better represents the inductive or stochastic nature of the logical evolution in natural systems.

Together with Chambers (1973) I have further illustrated that the minimax principle can be arranged to solve Von Neumann games (1944) operating on utilities as free energies. A game solved by dynamic play is a simulation of irreversible pattern formation or a spontaneous logistic.

We can use the foregoing intelligence to introduce a generalized idea of a phase transition as a transformation of pattern. The sets or classes involved in equilibrium transitions usually have a group character as exhibited by a symmetry change. This is also the case for transitions identified in non-linear bifurcation theory. Indeed, the term "symmetry-breaking" is often introduced. In the case of the pattern formations identified here the macroscopic pattern sets have no group attributes, or at best a groupoid character (no inverse), underlining the irreversible, macroscopic nature of the structure. Nonetheless, we believe that they deserve, because of their logistical or class theoretical origins, to be either identified as phase transitions or be subject to explicit technical exclusions in terms of group theory.

It is appropriate here to mention the role of fractals, a mathematical discipline which has captured the imagination of many workers in the field of pattern formation (Mandelbrot, 1983). In the present context the theory refers solely to the syntax of the underlying logistic. It is accordingly diagnostic rather than predictive, a kind of parsing which possesses artistic rather than scientific merit. After all, a tree is a fractal, but what have we learned if we leave out the dynamics or inference of the life process? A sea coast is also a fractal, but what significance does this possess if the irreversible processes of erosion, uplift or subsidence are left out of the description?

Of more scientific significance are the related studies pertaining to "deterministic chaos" (Schuster, 1984). Here the dynamics or inferential elements are recognized through the Kolmogoroff entropy (production).

EPISTEMOLOGICAL IMPLICATIONS

One can fairly ask whether the logistical approach to pattern formation is more fundamental than that through the science of irreversibility. We suggest that at this point in the development of thermologistics it is best to accord the two approaches equal footing, each enriching the other in a search for linguistic equivalences. Consider experimental physics, which is an irreversible process of pattern formation, and in the words of Niels Bohr (1958), "closed in the sense that...observation is based on registrations..with irreversible functioning". If, as we maintain, such an irreversible process of pattern formation always possesses a logistical equivalent, then for any experiment a logical or mathematical structure can always be elucidated, which is of course an empirical fact. We call this process of elucidation, which imposes its own characteristic irreversible propensities, theoretical physics. Relativity and quantum theory make quite clear the necessary symbiosis between system, experimenter and interpretation. Indeed, modern cosmologists and astrophysicists have been forced to be explicit about this in what is known as the "anthropic principle". That is to say, a universe which is capable of examining itself (through human propensities) must have certain evolutionary characteristics which connect ultimately with the fundamentals of physics (Carr and Rees, 1979).

Now if a logistical character is ubiquitous to the irreversible universe then the universe must possess elements of duality, analyticity, consistency and completeness. In turn, these must be reflected in the theories of the world in the form of duality, analyticity, differentiability and embeddability, etc., which again is an empirical fact. One here recalls Fermi's aphorism: "If in doubt exploit a Taylor Expansion".

Consider, for example, a generalized dissipation function

$$\dot{S} = JX \qquad (13)$$

where J and X are appropriate functionals describing composites of fluxes and forces. This bilinear function of functionals is differentiable in the sense of complex variable theory since it satisfies Laplace's equation. Its complementary function in terms of the variable $J + iX$ is

$$\pi = \frac{1}{2}(J^2 - X^2) \qquad (14)$$

which has the form of a Lagrangian, and a meaning equivalent to the Helmholz Free Energy. Figure 26 represents the syntactical (non-dynamic) structure schematically. While not in the precise conventional language this diagram together with imposed constraints and a time coordinate (dynamics) is inclusive of classical and irreversible thermodynamics, including the minimax. For example, path A → B represents a reversible process (infinitely slow force relaxation at zero flux) and path C → D represents a superconduction process (infinitely slow flux relaxation at zero force). A constrained bivariate non-zero stability point is defined by $dJ = -dX$ or $J = X$, and with the temporal path defined in the direction of the origin this represents a minimax (paths ES and FS).

Consider as a second example of the powers of a principle of analyticity a dual field $\vec{E}$, $\vec{H}$ where the latter form the complex quaternion $\vec{E} + i\vec{H}$. Suppose this is an analytic or differentiable function of the complex quaternion $\vec{r} + it$. The generalized Cauchy-Riemann differential equations invoking quaternion multiplication yield the pattern which we call Maxwell's equations (Kirkaldy, 1987). Because the latter are syntactical (no dynamics or inference) they are necessarily time reversible. We have found an analogous structure pertaining to Schroedinger's equation. Such a result has a direct bearing on quantum gravity, one of the few remaining unsolved problems of theoretical physics.

COSMOLOGICAL IMPLICATIONS

The universe as a whole is an irreversible pattern-forming system. Einstein's

Fig. 26. Schematic representation of the analytic dissipation functions $\dot{S}$ and π.

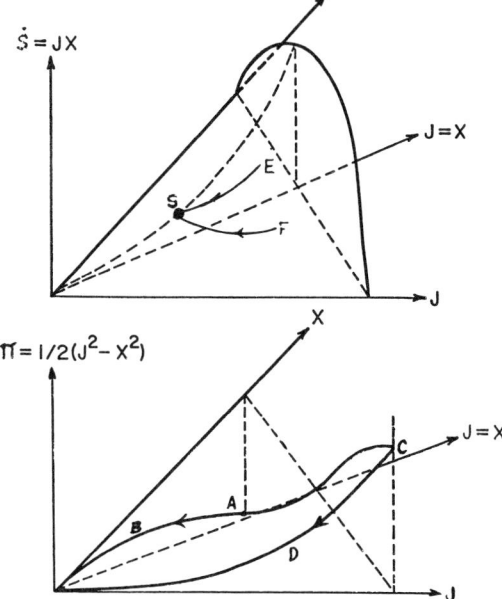

equations of general relativity like Maxwell's equations represent an analytic manifold of great power, and when solved subject to the Big Bang hypothesis and the approximation of adiabatic (isentropic) conditions predict or explain many of the features within present observational astronomy. By contrast, the irreversible nature of pattern formation has not been recognized, so further predictions appear possible within this hypothesis (Kirkaldy, 1987).

Here we assume that the Big-Bang can be represented by a single large fluctuation or singularity in the dissipation at time zero, $\Delta\dot{S}(0)$, which regresses continuously according to the Second Law and Eq. 3. Now time itself is deemed to be the order parameter. We undertake to establish the most probable monotone regression path subject to the Second Law requirement that

$$\Delta\dot{S}(\infty) = 0 \tag{15}$$

This asymptotic limit together with analyticity suggests via Eqs. 1,2 and 15 that C also be asymptotic. Consider from the entire set of monotonic C paths a possible regressing path which is decaying monotone to a minimal asymptote C_∞. In relation to Eq. 3 the solution which has the maximum probability for all path times will be that for which $C = C_\infty$ = constant. Next, consider a possible regressing path from the entire set C which is increasing monotone from an initial value C_0 and therefore minimal. The solution of Eq. 3 which has the maximum probability for all path times will be for $C = C_0$ = constant. Accordingly, all solutions which correspond to maximum path probability will be for $dC/dt = 0$. Thence ΔS satisfies the ordinary differential equation

$$\frac{d(\Delta S \Delta \dot{S})}{dt} = 0 \tag{16}$$

which is identical to the equation satisfied by the universal scale factor, R, in early universe cosmology.

If we adopt the conventional intelligence that the entropy is zero at time zero then

$$\Delta\dot{S} = 1/\sqrt{\kappa t} \tag{17}$$

and

$$\Delta S = 2\sqrt{t/\kappa} \tag{18}$$

where κ is an integration constant. The latter explicitly expresses the entropy as "times arrow".

The entropy density of the universe is supposed to be given very closely by the 3°K background radiation so one can calculate the current entropy within the light horizon (which yields κ) and thence the current dissipation rate. We obtain for the latter, 1.4×10^{54} ergs/Ks. Identifying the main current entropy-producing process as absorption of starlight within the interstellar dust of space we have obtained the empirical estimate 1.1×10^{54} ergs/Ks which is in good agreement with the theoretical value.

A second point of closure between theory and experiment can be obtained within the conventional assumption that the universe has remained within effective blackbody equilibrium throughout most of its evolution (equivalent to the adiabatic assumption). Thus we can conjecture that the instantaneous rate of formation of mass-energy is equal to its immediate dissipation rate, i.e.,

$$\frac{dM}{dt} = T \Delta \dot{S} \tag{19}$$

That is, the universe starts out with zero entropy and mass-energy. This is equivalent to a well-known expression used to calculate the entropy of a black hole. Conventional theory together with the adiabatic assumption yields $T \propto 1/\sqrt{t}$ which formula can be calibrated to the Planck temperature in quantum gravity, $T_p = 3.6 \times 10^{32}$ K at $t \simeq 0$. Thus Eq. 19 integrates for a time equal to that of radiation decoupling ($\sim 10^{13}$ s) to a universal mass-energy of $\sim 10^{79}$ baryons (protons and neutrons) which is in good accord with other theoretical and empirical estimates. These considerations do not conflict with the idea of an isentropic, constant mass-energy universe on any scale which is relevant to the human observer. They are also consistent with the idea of a marginally open universe which is favoured by many cosmologists.

If one examines the solutions of Eq. 16 and its scale factor analogue in the complex plane and connects these with some ideas of quantum gravity, inflation and thermologistics then one can infer that the ultimate age of the universe is 10^{100} years (Kirkaldy, 1987). This well exceeds the theoretical lifetime of baryons (neutrons and protons) and is about the same as the lifetime of the black holes which might have accreted at the core of the galaxies, and so reaching a black body configuration or pattern empty set the universe may be deemed to have ended.

Modern cosmology is already on a convergent path towards a dynamics which connects with the theory of phase transformations. Indeed, Guth's current popular model of the "inflationary universe" (Guth and Steinhardt, 1984), which was designed to avoid some anomolies of distribution and contradictions in the particle physics of the present and early universe, invokes the idea of supercooling of the rapidly chilled universe through expansion and parallel precipitation of bubble-like sub-universes which exclude the eventualities observationally missing (e.g., magnetic monopoles). One can forgive particle physicists for choosing in the first instance the analogue of the Wilson cloud chamber or the bubble chamber as their model, for conceivably it is the only phase transformation with which they are strongly familiar. Not surprisingly, the advanced model known as the "new inflationary universe" has much in common with spinodal decomposition. The theory of condensation of galaxies and stars seems also to have many fluid spinodal features including the fluctuation scale invariance attributed to near critical phenomena and a fractal geometry.

TECHNOLOGICAL IMPLICATIONS

We have already stated our conviction that a well-articulated science of thermologistics has a direct bearing on the development of the technology of machine intelligence (AI) and on the understanding of terrestrial evolution (von Neumann, 1956; 1966). This is no coincidence, for AI learning systems based on the predicate calculus are often patterned after evolutionary models (Yazdani, 1986). This is in preference to patterning from the human brain which is less well understood. On the other hand there is a mapping between the functioning of evolution and the brain (Kirkaldy, 1965a; 1965b). Adaptive switching models with a minimax analogue have already been used by Huberman and coworkers to explore the inductive process of machine learning (Huberman and Hogg, 1984; Humieres and Huberman, 1984; Kirkaldy, 1984). We thus see thermologistics as the synthesis of all the mathematics and physics which will ultimately qualify AI as a mature physical science and a potential source of technological innovation. For example, the connection between the minimax in games played dynamically with utility payoffs and the dissipation minimax suggests potential applications in the field of economics, if not broadly within the social sciences (Chambers, 1973; Kirkaldy and Black, 1972). Furthermore, its contributions to the general theory of evolution may be reflected in the new technologies of genetic engineering. The question as to how the genotype irreversibly determines the phenotype is still very much an open one. Here explicit continuity between genotype and phenotype is known to accelerate natural selection (Lerner, 1950).

The idea of a minimax optimum is already endemic to engineering design practice and often expressed as "maximum production (with a necessary parallel in the dissipation) and minimum energy use (with a necessary parallel in the dissipation)". The evolution in design of highway interchanges illustrates very graphically this process of evolutionary pattern formation in engineering practice (Fig. 27).

The engineering of robotics and pattern recognition can undoubtedly benefit from the thermologistic synthesis. One of the most important unsolved problems has to do with pattern or image formation in the human brain (otherwise known as the mind-matter problem). A logistical interpretation of the universe suggests that the generalized concept of space is the syntax of the universe while the concept of time is its inference. Thus the irreversible electrochemical, Boolean switching system which defines the process of the human brain translates the so-called external world directly and the mind-matter problem disappears. This idea reflects directly on the technology of analog to digital translation known to electrical engineers as "quantization". Significantly, the most powerful mathematical tool used in this discipline is the "minimax" of game theory (Morris and Vandelinde, 1985).

We have long contemplated and explored the possibility that there is a potential for an explicit hardware relevant to AI and to pattern formation and pattern recognition which represents a symbiosis of the minimax, stochastic propensities of natural systems and the digital computer (Kirkaldy, 1978). Herein would lie

(a) Cross-road   (b) Traffic circle

(c) Cloverleaf   (d) Freeway interchange

Fig. 27. Evolution of an engineering sturcture; the highway traffic interchange.

the attributes of a truly creative computer. Fluidic devices (McCloy and Martin, 1985) in their unstable modes have just such creative tendencies, and as is often the case in real life, these are suppressed in favour of discrete action. We have speculated that digital microelectronic devices combined symbiotically with patterned plasma instabilities could serve as the basis for a comprehensive modelling of the creative human brain. (Kirkaldy, 1984).

REFERENCES

Ashby, W.R. (1956). In C.E. Shannon and J. McCarthy (Eds.), Automata Studies, Princeton University Press.
Banerjee, S., J-S. Chang, R. Girard and V.S. Krishnan (1983). J. Heat Transfer, 105, 719-727.
Bohr, N. (1958). Atomic Physics and Human Knowledge, John Wiley, New York, p.73.
Boole, G. (1958). An Investigation of the Laws of Thought, Dover Publishing, New York.

Bourret, J.A., R.G. Lincoln and B.H. Carpenter (1969). Science, 166, 763-764.
Cahn, J.W. (1959). Acta Met. 7, 18-28.
Carr, B.J., and M.J. Rees, (1979). Nature 278, 605-612.
Chambers, D.B. (1973). Thermodynamics of Self-organizing Systems, Ph.D. Thesis, McMaster University.
Cole, G. (1963). Ph.D. Thesis, University of Toronto.
Copi, I. (1967). Symbolic Logic, The McMillan Company, New York.
Darken, L.S., and R.W. Gurry (1953). Physical Chemistry of Metals, McGraw-Hill, New York.
Feldman, S.E. (1978). In D.V. Doane and J.S. Kirkaldy (Eds.), Hardenability Concepts with Applications to Steel, AIME, Warrendale, PA.
Fisher, R.A. (1930). The Genetical Theory of Natural Selection, Oxford University Press.
Forsyth, R. (1986). In M. Yazdani (Ed.), Artificial Intelligence: Principles and Applications, Chapman and Hall, London, p.218 et seq.
Girard, R. (1985). Ph.D. Thesis, McMaster University, Hamilton, Canada.
Glansdorff P., and I. Prigogine (1971). Structure, Stability and Fluctuations, Wiley Interscience, London.
Guth, A.H. and P.J. Steinhardt (1984). Scientific-American, 250, 116-128.
Haken, H. (1983). Advanced Synergetics: Instability Hierarchies of Self-organizing Systems, Springer-Verlag, Berlin.
Haken, H. (1983). Synergetics: An Introduction; Non-Equilibrium Phase Transitions and Self-organization in Physics, Chemistry and Biology, Springer-Verlag, Berlin.
Huberman, B.A., and T. Hogg (1984). Phys. Rev. Lett., 52, 1048-1051.
Humieres, D., and B.A. Huberman (1984). J. Phys., 34, 361-379.
Hillert, M. (1957). Jernkontorets Ann., 141, 757-789.
Iguchi I., and Y. Suzuki (1983). Phys. Rev., B, 28, 4043-4045.
Jackson, K.A., and J.D. Hunt (1966). Trans. AIME, 236, 1129-1142.
Kirkaldy, J.S. (1959). Can. J. Phys., 37, 739-754.
Kirkaldy, J.S. (1964). Can. J. Phys., 42, 1437-1446.
Kirkaldy, J.S. (1965). Biophysical Journal, 5, 965-979.
Kirkaldy, J.S. (1965). Biophysical Journal, 5, 981-986.
Kirkaldy, J.S. (1978). Life, Logic and Bootstrap Physics, Jasak, Ancaster, Ontario.
Kirkaldy, J.S., and D.B. Black (1972). Social Reporting and Educational Planning, Ontario Ministry of Public Services, Toronto, p.203 et seq.
Kirkaldy, J.S. (1973). Met. Tran., 4, 2327-2333.
Kirkaldy, J.S. (1984). Phys. Rev., 31, 3376-3390.
Kirkaldy, J.S. (1986). Computers in Materials Technology, Institute of Technology, Linkoping, Sweden, June 4-5.
Kirkaldy, J.S., L.R.B. Patterson and J. Hubert (1987). J. Phys. C: Solid State Phys., 20, 1393-1411.
Kirkaldy, J.S., and S.C. Lonergan (1987). Unpublished research.
Kirkaldy, J.S. (1987). Unpublished research.
Langer, S.K. (1967). An Introduction to Symbolic Logic, Dover Publications, New York.
Lerner, I.M. (1950). Population Genetics and Animal Improvement, Cambridge University Press.
Liesegang, R.E. (1896). Naturwiss, Wochenschr., 11., 353-360.
Lotka, A.J. (1922). Proc. Nat. Acad. Sci., 8, 147-153.
Lotka, A.J. (1945). Human Biology, 17, 167-194.
Lotka, A.J. (1956). Mathematical Biology, Dover Publications.
Mandelbrot, B.B. (1983). The Fractal Geometry of Nature, W.H. Freeman, New York.
McCloy D., and H.R. Martin (1980). Control of Fluid Power: Analysis and Design, John Wiley and Sons, New York.
Morris, J.M., and V.D. Vandelinde (1985). In P.F. Swaszek (Ed.), Quantization, Van Nostrand, New York.

Nicolis, G., and I. Prigogine (1977). *Self-organization in Nonequilibrium Systems*, Wiley, New York.
Onsanger, L. (1931). *Phys. Rev.*, 37, 405-426; 38, 2265-2279.
Paltridge, G.W. (1981). *Quarterly J.R. Met. Soc.*, 107, 531-547.
Patterson, L.R.B. (1987). *Low Temperature Physics*, 31, 99-113.
Prigogine, I., and J.M. Wiame (1946). *Experientia* 2, 451-454.
Prigogine, I. (1955). *Introduction to Thermodynamics of Irreversible Processes*, Charles C. Thomas, Springfield, Ill.
Prigogine, I., G. Nicolis, and A. Babloyantz (1972). *Thermodynamics of Evolution*, *Physics Today*, 23-28 (Nov.), 38-44 (Dec.).
Purdy, G.R. (this volume).
Russell, B. (1927). *An Outline of Philosophy*, Meridan, London, p.27.
Sawada, Y. (1981). *Progress Theor. Phys.*, 66, 68-76.
Schuster, H.G. (1984). *Deterministic Chaos*, Physik-Verlag, Weinheim, FRG.
Shannon, C.E. (1949). *Bell System Tech. J.*, 28, 59-98.
Shannon, C.E., and W. Weaver (1949). *The Mathematical Theory of Commuinication*, University of Illinois Press, Urbana.
Thom, R. (1972). *Stabilite Structurelle et Morphogenese*, Benjamin, New York.
Thompson, D'A. (1917). *On Growth and Form*, Cambridge University Press, First Edition, (1917) Second Edition, (1941).
Turing, A.M. (1952). *Phil. Trans. Roy. Soc.*, B237, London, 37-72.
Turnbull, D. (1955). *Acta Met.*, 3., 55-63.
Tykodi, R.J. (1967). *Thermodynamics of Steady States*, Macmillan, New York.
Veil, S. (1934). *Les Phenomenes Periodiques de la Chimie*, Herman et Cie, Paris.
Venugopalan, D., and J.S. Kirkaldy (1984). *Acta Met.*, 32, 893-906.
Venugopalan, D. (this volume).
Vilella, R. (1962). In V.F. Zackay and H.L. Aaronson (Eds.), *Decomposition of Austenite by Diffusional Processes*, reproduced by L.S. Darken and W.C. Leslie, Interscience, New York.
Volterra, V. (1936). *Lecons sur la Theorie Mathematique de la Lutte sur la Vie*, Gauthier-Villars, Paris.
von Neumann, J., and O. Morgenstern (1944). *Theory of Games and Economic Behavior*, Wiley, New York.
von Neumann, J. (1956). In C.E. Shannon and J. McCarthy (Eds.), *Automata Studies*, Princeton University Press.
von Neumann, J. (1966). *Theory of Self-Reproducing Automata*, University of Illinois, Urbana.
Wilson, K.G. (1975). *Rev. Mod. Phys.*, 47, 773-840.
Wilson, K.G. (1979). *Sci. Am.*, 241, 158-179.
Wolfe, J.P. (1983). Private communication.
Wright, S. (1949). *Adaptation and Selection in Genetics, Paleontology and Evolution*, Princeton University Press.
Yazdani, M. (1986). *Artificial Intelligence: Principles and Applications*, Chapman and Hall, London.
Young, D.J. (this volume).
Zhabotinskii, A.M. (1974). *Concentrated Auto-oscillations*, Nauka, Moscow.
Zener, C. (1946). *Trans. AIME*, 167, 550-595.

KEYWORDS	Page
"A" segregate	101
alloy partition	1, 25
alloy steels	1, 25
amorphous alloys	166, 178
C_2H_4/graphite	145
carbide precipitation at $\alpha:\gamma$ boundaries	31
catalysts	116
cavity distribution	78
cellular phase transformation	116
channel	101
chemically induced grain boundary migration	204
chemical potential gradients	1
coherency strains	204
creep	221
critical exponents	145
damage	247
dealloying	116
diffusion	193
diffusional transformations	1
diffusion couples	69
Dirichlet Tesselation	247
discontinuous precipitation	204
dislocations	231
disordered boundaries	23, 33
displacement reactions	221
elastic energy	205
fcc crystal	78
ferrites	154
ferro-electrics	154
free boundary problem	90
freckle	101
glass forming range	166
glass transition	154
glassy alloys	166
grain boundary ferrite allotriomorphs	20
growth kinetics	21, 151
impingement	1
internal oxidation	221
isoteric heats of absorption	145
ledge mechanism	33
Lennard-Jones solid	78
Lillie number	154
liquid film migration	204
local equilibrium model	1, 21
macrosegregation	101
massive transformation	204
mechanical alloying	166
mechanical grinding	166
melting	78

KEYWORDS	Page
metastability	131
microstructure	247, 263
misfit strain	204
molecular dynamics	78
multicomponent phase equilibria	37, 52, 68
multicomponent diffusion	1, 68
multilayers	178
no partition	1
non-equilibrium	154
non-equilibrium interface	90
nucleation	78, 51
oxidation	131
paraequilibrium	1, 21, 204
partial coherency	27,
patterns	231, 247, 263
persistent slip bands	231
perturbation	101
phase diagrams	1, 37, 75, 193
phase decomposition	68
phase transition	1, 75, 151
phase transformation	1, 20, 68, 116, 193, 204, 221
plasticity	231
point of no return	78
pore measurements	116
rapid quenching	78
Rayleigh number	101
recrystallisation	247
rejector plate mechanism	29
ripening	116
selective dissolution	116
self-organization	231, 263
short-range ordering	154
solid-state reactions	1, 20, 68, 131, 193, 204, 221
solidification	90, 101
solidification cells	90
solute drag-like effect	34
square-root diffusivity	71
stress	193
sulfidation	131
ternary diffusion	1, 68
thermodynamics	1, 37, 52, 193, 204, 221
thermodynamic database	52
thermosolutal convection	101
transition metals and alloys	131
unstability	151
uphill diffusion	1
wetting transition	145